Evolution and Speciation
Essays in Honor of M. J. D. White

Michael J. D. White (Photograph by David Lamb-Jenkins)

Evolution and Speciation

ESSAYS IN HONOR OF M. J. D. WHITE

EDITORS:

WILLIAM R. ATCHLEY
University of Wisconsin, Madison

DAVID S. WOODRUFF
University of California, San Diego

CAMBRIDGE UNIVERSITY PRESS

Cambridge

London New York New Rochelle

Melbourne Sydney

Published by the Press Syndicate of the University of Cambridge
The Pitt Building, Trumpington Street, Cambridge CB2 1RP
32 East 57th Street, New York, NY 10022, USA
296 Beaconsfield Parade, Middle Park, Melbourne 3206, Australia

First published 1981

Printed in the United States of America
Typeset by Science Press, Ephrata, Pennsylvania
Printed and bound by The Murray Printing Co., Westford, Massachusetts

Library of Congress Cataloging in Publication Data
Main entry under title:
Evolution and speciation.
Includes index.
1. Evolution – Addresses, essays, lectures.
2. Species – Addresses, essays, lectures.   3. White,
Michael James Denham, 1910 –, – Addresses, essays,
lectures.   I. White, Michael James Denham, 1910 –.
II. Atchley, William R.   III. Woodruff, David S.
IV. Title: Speciation.
QH366.2.E854   575   80-24876
ISBN 0 521 23823 4

Contents

Contributors

W. R. Atchley, Department of Entomology, University of Wisconsin, Madison, WI 53706, USA

N. H. Barton, School of Biological Sciences, University of East Anglia, Norwich, NR4 7TJ, UK

R. Blackith, Department of Zoology, Trinity College, Dublin 2, Ireland

G. L. Bush, Department of Zoology, University of Texas at Austin, Austin, TX 78712, USA

H. L. Carson, Department of Genetics, University of Hawaii, Honolulu, HI 96844, USA

C. R. Carter, School of Biological Sciences, University of Sydney, Sydney, N.S.W. 2006, Australia

R. H. Crozier, School of Zoology, University of New South Wales, Kensington, N.S.W. 2033, Australia

E. S. Dennis, Division of Plant Industry, CSIRO, P. O. Box 1600, Canberra A.C.T. 2601, Australia

O. H. Frankel, Division of Plant Industry, CSIRO, P. O. Box 1600, Canberra A.C.T. 2601, Australia

D. L. Hayman, Department of Genetics, University of Adelaide, Adelaide, S.A. 5001, Australia

G. M. Hewitt, School of Biological Sciences, University of East Anglia, Norwich NR4 7TJ, UK

A. J. Hilliker, Division of Plant Industry, CSIRO, P. O. Box 1600, Canberra, A.C.T. 2061, Australia

B. John, Department of Population Biology, Research School of Biological Sciences, Australian National University, P. O. Box 475, Canberra, A.C.T. 2601, Australia

M. King, Department of Population Biology, Research School of Biological Sciences, Australian National University, P. O. Box 475, Canberra, A.C.T. 2601, Australia

B. T. O. Lee, Department of Genetics, University of Melbourne, Parkville, Victoria 3052, Australia

M. J. Littlejohn, Department of Zoology, University of Melbourne, Parkville, Victoria 3052, Australia

J. Martin, Department of Genetics, University of Melbourne, Parkville, Victoria 3052, Australia

J. H. Oliver, Jr., Department of Biology, Georgia Southern College, Statesboro, GA 30460, USA

P. A. Parsons, Department of Genetics and Human Variation, La Trobe University, Bundoora, Victoria 3083, Australia

W. J. Peacock, Division of Plant Industry, CSIRO, P. O. Box 1600, Canberra, A.C.T. 2601, Australia

A. J. Pryor, Division of Plant Industry, CSIRO, P. O. Box 1600, Canberra, A.C.T. 2601, Australia

D. D. Shaw, Department of Population Biology, Research School of Biological Sciences, Australian National University, P. O. Box 475, Canberra, A.C.T. 2601, Australia

S. Smith-White, Department of Botany, School of Biological Sciences, University of Sydney, Sydney, N.S.W. 2006, Australia. (Retired)

J. A. Thomson, School of Biological Sciences, Department of Genetics, University of Sydney, Sydney, N.S.W. 2006, Australia

D. S. Woodruff, Department of Biology C-016, University of California, San Diego, La Jolla, CA 92093, USA

Preface

Throughout his distinguished career Professor M. J. D. White has made significant contributions to many aspects of evolutionary biology including cytology and cytogenetics, speciation theory and systematics. These contributions are reflected in this collection of essays prepared by his friends to mark the occasion of his seventieth birthday. It is a pleasure to report that Michael White's field and laboratory research activities continue undiminished. His work over the past decade on the genetic systems of insects must rank among his best. May he long continue to throw light on Darwin's mystery of mysteries – the origins of species.

The essays in this book were prepared in early 1979 and the lists of Michael White's honors and publications are accordingly incomplete. We thank Constance MacDonald for her superb editorial assistance. We thank the staff and editors of Cambridge University Press for making possible this tribute to a great scientist.

W.R.A.
D.S.W.

PART I

INTRODUCTION

1   M. J. D. White: the scientist and the man

WILLIAM R.ATCHLEY

Michael James Denham White was born in the Chelsea district of London, England, on August 20, 1910, the first of three children of James Kemp White and Una Chase White. He was named Michael after the day on which his parents were married (Saint Michael's day), James after his father, and Denham after an ancestor, Admiral Henry Mangles Denham, F.R.S. Admiral Denham was the hydrographer on HMS *Herald,* which charted the coasts of Australia. A town, a sound, a river, and a mountain range in Australia are also named after Denham.

The original White home on Cheyne Walk, located on the fringe of Chelsea's artistic bohemia, was filled with books, pictures, and souvenirs from James's travels over the continent of Europe. James White was a tutor who taught Latin, Greek, algebra, and trigonometry to young English gentlemen attempting to enter Oxford University, Cambridge University, or the upper echelons of the British Civil Service.

Michael's mother, Una Chase, was an aspiring artist who had left the restricting confines of a Cornish vicarage to study art in London. In 1908 James White and Una Chase met in an art class in Chelsea, and in 1909 they were married by the father of the bride.

Soon after Michael was born, the family moved to a country house in Norfolk on the east coast of England. Here Michael's father had more room for tutoring, and the number of students increased. Unfortunately, however, when World War I broke out in 1914, all of his students went into the army, and he lost his primary source of support. As a result of the war, Michael's parents took the family to Italy. Michael was to remain in Italy and France until 1927 when he returned to England to attend University College.

Michael's early schooling can best be characterized as informal and extremely intermittent. From his father he learned Latin, classical Greek, algebra, geometry, and trigonometry, but he was never taught history, geography, or science. Rather, his father encouraged him to read about history and geography on his own. Not until he entered University College did Michael attend a formal school.

It is not clear exactly when Michael's great interest in biology began. He seemed always to be interested in natural history. When he was very young, he was usually rewarded with a new natural history book for enduring extractions of his milk teeth. At a very early age he had available many natural history books written in (or translated into) Italian. These included the Giglio-Tos butterfly book, L. Figuier's *Gli Insetti,* Arcangeli's *Compendio della Flora Italiana,* and other similar works. His father provided him with a home-made butterfly net, pinning box, killing jar, and spreading board so he could make successful forays against the local insect fauna. One of the early critical stages in his scientific education occurred on his seventh birthday when the entire family went on a picnic to a place called Tre Fossi (Three Creeks). In one day he learned about sexual dimorphism in dragonflies, gill respiration in freshwater crabs, and periodic molting of the arthropod cuticle. By the end of his birthday he had collected a box full of insects but, more important, had had his head filled by the exposure to many new biological ideas and concepts.

Michael seems to have been quite close to his father because James White was more than a father – he was Michael's primary source of education. Thus Michael endured a heavy psychological blow at age 13 when his father died of lung cancer. Before he died, the elder White wrote down his wishes for Michael's education. He was not to be sent to an English public school but rather was to study with an institution in Cambridge called the University Correspondence College. This institution was to prepare Michael for the entrance examinations for a British university. Once a week he would receive reading assignments, essay topics, and exercises, and once a week he would return his work to be corrected. Although he never saw his instructors or knew their names, he soon began to know what pleased each one as well as what would incur their wrath.

Michael studied with the University Correspondence School until 1927, and then returned to London to sit for the London Matriculation Examination. He was 16, and this was the first examination he had ever taken. In spite of his highly unorthodox education, Michael passed the examination and entered University College in September of 1927. Thus he was to begin an association with University College that would extend to 1946 (the "London period"). This university was strong in the sciences and, unlike most British universities, had no departments of divinity. In fact, it had a reputation for liberal agnosticism, which seemed to better fit Michael's temperament.

Michael began his university career with the intention of becoming a botanist. Indeed, a few years earlier he had made a special herbarium of orchids and had studied the alpine species of orchids while mountain climbing, which was one of his favorite hobbies as a teen-ager. He was particularly interested in the natural hybrids of *Gymnadenia odoratissima, G. conopsea,* and various species of *Orchis.*

Unfortunately, the botany professor at University College, Edward Salisbury, was unimpressive as a lecturer and a disappointment to an aspiring young botanist. On the other hand, Professor D. M. S. Watson, a vertebrate paleontologist in the zoology department, was a much more dynamic lecturer and inspired students with his confidence and self-assurance. It was Watson's influence that led Michael to become an evolutionary zoologist rather than a botanist or forester.

The first year at University College was a difficult one for Michael. Being thin, shy, foreign, and with only modest financial support, he found it difficult to establish a rapport with his colleagues who were better endowed financially. Lack of money and social rejection stimulated Michael to study much harder than he might otherwise have done. As a result, he began to do well in all of his examinations except physics. Physics was the most difficult of his subjects (as it is to many biologists), but his diligence began to yield results, and eventually he passed the course.

The second and third years at University College were more enjoyable. He was near the top of his class in zoology and many students who had previously scorned him began to ask for help with their studies. During his third year, Michael selected entomology as his "special subject." Because there was no course in entomology at University College, Michael traveled to the Royal College of Science to study entomology with F. Balfour-Brown, J. W. Munro, and O. W. Richards. It was Richards's lectures in entomology that influenced Michael to devote his career to the study of insects.

In 1931, Michael received a B.Sc. with First Class Honors in Zoology and Human Physiology, which was followed by an M.Sc. in 1932. Pursuit of a Ph.D. was difficult in the British university system during this time for individuals without considerable independent financial resources because students were required to be enrolled full-time, and part-time employment was precluded. Lacking any form of outside support, Michael took the position of assistant lecturer in zoology at University College in 1932, which permitted him to pursue a degree of doctor of science. The D.Sc. in the British university system is a doctoral degree reflecting scientific achievement in terms of scholarly publications rather than a specific course of study and a dissertation. Michael was awarded the D.Sc. from the University of London in 1940. During this period, he was promoted through the academic ranks at University College, achieving lecturer rank in 1935 and reader in 1946.

In 1937, Michael went to Columbia University as a Rockefeller research fellow to work in cytogenetics with Franz Schrader and Sally Hughes-Schrader. One of the most important things he learned during his time with the Schraders was rigid standards of intellectual honesty in observation and interpretation. It was also during this period that he became close friends with Theodosius Dobzhansky, who was to have a great influence on him. The two remained close friends until Doby's death.

During his years in London, Michael associated with many well-known biologists, including J. B. S. Haldane, C. D. Darlington, and R. A. Fisher. From Haldane he learned the importance of subjecting one's hypotheses to tests of significance. Darlington's 1932 book *Recent Advances in Cytology* was an influence in that it defined the field of cytogenetics as a discipline in its own right. Interestingly, however, these luminaries seemed to have little influence on his overall scientific development; rather he was stimulated by lesser-known figures such as Watson, Richards, and Lancelot Hogben, who was the first person to interest Michael in the process of speciation.

Knowledge of Michael's relations with C. D. Darlington is hazy. Both men were primarily responsible for the development of the field of cytogenetics – Darlington by his work on plants and White through his efforts on animals. There was obviously some degree of antagonism between the two, although its origins are unclear. Darlington wrote scalding and quite unfair reviews in *Nature* of the second and third editions of White's *Animal Cytology and Evolution*. He sarcastically described White as "out of his depth when he writes about chromosomes and evolution." Whether such erroneous comments were due to Darlington's own insecurities or to a lack of scientific judgment is unclear. History has shown that White knew a great deal about chromosomes and evolution, and his knowledge of the field has at least equaled Darlington's.

It is intriguing to note that in spite of Michael's interest in entomology, his first scientific publication, in 1932, dealt with the chromosomes of the domestic chicken. Working on the chromosomes of vertebrates must have been an unsatisfying experience for Michael because, with the exception of a single paper on spermatogenesis in hybrid urodele species (White, 1946c), he never again published on the cytology of a vertebrate (except in review papers).

Two publications came out of the London period that had a significant impact on the fields of chromosomal cytology and evolutionary biology. First was a small book entitled *The Chromosomes*, published in 1937. This little book not only provided a description of chromosome behavior in mitosis and meiosis but also included a terse but highly lucid discussion of chromosomes and evolution. In this latter chapter, Michael noted that it is difficult to formulate general cytogenetic laws of evolution because the chromosomal mechanisms can vary greatly from group to group. This motif would be developed extensively in his later publications. *The Chromosomes* has since undergone six revisions and has been translated into Spanish, French, and Italian. This small volume provided many scientists with their introduction to chromosomal cytology.

Animal Cytology and Evolution, published in 1945 by Cambridge University Press, was the second major publication from the London period. This work, the first critical account of cytology since E. B. Wilson's 1925 monograph on *The Cell in Development and Heredity*, provided an encyclopedia of information on animal chromosomal cytology. However, it would be highly misleading to label *Animal Cytology* as simply an enormous compilation of cytological observations, although it obviously was that. Rather, the most important aspect of this book was that it established evolutionary cytology as a discipline in its own right. *Animal Cytology*, along with Dobzhansky's *Genetics and the Origin of Species*, Huxley's *Evolution, the Modern Synthesis*, Mayr's *Systematics and the Origin of Species*, and Simpson's *Tempo and Mode in Evolution*, brought together aspects of cytology, genetics, evolution, systematics, and paleontology and integrated them into what could truly be called a synthetic theory of evolution.

Animal Cytology was strictly cytological in approach, but it ranged over an extraordinarily wide field. It placed in an evolutionary perspective the chromosomal mechanisms and variations of animal groups other than *Drosophila* as well as providing critical discussions on the evolution of meiosis, karyotypes, sex determination, and parthenogenesis. Further, the field of chromosomal cytology was dominated at this time by botanists, such as C. D. Darlington; and one of Michael's contributions at this point was to demonstrate clearly that the evolution of meiosis in animals has involved a far greater range of variant cycles than that found in plants.

Animal Cytology and Evolution is regarded by many biologists as

White's greatest contribution to the field of cytogenetics and evolutionary biology. This book is now in its third revised edition, and its major thesis remains threefold: (1) Evolution is essentially a cytogenetic process because evolution ultimately involves a complex sequence change in chromosomal DNA. Thus, reliance on morphological, ethological, and ecological studies and distribution data alone without critical cytogenetic studies will provide an inadequate explanation of major evolutionary phenomena. (2) Understanding the role of genetic systems in evolution requires both population genetics and cytological data. This aspect has become increasingly obvious in recent studies on the genetics of the speciation process. (3) The chromosomal mechanism itself is highly antitypological and varies in a number of ways from taxon to taxon within the taxonomic hierarchy. Different species of animals vary considerably in the mechanisms of meiosis and in groups such as insects may include bizarre chromosomal cycles. Further, the system of chromosomal polymorphisms can vary widely in different groups of organisms, as is evidenced by the insects. There is a preponderance of paracentric inversions in *Drosophila,* pericentric inversions in the Orthoptera, translocations in roaches and scorpions, Robertsonian fusion polymorphisms in various beetle groups, and so on.

The initial completion of *Animal Cytology* must have been very difficult because much of the compilation of literature sources and writing occurred during the early years of World War II. Michael was with the British Ministry of Food from 1940 to 1945, so that any critical work in cytology was most difficult. The necessary reference works and periodicals needed to compile a large reference book were accessible only under the most trying of circumstances.

In 1947, Michael left University College and moved to the United States. This move was precipitated by a general unhappiness with the intellectual atmosphere in British science after the war. The move to the United States marks the beginning of the "Texas period." After spending a few months as a guest investigator in the Department of Genetics in the Carnegie Institution of Washington, Michael accepted a position as professor of zoology at the University of Texas at Austin. The University of Texas was an intellectually stimulating place, the faculty including such diverse personalities as T. S. Painter, Wilson Stone, J. T. Patterson, and others.

It was during this period that he began to publish on the chromosomal behavior of organisms with aberrant chromosome cycles. He focused his attention on the unique and bizarre chromosome cycles in midges of the family Cecidomyiidae. In these insects there are a large number of chromosomes (E chromosomes) which are present in the cells of the germ line but absent from the somatic cells. These E chromosomes are eliminated from the early embryonic stages, and only a small number of so-called S chromosomes are retained in the soma. In most species, the sperm transmits only a haploid set of S chromosomes, with the egg contributing the other haploid set of S's and all of the E chromosomes.

Michael's work on the cecidomyiids clarified their highly unusual chromosomal cycle; however, the evolutionary significance of this extraordinary chromosomal dualism has remained enigmatic. Such bizarre chromosome cycles occur in other Diptera (e.g., the Sciaridae and Chironomidae), scale insects of the superfamily Coccoidea, and the lice (Mallophaga and Anoplura). These chromosomal cycles are "successful" in an evolutionary sense because they occur in a large number of species in divergent lineages and are, therefore, obviously very old in an evolutionary sense. However, the adaptive significance of such genetic systems clearly awaits further clarification.

A humorous story arises from Michael's work on the gall midges. It seems that he had stopped along the highway leading from Austin one day and was collecting flies from a patch of sunflowers. He placed the flies into coffee cans and then stored them under the plants where it was cooler. Unfortunately, at about the same time a robbery occurred in Austin, and the culprits fled the city using the same route and model of car as Michael's. Michael soon found himself surrounded by the local constabulary, who must have thought he was hiding the loot from the robbery in the coffee cans around the sunflowers. Only after considerable explanation was he able to convince the police that he was really a professor from the University of Texas collecting insects for his research and totally uninvolved in any robbery. One can only imagine the look on the face of the officer who first opened the cans and found flies rather than loot. The constabulary gave him a stern lecture about "acting suspicious," the lecture being given in a manner that can only be appreciated by someone who has lived in Texas.

Another major line of research pursued by Michael during the Texas period involved the population cytology of the North American trimerotropine grasshoppers. These grasshoppers exhibit a great deal of structural heterozygosity for pericentric inversions as well as a relatively high frequency of supernumerary chromosomes. Much of Michael's work on these grasshoppers involved deducing the adaptive significance of the high levels of inversion polymorphism. Recently, several groups of cytogeneticists have started to work again on these interesting organisms.

The studies on gall midges and trimerotropine grasshoppers firmly established Michael as a pioneer in the field of population cytology along with Sturtevant and Dobzhansky. Michael consistently, and in a systematic manner, studied the cytology of whole samples from many localities and analyzed their cytological variation. Like Dobzhansky and Sturtevant, he truly appreciated the population dynamics of chromosomal polymorphisms, but, unlike his contemporaries, he attacked a wide variety of cytological problems with this approach and integrated his results with comparative information from other, often unrelated organisms. He used a population approach to study problems ranging from the evolution of chromosomal cycles in gall midges to rearrangement polymorphisms in grasshoppers.

In 1953, Michael left the University of Texas to take a position as senior research fellow with the CSIRO Division of Plant Industry in Canberra, Australia, where he remained until 1956. He spent a brief period from January 1957 until June of 1958 as professor of zoology at the University of Missouri before returning to Australia to assume the chair in zoology at the University of Melbourne. He was the professor of zoology at Melbourne from 1958 until 1964 and during this period was the prime mover in establishing a new Department of Genetics at Melbourne. After the genetics department was established, he became its first professor and occupied the chair in genetics from 1964 until his retirement from the University of Melbourne in 1975. Since retirement, he has been a visiting fellow in the Department of Population Biology at the Australian National University in Canberra.

Although he was employed in the Division of Plant Industry when he first moved to Australia, Michael never worked on plant genetics, but instead began working immediately on the Australian grasshoppers. During the first year or two he examined the cytology of the grasshopper genera *Atractomorpha* (Pyrogomorphidae) and *Cryptobothrus* (Acrididae). Only brief references to this work were ever published except in the context of general cytological topics. However, this early work has since been expanded by geneticists such as Bernard John and Neil Nankivell.

The year 1956 marked a pivotal point in Michael's scientific career because at this time he began work on the morabine grasshoppers. The Morabinae are a large and very distinctive group of endemic, wingless Australian grasshoppers within the orthopteran family Eumastacidae. It is for his numerous and imaginative studies on this group of insects that he is probably best known to noncytologists. The work on the morabines has involved three major projects: (1) population cytology of *Keyacris* (formerly *Moraba*) *scurra;* (2) analysis of raciation and speciation in the

Vandiemenella (formerly *Moraba*) *viatica* complex and the subsequent development of the stasipatic speciation model; and (3) studies on genetic variability and the evolution of parthenogenesis in *Warramaba* (formerly *Moraba*) *virgo* and its bisexual relatives.

One of the most carefully documented studies of chromosomal polymorphism in natural populations outside of the *Drosophila* has been White's work on *Keyacris scurra* in southern Australia. He examined the adaptive role played by naturally occurring pericentric inversions in this insect and the causes of genetic equilibrium. A complex epistatic interaction in the determination of viability exists between the inversion polymorphisms on two nonhomologous chromosomes. Studies over a 10-year period showed that the inversion frequencies in natural populations were in genetic equilibrium; that is, temporal changes in the frequencies were statistically nonsignificant. Further, it was noted that demes separated by as little as 5 miles differed considerably in cytogenetic composition.

In collaboration with R. C. Lewontin, Michael computed a series of adaptive topographies for the various inversion frequencies. The surprising result of studies on adaptive topographies was the finding that the populations were not located at or near an adaptive peak but were always in the saddle between the peaks. Lewontin and White suggested that the equilibria in inversion frequencies are most easily explained by assuming that the selective values of the various karyotypes are frequency-dependent, declining in the case of each genotype as it becomes more common in the population. They suggested that such a system of frequency-dependent selective values could result from a situation in which each genotype was specialized to exploit a particular type of ecological niche, its fitness declining if it were forced by competition to exist outside that niche.

The *K. scurra* data have been examined in detail by many other geneticists, and several possible explanations have now been proposed to account for the variation seen in this species. R. C. Allard and C. Wehrhahn have reexamined the *scurra* data to assess the effect of inbreeding, whereas J. R. G. Turner has examined the methods for determining relative variabilities. Most recently, Sewell Wright has examined the *K. scurra* story in considerable detail in vol. 4 of his treatise on *Evolution and the Genetics of Populations* and has provided a very plausible interpretation for the rather paradoxical results of Lewontin and White.

The work on *K. scurra* remains the classic account on the correlation between morphological variation and chromosomal inversions. Biometrical analyses indicated that there has been a differential accumulation of size-increasing alleles on the mutually inverted cytological sequences. These effects are stable between demes and chromosomal races over a large geographic area.

After working on *K. scurra* for several years, Michael turned his attention to the morabine genus *Vandiemenella* (formerly *Moraba*). This genus, which has become well-known as the *viatica* species group, includes seven species, of which four have at least two chromosomal races. These grasshoppers are wingless, they exhibit low vagility, and the taxa have parapatric distributions along the coast of southeastern Australia. Trying to explain the patterns of chromosomal evolution in the *viatica* group caused Michael seriously to question contemporary dogma about speciation theory.

Michael was among the many evolutionary biologists who had generally reached the consensus by the mid 1960s that new species of sexually reproducing animals probably arose exclusively by geographic speciation. In fact, in 1962 (*Aust. J. Sci.* 25:183) he wrote that: "The morabine grasshoppers, with their restricted powers of locomotion and dispersal seem to provide a classic example of geographic differentiation of races and species. There does not seem to be the slightest evidence in this group for the concept of sympatric speciation nor even the formation of sympatric ecological races."

These comments were obviously written before he tried to account for the observed chromosomal evolution in the *viatica* group by the classic allopatric model of speciation. Based on his experiences with the *viatica* group, Michael published a seminal paper in 1968 entitled "Modes of speciation," which indicated that his thoughts on geographic speciation had undergone some radical changes. He proposed that in plants and animals with low vagility, chromosomal rearrangements could result in the rapid origin of new species. He suggested that a chromosomal rearrangement could arise that would form the basis of a cytogenetic isolating mechanism. The chromosomal rearrangement would be spread throughout a substantial part of the range of an already existing species. The model, called *stasipatric speciation*, combined two important points. First, it involved a chromosomal rearrangement as a postmating isolating mechanism, and second, it implied that geographic isolation was not always a necessary prerequisite for this mode of speciation to occur.

The idea of stasipatric speciation was met with skepticism in many quarters. By this time the universality of geographic speciation had become part of evolutionary dogma. However, other papers, such as John

Maynard Smith's important 1966 work on sympatric speciation, began to appear that broke new evolutionary ground and offered several alternatives to the geographic speciation model. As a result, by the mid-1970s many evolutionary biologists began questioning the conventional wisdom on geographic speciation.

In 1978, Michael published his most recent book, appropriately entitled *Modes of Speciation*. As has been the case with *Animal Cytology and Evolution*, he gathered extensive data from a wide variety of organisms to evaluate the subject – in this case the various speciation models proposed over the years. His conclusions lend strong support to the thesis that the modes of speciation in animals and plants are diverse and that geographic speciation represents only one of several ways new species can arise. This latest book has stimulated many students interested in evolution to reexamine their views on speciation. The subject itself has been revitalized after a period in the doldrums during the 1960s when speciation theory was considered resolved and no longer interesting.

Most recently, Michael has devoted most of his efforts to the study of the Australian parthenogenetic grasshopper *Warramaba virgo* and its bisexual relatives. This thelytokous species arose by hybridization between two bisexual species possibly as long as a half million years ago. Studies by White and co-workers have described the meiotic mechanism in *virgo,* and documented its hybrid origin and extensive cytogenetic diversity. As a result of his endeavors, *virgo* has become the most extensively studied thelytokous species. Much of the work on this species is summarized in Chapter 18 of this volume.

Although he formally retired from academic life in 1975, Michael shows no evidence of slowing his research output. In fact, the tremendous capacity for sheer hard work that has always characterized him has continued unabated since his supposed retirement. Since 1975, he has published his book on modes of speciation, a monograph on orthopteran cytogenetics, and 15 papers. He is currently working on a book on parthenogenesis and continues a full research program supported by outside funding.

As further evidence that retirement from academic duties is not diminishing his influence, consider the following data from *Science Citation Index*. For the 5-year period 1970–4, there were 547 references to fully 121 of his then-published 128 papers. The rate of citation was an average of nine per month. In 1978, 2 years after supposed retirement, his work was being cited 16 times a month, and during the first half of 1979 it was cited 18 times a month.

His stature in the international scientific community is further reflected by his membership in scientific academies. In 1955, he was elected a fellow of the Australian Academy of Science and in 1961 elected fellow of the Royal Society of London. He is an honorary foreign member of the American Academy of Arts and Science (elected 1963), a foreign member of the American Philosophical Society (elected 1978), and Socio Straniero, Accademia Nazionale dei Lincei (elected 1978). In 1962, he was elected fellow of University College, London.

Michael married Isobel Mary (Sally) Lunn in 1939. Sally is an anthropologist who has done significant research on the Australian aborigines. In spite of her own independent scientific career, she has remained a major source of support to Michael as well as a companion on many field trips. They have three children: Charlotte, a physician in private practice; Jonathan, a senior lecturer in Shakespearian literature at the University of Essex, and Nicholas, who is employed by the Biological Standards Laboratory at the Australian Commonwealth Department of Health.

I am deeply indebted to Michael White for sharing many reflections and stories with me over the past 10 years. I am also indebted to Sally White, K. H. L. Key, R. C. Jackson, Bernard John, Ernest Mayr, Guy L. Bush, Hampton Carson, and David Woodruff for helpful comments and assistance in preparing this account. However, I accept responsibility for the conclusions reflected in it.

Publications of M. J. D. White

1932. The chromosomes of the domestic chicken. *J. Genet. 26:*345–50.

1933. Tetraploid spermatocytes in a locust, *Schistocerca gregaria. Cytologia 5:*135–9.

1934. The influence of temperature on chiasma frequency. *J. Genet. 29:*203–15.

1935a. Eine neue Form von Tetraploidie nach Roentgenbestrahlung. *Naturwissenschaften* 23:390–1.

1935b. The effects of X-rays on mitosis in the spermatogonial divisions of *Locusta migratoria* L. *Proc. R. Soc. Lond. B 119:*61–84.

1936a. Chiasma localization in *Mecostethus grossus* L. and *Metrioptera brachyptera* L. *Z. Zellforsch. 24:*128–135.

1936b. The chromosome cycle of *Ascaris megalocephala. Nature (Lond.) 137:*783.

1937a. The effect of X-rays on the first meiotic division in three species of Orthoptera. *Proc. R. Soc. Lond. B 124:*183–96.

1937b. *The Chromosomes.* London: Methuen, viii + 128 pp.

1938. A new and anomalous type of meiosis in a mantid, *Callimantis antillarum* Saussure. *Proc. R. Soc. Lond. B 125:*516–23.

1940a. The heteropycnosis of sex chromosomes and its interpretation in terms of spiral structure. *J. Genet. 40:*67–82.

1940b. The origin and evolution of multiple sex chromosome mechanisms. *J. Genet. 40:*303–36.

1940c. A translocation in a wild population of grasshoppers. *J. Hered. 31:*137–40.

1940d. Evidence for polyploidy in the hermaphrodite groups of animals. *Nature (Lond.)* *146*:132.

1941a. Chromosomal evolution and the mechanisms of meiosis in praying mantids. *Proc. VII Int. Genet. Congr.*, p. 313.

1941b. The evolution of the sex chromosomes. I. The X0 and X_1X_2Y mechanisms in praying mantids. *J. Genet. 42*:143–72.

1941c. The evolution of the sex chromosomes. II. The X-chromosome in the Tettigoniidae and the Acrididae and the principle of "evolutionary isolation" of the X. *J. Genet. 42*:173–90.

1942a. *The Chromosomes*, 2nd ed. London: Methuen, x + 124 pp.

1942b. Nucleus, chromosomes and genes. Chapter 5 in *Cytology and Cell Physiology*. London: Oxford University Press.

1943. Amount of heterochromatin as a specific character. *Nature (Lond.) 152*:536–7.

1945. *Animal Cytology and Evolution*. Cambridge: Cambridge University Press, vii + 375 pp.

1946a. The cytology of the Cecidomyidae (Diptera). I. Polyploidy and polyteny in salivary gland cells of *Lestodiplosis* spp. *J. Morphol. 78*:201–19.

1946b. The cytology of the Cecidomyidae (Diptera). II. The chromosome cycle and anomalous spermatogenesis of *Miastor. J. Morphol. 79*:323–70.

1946c. The spermatogenesis of hybrids between *Triturus cirstatus* and *T. marmoratus* (Urodela). *J. Exp. Zool. 102*:179–207.

1946d. The evidence against polyploidy in sexually-reproducing animals. *Am. Nat. 80*:610–18.

1947a. The cytology of the Cecidomyide (Diptera). III. The spermatogenesis of *Taxomyia taxi. J. Morphol. 80*:1–24.

1947b. Chromosome studies on gall midges. *Yearb. Carnegie Inst.*, pp. 165–9.

1947c. *Los Cromosomas* (Spanish translation of *The Chromosomes* by F. A. Saez). Madrid: Espasa-Calpe, 146 pp.

1948a. The chromosomes of the parthenogenetic mantid *Brunneria borealis. Evolution 2*:90–3.

1948b. The cytology of the Cecidomyidae (Diptera). IV. The salivary gland chromosomes of several species. *J. Morphol. 82*:53–80.

1948c. The cytogenetic system of the gall midges (abstr.) *Anat. Rec. 101*:21–2.

1949a. A cytological survey of wild populations of *Trimerotropis* and *Circotettix* (Orthoptera, Acrididae). I. The chromosomes of twelve species. *Genetics 34*:537–63.

1949b. Cytological evidence on the phylogeny and classification of the Diptera. *Evolution 3*:252–60.

1949c. Cytological polymorphism in wild populations of western grasshoppers (abstr.). *Rec. Genet. Soc. Am. 18*:119.

1949d. Chromosomes of the vertebrates (essay review of R. Matthey: *Les Chromosomes des Vertebres*). *Evolution 3*:379–81.

1950a. Cytological studies on gall midges. *Univ. Tex. Publ. 5007*:1–80.

1950b. Cytological polymorphism in natural populations of grasshoppers. *Yearb. Am. Phil. Soc. 1949*:183–5.

1950c. *I Cromosomi* (Italian translation of *The Chromosomes* by C. Winspeare). Einaudi, 144

1951a. Nucleus, chromosomes and genes. Chapter 5 in *Cytology and Cell Physiology*, 2nd ed. London: Oxford University Press.

1951b. Evolution of cytogenetic mechanisms in animals. Chapter 16 in *Genetics in the Twentieth Century*. Macmillan.

1951c. Cytological polymorphism and racial differentiation in grasshopper populations. *Yearb. Am. Phil. Soc. 1950*:158–60.

1951d. A cytological survey of wild populations of *Trimerotropis* and *Circotettix* (Orthoptera, Acrididae). II. Racial differentiation in *T. sparsa. Genetics 36:*31–53.

1951e. Cytogenetics of orthopteroid insects. *Adv. Genet. 4:*267–330.

1951f. Structural heterozygosity in natural populations of the grasshopper *Trimerotropis sparsa. Evolution 5:*376–94.

1951g. White, M. J. D. and Nickerson, N. H. Structural heterozygosity in a very rare species of grasshopper. *Am. Nat. 85:*239–46.

1951h. Supernumerary chromosomes in the trimerotropine grasshoppers (abstr.) *Rec. Genet. Soc. Am. 20:*130–1.

1951i. *Citologia Animal y Evolucion* (Spanish translation of *Animal Cytology and Evolution* by F. A. Saez). Madrid: Espasa-Calpe, 512 pp.

1952. Review of H. F. Barnes: *Gall Midges of Economic Importance,* vol. 5. *Q. Rev. Biol. 27:*219–20.

1953a. Multiple sex chromosome mechanisms in the grasshopper genus *Paratylotropidia. Am. Nat. 87:*237–44.

1953b. Review of J. A. G. Rehn: *The Grasshoppers and Locusts (Acridoidea) of Australia,* vol. 1, *Families Tetrigidae and Eumastacidae. Q. Rev. Biol. 28:*184–5.

1954a. *Animal Cytology and Evolution,* 2nd ed. Cambridge: Cambridge University Press, xiv + 454 pp.

1954b. An extreme form of chiasma localization in a species of *Bryodema* (Orthoptera, Acrididae). *Evolution 8:*350–8.

1954c. Review of J. A. G. Rehn: *The Grasshoppers and Locusts (Acridoidea) of Australia,* vol. 2, *Family Acrididae (Subfamily Pyrgomorphinae). Q. Rev. Biol. 29:*376–7.

1955a. Patterns of spermatogenesis in grasshoppers. *Aust. J. Zool. 3:*222–6.

1955b. White, M. J. D., and Morley, F. H. W. Effects of pericentric rearrangements on recombination in grasshopper chromosomes. *Genetics 40:*604–19.

1956a. Adaptive chromosomal polymorphism in an Australian grasshopper. *Evolution 10:*298–313.

1957a. An interpretation of the unique sex-chromosome mechanism of the rodent, *Ellobius lutescens* Thomas. *Proc. Zool. Soc. Calcutta, Mookerjee Memor. Vol.,* pp. 113–14.

1957b. Some general problems of chromosomal evolution and speciation in animals. *Surv. Biol. Progr. 3:*109–47.

1957c. Cytogenetics and systematic entomology. *Annu. Rev. Entomol. 2:*71–90.

1957d. White, M. J. D., and Key, K. H. L. A cytotaxonomic study of the *pusilla* group of species in the genus *Austroicetes* Uv. (Orthoptera: Acrididae). *Aust. J. Zool. 5:*56–87.

1957e. Cytogenetics of the grasshopper *Moraba scurra.* I. Meiosis of interracial and interpopulation hybrids. *Aust. J. Zool. 5:*285–304.

1957f. Cytogenetics of the grasshopper *Moraba scurra.* II. Heterotic systems and their interaction (with a statistical appendix by B. Griffing). *Aust. J. Zool. 5:*305–37.

1957g. White, M. J. D., and Chinnick, L. J. Cytogenetics of the grasshopper *Moraba scurra.* III. Distribution of the 15- and 17-chromosome races. *Aust. J. Zool. 5:*338–47.

1957h. Cytogenetics of the grasshopper *Moraba scurra.* III. Heterozygosity for "elastic constrictions." *Aust. J. Zool. 5:*348–54.

1957i. Genetic interaction of heterotic systems in grasshopper populations (abstr.). *Rec. Genet. Soc. Am.* in *Genetics 42:*402.

1957j. Review of H. B. Johnston: *Annotated Catalogue of African Grasshoppers. Q. Rev. Biol. 32:*188.

1958. Restrictions on recombination in grasshopper populations and species. *Cold Spring Harbor Symp. Quant. Biol. 23:*307–17.

1959a. Speciation in animals. *Aust. J. Sci. 22:*32–9.

1959b. Telomeres and terminal chiasmata – a reinterpretation. *Univ. Tex. Publ. 5914:* 107–11.

1959c. Review of J. A. G. Rehn: *The Grasshoppers and Locusts (Acridoidea) of Australia*, vol. 3. *Q. Rev. Biol. 34:*158.

1960a. Reply to Dr. Brown. *Aust. J. Sci. 22:*298–9.

1960b. Lewontin, R. C., and White, M. J. D. Interaction between inversion polymorphisms on two different chromosome pairs in the grasshopper *Moraba scurra. Evolution 14:*116–29.

1960c. White, M. J. D. and Andrew, Lesley E. Cytogenetics of the grasshopper *Moraba scurra.* V. Biometric effects of chromosomal inversions. *Evolution 14:*284–92.

1960d. Are there no mammals with X0 males – and if not, why not. *Am. Nat. 94:*301–4.

1961a. Prof. J. T. Patterson (obituary). *Nature (Lond.) 189:*709.

1961b. Chromosome. In *Encyclopedia of the Biological Sciences* (ed. Gray, P.), pp. 230–2. New York: Reinhold.

1961c. *The Chromosomes*, 5th ed. London: Methuen, viii + 188 pp.

1961d. The role of chromosomal translocations in urodele evolution and speciation in the light of work on grasshoppers. *Am. Nat. 95:*315–21.

1961e. Cytogenetics of the grasshopper *Moraba scurra.* VI. A spontaneous pericentric inversion. *Aust. J. Zool. 9:*784–90.

1962a. White, M. J. D., and Andrew, L. E. Effects of chromosomal inversions on size and relative viability in the grasshopper *Moraba scurra.* In *Evolution of Living Organisms*, Melbourne: Melbourne University Press. pp. 94–101.

1962b. Cell division. Chromosome. In *Chambers Encyclopedia.*

1962c. A unique type of sex chromosome mechanism in an Australian mantid. *Evolution 16:*75–85.

1962d. A parthenogenetic species of grasshopper feeding on a rare species of *Acadia. Aust. J. Sci. 25:*63–4.

1962e. Genetic adaptation (presidential address to Section D, Aust. N.Z. Assoc. Adv. Sci.). *Aust. J. Sci. 25:*178–86.

1963a. White, M. J. D., Cheney, J., and Key, K. H. L. A parthenogenetic species of grasshopper with complex structural heterozygosity (Orthoptera: Acridoidea). *Aust. J. Zool. 11:*1–19.

1963b. Cytogenetics of the grasshopper *Moraba scurra.* VIII. A complex spontaneous translocation. *Chromosoma (Berl.) 14:*140–5.

1963c. White, M. J. D., Lewontin, R. C., and Andrew, L. E. Cytogenetics of the grasshopper *Moraba scurra.* VII. Geographic variation of adaptive properties of inversions. *Evolution 17:*147–62.

1964a. Cytogenetic mechanisms in insect reproduction. *Insect Reproduction* (Symp. 2 of R. Entomol. Soc.), pp. 1–12.

1964b. *Les Chromosomes* (French translation of *The Chromosomes* by A. Berkaloff). Paris: Dunod.

1964c. Principles of karyotype evolution in animals. *Proc. XI Int. Congr. Genet. 2:*391–7.

1964d. White, M. J. D., Carson, H. L., and Cheney, J. Chromosomal races in the Australian grasshopper *Moraba viatica* in a zone of geographic overlap. *Evolution 18:*417–29.

1964e. The next stage in evolution studies. *Probe* (Witwatersrand University, Johannesburg) no. 3: 59–62.

1965a. Chiasmatic and achiasmatic meiosis in African eumastacid grasshoppers. *Chromosoma (Berl.) 16:*271–307.

1965b. Sex chromosomes and meiotic mechanisms in some African and Australian mantids. *Chromosoma (Berl.) 16:*521–47.

1965c. J. B. S. Haldane. *Genetics 52:*1–7.

1966a. Further studies on the cytology and distribution of the Australian parthenogenetic grasshopper *Moraba virgo. Rev. Suisse Zool. 73:*383–98.

1966b. White, M. J. D. and Cheney, J. Cytogenetics of the *Cultrata* group of morabine grasshoppers. I. A group of species with XY and X_1X_2Y sex chromosome mechanisms. *Aust. J. Zool. 14:*821–34.

1966c. A case of spontaneous chromosome breakage at a specific locus occurring at meiosis. *Aust. J. Zool. 14:*1027–34.

1967a. White, M. J. D., Blackith, R. E., Blackith, R. M., and Cheney, J. Cytogenetics of the *viatica* group of morabine grasshoppers. I. The coastal species. *Aust. J. Zool. 15:*263–302.

1967b. Karyotypes of some members of the grasshopper families Lentulidae and Charilaidae. *Cytologia 32:*184–9.

1967c. White, M. J. D., Mesa, A., and Mesa, R. Neo-XY sex chromosome mechanisms in two species of Tettigonioidea (Orthoptera). *Cytologia 32:*190–9.

1968a. A gynandromorphic grasshopper produced by double fertilization. *Aust. J. Zool. 16:*101–9.

1968b. Models of speciation. *Science 159:*1065–70.

1968c. White, M. J. D., and Webb, G. C. Origin and evolution of parthenogenetic reproduction in the grasshopper *Moraba virgo* (Eumastacidae: Morabinea). *Aust. J. Zool. 16:*647–71.

1968d. Karyotypes and nuclear size in the spermatogenesis of grasshoppers belonging to the subfamilies Gomphomastacinae, Chininae and Biroellinae (Orthoptera: Eumastacidae). *Caryologia 21:*167–79.

1968e. Sex. *Encyclopaedia Brittanica 20:*287–93.

1969a. White, M. J. D., Key, K. H. L., Andre, M., and Cheney, J. Cytogenetics of the *viatica* group of morabine grasshoppers. II. Kangaroo Island populations. *Aust. J. Zool. 17:*313–28.

1969b. Chromosomal rearrangements and speciation. *Annu. Rev. Genet. 3:*75–98.

1969c. The infinite variety of Haldane. *Science 164:*678–80.

1969d. Chromosomes and human personality. *Aust. Rationalist 1(3):*13–16.

1970a. Cytogenetics. Chapter 3 in *The Insects of Australia,* pp. 72–82. Melbourne: Melbourne University Press.

1970b. Heterozygosity and genetic polymorphism in parthenogenetic animals. In *Essays in Evolution and Genetics in Honor of Theodosius Dobzhansky* (ed. Hecht, M. K., and Steere, W. C.), pp. 237–62. New York: Appleton-Century-Crofts.

1970c. Asymmetry of heteropycnosis in tetraploid cells of a grasshopper. *Chromosoma (Berl.) 30:*51–61.

1970d. Karyotypes and meiotic mechanisms of some eumastacid grasshoppers from East Africa, Madagascar, India and South America. *Chromosoma (Berl.) 30:*62–97.

1970e. Cytogenetics of speciation. *J. Aust. Entomol. Soc. 9:*1–6.

1970f. Webb, G. C., and White, M. J. D. A new interpretation of the sex determining mechanism of the European earwig, *Forficula auricularia. Experientia 26:*1387–9.

1971a. The chromosomes of *Hemimerus bouvieri* Chopard (Dermaptera). *Chromosoma (Berl.) 34:*183–9.

1971b. Review of *Studies in Evolution and Genetics in Honor of Theodosius Dobzhansky. Search 2:*175.

1972a. The value of cytology in taxonomic research on Orthoptera. *Proc. Int. Study Conf. Curr. Future Probl. Acridology,* pp. 27–33.

1972b. The chromosomes of *Arixenia esau* Jordan (Dermaptera). *Chromosoma (Berl.) 36:*338–42.

1972c. White, M. J. D., and Cheney, J. Cytogenetics of a group of morabine grasshoppers with XY and X_1X_2Y males. *Chromosomes Today 3:*177–96.

1973a. Chromosomal rearrangements in mammalian population polymorphism and speciation. In *Cytotaxonomy and Vertebrate Evolution* (ed. Chiarelli, B., and Capanna, E.), pp. 95–128. London: Academic Press.

1973b. *Animal Cytology and Evolution,* 3rd ed. Cambridge: Cambridge University Press, viii + 961 pp.

1973c. White, M. J. D., Webb, G. C., and Cheney, J. Cytogenetics of the parthenogenetic grasshopper *Moraba virgo* and its bisexual relatives. I. A new species of the *virgo* group with a unique sex chromosome system. *Chromosoma (Berl.) 40:*199–212.

1973d. *The Chromosomes,* 6th ed. London: Methuen [Chapman and Hall].

1974a. White, M. J. D. (ed.). *Genetic Mechanisms of Speciation in Insects* (papers presented at two symposia, 14th Int. Entomol. Congr. Canberra, 1972). Sydney: Australia and New Zealand Book Co.

1974b. Speciation in the Australian morabine grasshoppers – the cytogenetic evidence. In *Genetic Mechanisms of Speciation in Insects,* pp. 57–68. Sydney: Australia and New Zealand Book Co.

1974c. Cytogenetics. In supplementary volume to *Insects of Australia,* pp. 15–18. Melbourne: Melbourne University Press.

1975a. Chromosomal repatterning – regularities and restrictions (paper delivered at the 13th Int. Congr. Genet. Berkeley, August 1973). *Genetics, 79:*63–72.

1975b. Karyotypes in the genus *Biroella* and the origin of the Australian morabine grasshoppers. *J. Aust. Entomol. Soc. 14:*135–8.

1975c. Webb, G. C., and White, M. J. D. Heterochromatin and timing of DNA replication in morabine grasshoppers. In *Proceedings of Eukaryote Chromosome Conference* (ed. Brock, R. D., and Peacock, W. J.), pp. 395–408. Canberra: Australian National University Press.

1975d. An XY sex chromosome mechanism in a mantid with achiasmatic meiosis. *Chromosoma (Berl.) 51:*93–7.

1975e. Review of S. Makino: *Human Chromosomes. Med. J. Aust. 1975:*885–6.

1976. Blattodea, Mantodea, Isoptera, Grylloblattodea, Phasmodea, Dermaptera, Embioptera. In *Animal Cytogenetics* (ed. John, B.), vol. 3, *Insecta,* part 2. Stuttgart: Borntraeger, pp. v + 1–75.

1977a. White, M. J. D., Contreras, N., Cheney, J., and Webb, G. C. Cytogenetics of the parthenogenetic grasshopper *Warramaba* (formerly *Moraba*) *virgo* and its bisexual relatives. II. Hybridization studies. *Chromosoma (Berl.) 61:*127–48.

1977b. Karyotypes and meiosis of the morabine grasshoppers. I. Introduction and genera *Moraba, Spectriforma and Filoraba. Aust. J. Zool. 25:*567–80.

1977c. Acridology in the Uvarovian style. *Science 198:*1247–8.

1977d. *Animal Cytology and Evolution.* Paperback edition (with minor changes). Cambridge: Cambridge University Press.

1978a. *Modes of Speciation.* San Francisco: Freeman.

1978b. White, M. J. D., and Contreras, N. Cytogenetics of the parthenogenetic grasshopper *Warramaba* (formerly *Moraba*) *virgo* and its bisexual relatives. III. Meiosis of male "synthetic *virgo*" individuals. *Chromosoma (Berl.) 67:*55–61.

1978c. Webb, G. C., White, M. J. D., and Contreras, N. Cytogenetics of the parthenogenetic grasshopper *Warramaba* (formerly *Moraba*) *virgo* and its bisexual relatives. IV. Chromosome banding studies. *Chromosoma (Berl.) 67:*309–39.

1978d. Chain processes in chromosomal speciation. *Syst. Zool. 27:*285–98.

1978e. The karyotype of the parthenogenetic grasshopper *Xiphidiopsis lita* (Orthoptera, Tettigoniidae). *Caryologia 31:*291–7.

1979a. White, M. J. D., and Contreras, N. Cytogenetics of the parthenogenetic grasshopper *Warramaba* (formerly *Moraba*) *virgo* and its bisexual relatives. V. Interaction of *W. virgo* and a bisexual species in geographic contact. *Evolution 33:*85–94.

1979b. Karyotypes and meiosis of the morabine grasshoppers. II. The genera *Culmacris* and *Stiletta. Aust. J. Zool. 27:*109–33.

1979c. Speciation – is it a real problem? *Scientia 114*:455–68.

1980a. The present status of myriapod cytogenetics. In *International Myriapod Biology 1978* (ed. Carmatini, M.), New York: Academic Press.

1980b. Meiotic mechanisms in a parthenogenetic grasshopper species and its hybrids with related bisexual species. In *Animal Genetics and Evolution* (ed. Vorontsov, N. N.). The Hague: W. Junk.

1980c. Significato avattativo della partenogenesi negli insetti: *Atti Accad. Naz. Entomologia* (in press).

1980d. Modes of speciation in orthopteroid insects. *Boll. Zool.* (in press).

1980e. The genetic system of the parthenogenetic grasshopper *Warramaba virgo*. In *Insect Cytogenetics*, Xth Symposium of Royal Entomological Society of London (ed. Blackman, R. L., Ashburner, M. and Hewitt, G. M.), Oxford: Blackwell.

1980f. White, M. J. D., Webb, G. C., and Contreras, N. Cytogenetics of the parthenogenetic grasshopper *Warramaba* (formerly *Moraba*) *virgo* and its bisexual relatives. VI. DNA replication patterns of the chromosomes. *Chromosoma (Berl.)* (in press).

1981a. Karyotypes and meiosis of the morabine grasshoppers. III. The genus *Hastella. Aust. J. Zool. 29* (in press).

1981b. The chromosome architecture of the parthenogenetic grasshopper *Warramaba virgo* and its bisexual ancestors. In *Chromosomes Today*, Vol. 7 New York: Wiley.

PART II

CYTOLOGY AND CYTOGENETICS

2   Chromosome change and evolutionary change: a critique

BERNARD JOHN

It is grossly selfish to require of one's neighbour that he should think in the same way and hold the same opinions, why should he? If he can think, he will probably think differently.

Oscar Wilde

Most of the key problems of biology have been with us for many years, and, despite the fact that they are regularly resurrected, redefined or even rediscovered by each succeeding generation of biologists, they remain unresolved. Paramount among these problems is that of the relationship between chromosome change and evolutionary change, the subject of this essay. In dealing with this relationship it is tempting to ask why not let the facts speak for themselves? Unfortunately facts cannot and do not speak. Facts, however authoritative, always have to be spoken for. Much, therefore, depends not only on the facts but on who is speaking for them. Coupled with this there is a temptation to see all facts as being of equal significance, which they rarely are. Finally, many of the arguments relating to chromosome change and evolutionary change depend not on facts but on speculation. It is only when we ascend from rhetoric to realism that these arguments can be seen for what they are worth.

The immense variety of, and the striking variation in, the karyotypes of extant forms provide suggestive evidence that chromosome change has played, and continues to play, a major role in evolutionary change. Despite this many of the available texts on population genetics and evolution assume that chromosomes are kept unchanged except for point mutation and recombination and make little or no reference to the staggering range of chromosome variation now known to occur within and between populations of many eukaryotes that have been adequately examined. This variation exists at three principal levels:

1. Variation within populations due to the occurrence of chromosome polymorphisms where distinct karyomorphs exist at stable frequencies too high to be accounted for by recurrent mutation. Such polymorphisms thus result from segregational variation within a population.
2. Variation between populations within a single species where there can be two or more chromosome races differentiated by fixed differences in chromosome constitution giving rise to a state of polytypism.
3. Variation between species, where related species are differentiated by fixed chromosome differences.

To appreciate the difference between these three categories we need to give some consideration to the kinds of chromosome change that can occur and their genetical consequences.

Discussion
Types of chromosome change

The chromosomes of eukaryotes appear to be built of two quite distinct components: euchromatin and constitutive heterochromatin.

Euchromatin contains the bulk of the unique DNA sequences of an organism, that is, those DNA segments that are represented only once, or at best a few times, in the haploid genome. It also includes moderately repeated DNA sequences, which are interspersed with the unique DNA. Part of the unique DNA forms the material basis for the greater part of the structural gene system, that is, the system that transcribes the mRNA, which, when subsequently translated, gives rise to the enzymic and structural proteins of the organism. Not all the unique DNA, however, behaves in this way, and all eukaryotes contain a proportion of unique DNA whose function has not been adequately defined. Thus in sea urchins only some 10%–30% of the unique DNA appears to be transcriptively active, whereas in *Drosophila,* with its much smaller genome, the figure is on the order of 25% (Davidson et al., 1977). Additionally, with the exception of the transfer, ribosomal, and histone-DNA, the function of most of the moderately repeated DNA is also largely unknown. Even so, it is clear that genes in the classical sense are not the only, or necessarily the principal, components of eukaryotic chromosomes. Yet conventional evolutionary theory is predicated on the assumption that the chromosome is essentially nothing but a linear linked system of structural genes. This, clearly, is a gross oversimplification of the truth, one moreover that has led to the commonly stated and erroneous belief that evolution consists simply of changes in gene frequency.

Constitutive heterochromatin frequently includes the bulk of the DNA with very high repetition frequencies ($> 10^6$ copies) and especially the class of DNA that appears as a satellite fraction in a buoyant density equilibrium gradient (Skinner, 1977). This second structural component characterizes both members of a given homologous pair in the same way and to the same extent. Euchromatic regions may under special circumstances become facultatively heterochromatinized in such a way that the DNA component of the region is transcriptively silenced. Such a process, however, affects only one of a pair of homologues and even then only in certain specific cell types.

The visible differences between karyotypes, both within and between species, stem from four sources (John, 1976b):

1. Structural rearrangements, which invariably lead to modifications in the ordering of the unique DNA sequences within the euchromatin. All such rearrangements thus alter the linkage relationships of structural genes. Additionally, the amount, or else the position, of effective crossing-over between linked genes may be reduced, restricted, increased, or canalized.
2. Multiplication of whole genomes (polyploidy), frequently coupled either with subsequent structural rearrangement (Jones, 1977) or else with the loss or gain of whole chromosomes.
3. The addition of supernumerary chromosomes, which although derived from conventional members of a given chromosome set, have rapidly become differentiated from them, often by becoming heterochromatinized. As such they no longer pair homologously with their progenitors at meiosis although they often pair with one another when two or more are present.
4. Additions or losses of chromosome material including changes in the amount of constitutive heterochromatin as well as changes in the amount of euchromatin.

Among them these four processes account for all the known variants found in any eukaryote.

It is, of course, important to determine what part of the chromosome variation in a given population or a given species is functionally relevant in an evolutionary sense. This, as we shall see, is no easy task, being confounded by the fact that chromosome changes are not usually associated with obvious morphological changes, let alone morphological changes that are adaptive. It is customary to argue that to stand any chance of survival chromosome changes must lead neither to developmental disturbances, as a consequence of either mechanical inefficiency or epigenetic imbalance, nor to a reduction in fertility, as a result of an upset in the efficiency of meiotic pairing or meiotic segregation. In fact, as will become evident, there is compelling evidence to demonstrate that some chromosome changes become fixed despite the fact that they do indeed reduce fertility; for example, interchanges between telo or acrocentric chromosomes and tandem translocations. But, for the moment, let us consider the fate of those changes that can meet such exacting demands and that reduce neither viability nor fertility. Clearly much will depend on what genetic effects accompany such changes.

Chance, change, and choice

There has always been a school of thought that accords no adaptive significance whatsoever to chromosome change. Such an opinion rests predominantly on the fact that gross changes in chromosome constitution are accompanied, at best, by only minor morphological modifications in the organism itself. Even Dobzhansky initially held just such a view. Thus in 1961 he wrote that "since flies with different karyotypes are externally indistinguishable, many investigators, including the present author, believed these karyotypes to be adaptively neutral traits," and then, in a conversion that must surely rank second only to that of St. Paul, added, "This proved to be wrong." The proof came from the fact that both in nature and in the laboratory he was able to demonstrate that different karyotypes responded differentially to altered environmental conditions.

Neutrality, however, is a difficult concept to dispense with. How does one prove that something is indeed doing nothing? Neutrality is currently in vogue once more and in a variety of ways. From a purely chromosomal point of view the most radical statement is that of Ohno (1974:36):

> The very fact that two morphologically similar species belonging to the same genus can have such different karyotypes suggests that the karyotype can change drastically in a very short time without appreciable genetic consequences. Thus I am inclined to believe that karyotype changes in evolution are to be regarded as neutral changes which accompanied speciation not because they were advantageous but because they were harmless.

There is, of course, the possibility that some polymorphisms or some fixed differences are relics of past selection that, having outlived their usefulness to the species, nevertheless persist as a kind of vestigial variation and, therefore, do indeed behave in a neutral fashion (Thoday, 1975). But given that there are segments of the genome that are genuinely nonfunctional, it is not easy to accept the idea that so much of the chromosome variation observed in natural populations can be explained in the way Ohno suggests. Moreover, although chance establishment of a neutral chromosome change may indeed explain the abundance of particular chromosome variants in particular populations, it will not so easily explain the high and differential frequencies of similar variant types in different populations of the same species, especially when these variants fall into a clinal relationship or can be related in some other obvious way to environmental variation. One can, of course, always resort to the argument that the change in question is neutral but happens to be linked to some other genetic component that is not. But, again, there is a limit to the extent that one can use this argument to explain away the vast range of variation now known at the chromosome level. Moreover, although a chromosome change can become fixed by a combination of inbreeding and random drift, it is important to emphasize that fixation in itself is not enough. To persist after fixation the changed form must, unless neutral, sooner or later become subject to selection. For example, a polyploid does not originate by selection; it originates by mutation. Nor does it become fixed by selection; its fixation depends on some form of inbreeding. But, once it has become fixed, its persistence depends on its capacity to compete with its progenitor or progenitors, and this demands some form of selection. The same principle applies to any nonneutral change that is fixed by chance rather than choice. Sooner or later choice becomes inevitable. Indeed this fact was recognized by Sewall Wright himself, though it has been largely ignored by those who have used his principle of drift as a convenience. As Wright wrote:

> Many critics have seized on the concept of random drift and asserted that I advocated this as a significant alternative to natural selection. Actually I have never attributed any evolutionary significance to random drift except as a trigger that may release selection towards a higher selective peak through incidental crossing of a threshold.

That is, evolution often proceeds most rapidly when random components are coupled with selection, drift then triggering the initial change but with subsequent selection either stabilizing the change or else rejecting it.

Chance apart, at least some of the chromosome changes found in nature may well be there not because they are selectively favored but because selection is powerless to cope with the ruthless efficiency with which they propagate themselves. The most forceful of these cases is that presented by Nur (1977) in the grasshopper *Melanoplus differentialis*. Here Nur argues, and with considerable force, that the proportion of B chromosomes within a population reaches a stable equilibrium that is related to the extent to which the accumulation of the supernumerary, as a result of meiotic drive in the egg, is able to overcome the decrease in fitness, which can be measured in terms of the effect of the supernumerary on viability. That selection is unable to remove a chromosome with such deleterious effects on fitness is not difficult to understand. In boosting its own frequency within the female gametes, the B chromosome is simply capitalizing on a meiotic mechanism that has itself been selected specifically for ensuring that only one of its four products becomes a functional egg.

This raises the more general issue of whether an equivalent argument can be applied to other cases too. In considering this possibility we need not be disturbed if some aspects of fitness are indeed reduced by some types of chromosome change. The crucial question is, are all aspects of fitness affected in the same way and to the same extent? Although fitness is sometimes equated with adaptedness, the term is more sensibly employed to describe the contribution made by individuals to subsequent generations, that is, to adaptability. In this wider sense of the term, adaptedness is but one component of fitness, a component that rests on the pattern of development and metabolism – on what Kenneth Lewis and I have termed the exophenotype. A second fitness component relates not to epigenesis and exophenotype but to genesis and reproduction – to what we may call endophenotype (Lewis and John, 1963). Endophenotype is concerned with regulating the genetic composition of the gametes and the zygotes produced by an individual. The fact that chromosome changes do not produce alterations in the adaptedness of their possessors in no sense excludes them from influencing endophenotypic components of those same individuals, and there are clear grounds for arguing that chromosome change has been both restricted and promoted by its effects on meiosis.

The roles of chromosome change

There has been a persistent feeling among evolutionists that chromosome change ought to produce an obvious exophenotypic effect. This is not surprising when one recalls that changes in exophenotype represent the outward and, to many, the most obvious aspects of evolutionary change. Thus as recently as 1978 Yoon and Richardson could, without any supporting evidence that I am aware of, write that structural chromosome changes "may contribute to abrupt changes in behaviour, morphology and other genetic features of a species." In arriving at such an unwarranted point of view, much emphasis has been attached to the quite singular fact that in *Drosophila* the phenotypic expression of a given gene can, in certain experimental circumstances, be affected by its spatial relationship with neighboring genetic material. It was left to Goldschmidt

(1955) to extend this concept and to argue that, in nature too, chromosome rearrangements provide a source of new position effects and hence represent a means of producing new character combinations. Although no direct evidence has ever been produced to support this concept, it has, nevertheless, persisted and has recently been resurrected in a seemingly new form by Wilson and his colleagues (see, e.g., Wilson et al., 1977), who speculate that chromosome rearrangements and their resulting effects on gene regulation may account for the most profound differences between species.

This view assumes that the important distinguishing features of related species depend not on differences in their structural genes so much as on differences in their regulatory gene systems, and that the rearrangement of the order of genes can result in altered patterns of regulation. Because no one knows how any eukaryotic gene is regulated, this idea is patently unsupportable in terms of hard fact. Wilson's speculation is in reality built upon the earlier speculation of Britten and Davidson (1971) that moderately repetitive DNA exerts a fine-tuning regulatory control over the expression of individual structural genes. Not only are there no meaningful data to support this speculation, but some data clearly contradict it. Thus some 97%–98% of the genome of the fungus *Aspergillus nidulans* has been shown to consist of unique DNA, whereas the remainder can be accounted for by ribosomal cistrons (Timberlake, 1978). Evidently here regulation must be achieved in the absence of repetitive DNA. Equally difficult to explain in terms of a regulatory role is the fact that the amount of repetitive DNA may vary widely between closely related species that show only minor morphological differences.

In fact it has long been evident that there is no necessary, and certainly no consistent, relationship between chromosome change and morphological change. Sibling speciation based on chromosome change makes it clear that certain taxa can undergo chromosome repatterning with little or no accompanying genetic or morphological differentiation. Yet many of those who have considered the relationship between chromosome change and evolutionary change have assumed just such a relationship, and this assumption has had a number of unfortunate consequences. For example, in determining the direction that change has followed when considering chromosome evolution, or for that matter molecular evolution too, it has become standard practice to use morphological change as a basis for defining directionality. Two further irrationalities follow from such a practice: first, the assumption that "primitive" or ancestral karyotypes are most likely associated with "primitive" morphological character states

(Williams and Hall, 1976); and second, the argument that rates of chromosome evolution can be estimated by comparing the karyotype of one species with that of another related species whose time of divergence is known from morphological evidence. This, again, presupposes a simple relationship between the rate of chromosome change and the rate of change of other characters. The case has been well put by Hsu and Mead (1969), who used the following argument:

> Suppose for one reason or another one decides that karyotype A is the ancestral form of karyotype B. This does not necessarily mean that species M with karyotype A is more primitive than species N with karyotype B. Evolutionary processes may have taken place at various directions and levels other than changes in chromosome morphology.

Given that chromosome change is not sensibly related to exophenotype change, to what then can it be related? Many chromosome changes affect the course and the consequences of meiosis. In so doing they affect either the level of recombination, and hence the amount of variability, within (polymorphisms) or between (polytypisms) populations, or else they restrict the exchange of genetic material between populations and, as a result, may play a role in speciation. All these effects depend on the fact that meiosis involves four genetically significant events:

1. The pairing of homologous chromosomes or chromosome segments at zygotene of first prophase. This, in turn, depends on the genetic content of the segments in question. The rules of pairing determine that in diploids, pairs of homologues regularly associate as bivalents. In aneuploid or polyploid states, on the other hand, the pairing of complete homologues leads to the formation of multivalent associations involving three or more homologues, whereas in structurally rearranged complements involving interchange heterozygotes, fusion/fission heterozygotes, or tandem translocation heterozygotes the pairing of partially homologous members gives rise to the formation of multiple chromosome associations.

2. Crossing-over between homologously paired euchromatic sections at late pachytene of first prophase. This can lead to the recombination of linked genetic elements. It should be stressed that crossing-over does not occur in constitutively heterochromatic regions (John, 1976a; Hotta and Stern, 1979).

3. The orientation at first metaphase of those homologously paired associations that are maintained by the formation of chiasmata as a consequence of crossing-over. Where bivalents are present, this

orientation depends on homologous centromere pairs co-orienting to opposite spindle poles. Co-orientation is, of course, not possible in unpaired chromosomes. Consequently, with the exception of certain sex chromosome systems (XO or X_nO), univalents, however formed, invariably create a segregational problem (see next paragraph). In multivalent and multiple configurations there are a variety of possible orientation types, not all of which are necessarily co-oriented. Those that are not lead to irregular segregation.

4. The segregation of oriented configurations at first anaphase. Spontaneous univalents which cannot co-orient, and those multiples of multivalents that do not, invariably lead to nondisjunction and hence to the production of unbalanced aneuploid chromosome sets in the meiotic products. Alternatively, univalent chromosomes may lag on the first division spindle and, in consequence, impede cytokinesis, thus giving rise to diploid restitution nuclei, which, if they divide normally, produce diploid meiotic products. Where univalents prevent cytokinesis at both first and second division, they produce tetraploid products, and in males they may give rise to macrospermatids. Univalents, by lagging, may also be excluded from the telophase groups and so form micronuclei, which, in males, subsequently produce microspermatids (John and Weissman, 1977). Neither micro- nor macrospermatids lead to successful fertilization. For this reason they are a potential source of infertility when present in substantial numbers. So, too, are the aneuploid gametes, which, although they fertilize, produce inviable progeny that die during embryogenesis. In mice, for example, at least 10 types of chromosome change are associated with male infertility (Evans, 1976), namely: (1) X–A translocations, (2) Y–A translocations, (3) insertional translocations, (4) spontaneous autosomal univalency, (5) failure of association of X and Y chromosomes in meiotic prophase, (6) sex chromosome trisomics of the XXY and XYY types, (7) tertiary trisomy, (8) single and multiple reciprocal translocations, (9) double heterozygosity for certain reciprocal translocations involving a common chromosome, and (10) combinations of two Robertsonian translocations that share a homologous arm. These various types of change show a wide variety of effects ranging from a complete absence of spermatogenic cells (XXY mouse) to the progressive decline of spermatogenesis, which may begin at any one of a number of stages in the meiotic cycle. There is no satisfactory explanation for most of these effects. Case (10) is especially difficult to understand because it is known that: (1) males singly

heterozygous for the Robertsonian translocations in question are fully fertile as far as the production of zygotes is concerned, and (2) multiple Robertsonian heterozygotes in which no two metacentrics share a common arm are similarly fertile.

Thus what chromosomes do at meiosis is fundamental to fertility and to sterility. It is also fundamental to the genetic composition of the next generation because chromosome changes may affect not only gametic quantity but also zygotic quality. Effects on meiosis thus have consequences that do not concern the individual in which these effects occur. Rather they concern the future, and the future, after all, is the only real concern of evolution. The important distinction is thus between aspects of the phenotype that affect the survival of the individual and those that have a meaning only for posterity.

Some chromosome changes unquestionably lead to alterations in the pattern of the amount of crossing-over. This applies to inversions (White and Morley, 1955; Lucchesi and Suzuki, 1968), fusions (Hewitt and John, 1972; Cattanach and Moseley, 1973; Craddock, 1973; John and Freeman, 1975), supernumerary chromosomes, and supernumerary segments (John, 1973). Taken individually any of these cases could be dismissed as a mere correlation, but, as Smith (1978) points out, corroborating evidence has much more significance in biological science than contrary evidence, and collectively these cases add up to a convincing argument that the regulation of recombination represents an important facet of chromosome change. Although these effects cannot and do not influence the adaptive properties of the individual in which they occur, they may, through their contributions to meiotic recombination, influence the adaptive properties of the progeny to which they contribute (Table 2.1). White (1979), in common with other authors, has tended to dismiss such second-order effects as "naive and unrealistic." By comparison with the unsupported speculations about unknown first-order effects, however, benefits to remote descendants assume a considerable degree of credence and respectability simply because they at least rest on a solid observational basis. Indeed if, as Thoday (1953) has argued, the fitness of contemporary individuals cannot be defined without reference to their remote offspring, then the genetic composition of these offspring cannot be unimportant. Moreover, differential viability (exophenotype) and differential fecundity or fertility (endophenotype) are quite different forms of fitness, and there are clear situations in which they do not go together. Thus hybridity may lead to vigor (exophenotypic effect) but also to sexual sterility (endophenotypic effect). Similarly, the absence of any effect on differential viability should not be taken to mean that fecundity or fertility will also be unaffected.

Compared to the occurrence of effects on crossing-over, which may play a causative role in the development of polymorphism, and in some cases at least (fusions) to polytypism, the development of mechanisms that restrict gene exchange and gene flow can be expected to lead to the breakup of a species because the essential criterion for reproductive isolation is lack of gene flow. It is important to recognize that mere interbreeding between taxa does not necessarily indicate an absence of reproductive isolation, because the hybrids produced may be sterile or may break down subsequent to their formation. Indeed, in a sense, the formation of inviable or sterile hybrids is often the most objective evidence for full species status. Added to this, the formation of species hybrids, whether natural or experimental, offers the only way of re-creating the heterozygous condition through which structural changes must have passed prior to their fixation. Even here, however, the analysis of hybrids may be confounded by the

Table 2.1. *Effects of B-chromosomes on progeny phenotype in the spring variety of* Secale cereale *from the Transbaikal*

| | | No. of Bs per ♀ parent | | | |
| | | OB | | 2B | |
Character	No. of Bs per seed	Variance	Coefficient of variation	Variance	Coefficient of variation
Seed weight	0	8.38	21.4	19.97	23.3
(mg)	2	9.29	24.5	24.17	27.7
	4	—	—	28.05	27.5
Germination	0	0.33	16.9	0.57	22.2
time (days)	2	0.62	22.6	8.88	67.7
	4	—	—	4.12	40.6
Plant height	0	11.59	11.03	16.25	13.07
(cm) 3 wk	2	11.50	10.95	32.92	18.94
after	4	—	—	32.15	17.45
germination					

Note that the presence of Bs in the female parent increases the variability of their progeny irrespective of the number of B-chromosomes present in the progeny themselves.

Source: Data from Moss (1966).

difficulty of distinguishing between genotypic and structural differences. Indeed the behavior of differences that are without doubt structural in basis is known to be influenced by the genotype (Cattanach and Moseley, 1973).

There are two clear cases in which we can, with some confidence, claim chromosome change as a causative factor in speciation. The first of these is polyploidy; the second is in the production of sibling species in cases where the interchange differences that distinguish the siblings involve telocentric chromosomes (John and Lewis, 1965). The lesson from both of these cases is clear. An organism must survive to maturity if it is to reproduce, and must reproduce if it is to leave progeny. To the extent that chromosome differences lead to consistent and extensive hybrid inviability, hybrid sterility, or hybrid breakdown, there is the possibility that they can play fairly decisive roles in speciation because they can be expected to function as strong primary, though of course postzygotic, reproductive isolating mechanisms. In hybrid inviability we are dealing directly with the exophenotype, but with hybrid sterility and hybrid breakdown we are very much concerned with endophenotype. Chromosome changes may thus produce hybrid inviability or sterility as a result of disturbances in pairing and segregation at meiosis. They are of particular interest because they are efficient at their inception, and their efficiency does not improve with contact. For this reason, they can arise in any situation – allopatric, sympatric, or parapatric.

Fixed and polymorphic differences

It is important to emphasize that many, if not all, of the chromosome changes that are successful in leading to a stable polymorphism, or even to a fixed polytypism, are not the same as those required for creating a species divergence or vice versa. It is true that some types of chromosome change can serve a variety of functions. In *Drosophila* paracentric inversions may exist either as stable polymorphisms or else as fixed differences between related species. This is also the case for interchanges. Even so, the interchanges that occur as polymorphisms or as fixed permanent heterozygotes invariably involve meta- or submetacentric chromosomes. Those that distinguish species and serve to isolate those species involve telocentric or acrocentric chromosomes, which are self-sterilizing (Lewis and John, 1963). It is, of course, precisely for this reason that the collapse of a stable interchange polymorphism cannot conceivably provide a basis for speciation.

Fusions, too, can exist in a polymorphic or a fixed state and, in the latter condition, can distinguish either races or species. Some fusions evidently do not lead to sterility. Although the relevant meiotic studies have not been made, this must be true in the case of the shrew *Sorex araneus,* in which populations are polymorphic for six fusions (Ford and Hamerton, 1970). A second, though rather different, example is found in sheep. Wild sheep range from Europe, through the xeric and subxeric regions of the Middle East, Soviet Turkmenistan and the Indian subcontinent, central Asia, Siberia, and North America. They show a diverse phenotypic variability throughout this extensive and extremely wide holarctic distribution, and various authors have recognized from one to nine species on morphological grounds. From a cytological point of view the genus *Ovis* can be divided into four groups distinguished by fusion differences (Bunch et al., 1976; see Table 2.2).

Wild sheep are particularly variable in Iran, where three species have been recognized. Cytologically, however, two principal types exist with $2n = 58$ (NE form) and $2n = 54$ (NW form). These two forms represent the basic chromosomal, geographic, and morphological extremes of Iranian sheep (Valdez et al., 1978). Where these two forms converge, a hybrid zone exists with $2n = 54, 55, 56, 57$, and 58 types. Of eight ewes in late gestation examined from one location in this hybrid zone, seven had fetuses with hybrid numbers. The high fetal rate of hybrid ewes and the presence of backcross hybrids in the sample indicate that the fusions in this case appear to have no effect on fertility. If this is so, then what have been referred to as distinct species are in reality only chromosome races. Extensive interbreeding between overlapping populations of chromo-

Table 2.2. *Robertsonian relationships in sheep (genus* Ovis*)*

58	56	54	52
Urial sheep (*O. vignei*)	Arkhar/Argali sheep (*O. ammon*)	Mouflon sheep (*O. musimon*) Red sheep (*O. orientalis*) Dall sheep (*O. dalli*) North American bighorn sheep (*O. canadensis*) Domestic sheep (*O. aries*)	Siberian snow sheep (*O. nivicola*)

somally characterized taxa, even ones with wide chromosome differences, has also been reported in the gerbil *Gerbillus pyramidum* (Wahrman and Gourevitz, 1973), the spiny mouse *Acomys caharinus* (Wahrman and Goitein, 1972), and the neotropical bat *Uroderma bilobatum* (Baker et al., 1975). None of these studies detected reduced gene flow between the different cytotypes, and the populations in question have all been considered conspecific.

Select inbred stocks of domestic sheep from New Zealand are known to carry three polymorphic fusions in addition to those represented in the $2n = 54$ type of Iran. Although they form only trivalents at meiosis when heterozygous, they nevertheless show a proportion of aneuploid meiotic products (Table 2.3). Despite this, there is no evidence from either blastocyst studies (Long, 1977) or mating data involving such fusions (Bruère, 1974; Chapman and Bruère, 1975) that aneuploid gametes lead to aneuploid zygotes in these sheep. It would appear, therefore, that either aneuploid metaphase-II cells degenerate during spermatogenesis or else they are selected against at fertilization.

Contrasting sharply with the state of affairs in sheep is that found in mice. Most European domestic mice (*Mus musculus*) and a majority of Asiatic and sub-Asiatic species of *Mus* have $2n = 40$ telocentrics. However, in Switzerland and Italy mouse populations homozygous for up to nine different fusions occur (Capanna et al., 1976, 1977). The boundaries of most of these fusion populations are still uncertain, but, in general, they appear as isolates in an otherwise continuous distribution of $2n = 40$ forms. The Alpine system includes populations in the val Bregaglia (3 fusions $2n = 34$), the val Mesolecina (6 fusions $2n = 28$), the val Poschiavo (7 fusions $2n = 26$), the Lombardy (8 fusions $2n = 24$), and the Orobian Alps (9 fusions $2n = 22$). The Apennine system includes a population on the Isle of Liparia (7 fusions $2n = 26$), the Laga Mountain population (8 fusions $2n = 24$), and the Abruzzi and Molise populations (each with 9 fusions $2n = 22$).

There is a hybrid zone in which the 22-chromosome Abruzzi population meets the 40-chromosome populations in the east and northeast of Rome. Here individuals heterozygous for fusions occur. When mice from Poschiavo are experimentally crossed with Abruzzi and Molise mice, spermatogenesis is completely blocked in the F_1 hybrids, and no sperm are present in the epididymis. Likewise, in Abruzzi and Molise intercrosses both male and female hybrids are completely sterile, and long-chain multiples are present at meiosis whose disjunction is irregular.

The mice from val Poschiavo are known under a separate specific name,

Table 2.3. *The frequency of gametic aneuploidy in three spontaneous fusions (t₁, t₂, and t₃) present in select New Zealand breeds of domestic sheep and in laboratory synthesized mouse hybrids.*

Sheep	% ♂ MII cells with chromosome arm nos. of						Total cells
	<28	28	29	30	31	>31	
Standard, 2n = 54	9.68	1.61	4.84	*83.87*	—	—	62
Fusion heterozygotes {2n = 53 (t₁)	3.13	3.75	20.63	*66.88*	5.63	—	160
{2n = 53 (t₂)	3.02	12.69	25.46	*54.38*	4.53	—	331
Fusion homozygotes {2n = 53 (t₃)	7.67	3.68	25.46	*54.99*	8.59	0.61	326
{2n = 52 (t₁t₁)	6.06	2.02	6.06	*85.86*	—	—	99
{2n = 52 (t₃ t₃)	—	—	8.70	*91.30*	—	—	23

Mice	% ♂ MII cells with arm nos. of					Total cells
	<19	19	20	21	>21	
2n = 40 parent	1.00	2.00	*96.50*	—	0.50	200
2n = 22 parent	1.16	0.58	*97.05*	0.58	1.16	170
F₁ = 31 = 22 × 40	27.09	19.30	*29.68*	14.99	8.94	347
2n = 40 parent	2.00	3.80	*92.80*	1.00	0.50	207
2n = 26 parent	1.70	4.30	*94.00*	—	—	233
F₁ = 33 = 26 × 40	10.60	19.10	*45.80*	14.80	8.80	455

	% MII cells with arm nos. of					Total cells
	<19	19	20	21	>21	
pos/+ { sp'cytes	8.0	20.0	*48.0*	19.0	5.0	300
{ oocytes	12.0	17.0	*32.0*	20.0	19.0	100
CD/+ { sp'cytes	3.5	20.0	*49.0*	22.0	5.5	400
{ oocytes	15.0	27.0	*23.0*	24.0	11.0	100
pos/CD oocytes	4	3	1	2	3	13

Nos. ♀ MII cells

Note pos. × CD Abruzzi 2n = 31, pos. × CB Molise, and CD Abruzzi 2n = 31 × CB Molise Abruzzi 2n = 22 hybrids all give completely sterile males in which the hybrid testes show spermatogenic arrest at different stages.

Source: For sheep – data of Bruère, 1974; Chapman and Bruère, 1975. For mouse hybrids – data of Capanna, 1976, for the F₁ = 31 Apennine hybrid; of Tettenborn and Gropp, 1970, for the F₁ = 33 *poschiavinus* hybrid; and of Capanna et al., 1976 for the *poschiavinus* × CD Abruzzi hybrid.

Mus poschiavinus. When hybridized with $2n = 40$ *Mus musculus* in the laboratory, they consistently give F_1s with 7 trivalents + 13 bivalents. An analysis of second metaphase cells in these hybrids indicates that in males more than 50% of these cells are aneuploid, resulting from unequal segregation of the trivalents at first anaphase (Table 2.3). A majority of these aneuploids develop into functional sperm because:

1. There is considerably greater variation in the DNA values of morphologically normal F_1 sperm than in either parent (Stolla and Gropp, 1974).
2. F_1 hybrid males when mated to 40-chromosome females generate 50% or more aneuploid zygotes which abort (Tettenborn and Gropp, 1970). More specifically, although all zygotes with unbalanced genomes survive up to the time of implantation, hypoploid $(2n - 1)$ embryos are rapidly eliminated after implantation, whereas hyperploid $(2n + 1)$ embryos, though they survive longer, also die before birth.

A similar story can be told for the $2n = 22$ Apennine mice of Italy homozygous for nine fusions. Here some 70% of the second division cells are aneuploid in male hybrids. Given that the products of female meiosis are similarly affected, then the actual reproductive fitness of the hybrid (i.e., the probability of forming euploid zygotes) corresponds to the square of the frequency of balanced gametes. This means it will be in the order of $0.5^2 = 25\%$ for 26 × 40 hybrids and $0.3^2 = 9\%$ for the 22 × 40 hybrids.

In contrast to this situation, Evans et al. (1967) studied a spontaneous fusion in the mouse at the actual time of its origin. Nondisjunction in the heterozygous male, as measured by second metaphase counts, was estimated to be 1.0%, a very much lower figure than obtained for the nondisjunction values of individual *poschiavinus* metacentrics against a predominantly *musculus* background in laboratory F_1 hybrids, in which estimates range from 10.7% to 35.5% per fusion heterozygote. The correspondingly low expected output of unbalanced gametes in this spontaneous *musculus* mutant was complemented by the fact that the postimplantation loss of embryos in matings of heterozygous males to normal females did not differ from that of normal control matings. Surprisingly, however, the preimplantation losses were much higher, 39.9% compared with 13.0% in controls. This effect could not have been due to unbalanced zygotes, and Ford and Evans (1973) concluded that it depends on a generalized reduced capacity of the sperm to fertilize effectively. This, in turn, implies that the total effect of heterozygosity for a fusion on fertility, and consequently its total selective disadvantage, may be greater than is expected from the frequency of irregular disjunction at meiosis.

The conclusion from these several examples of fusion is inescapable: Whether fusion trivalents behave in a regular or an irregular manner is unpredictable. Added to this, malorientation need not necessarily lead to infertility, at least in domestic inbred lines of sheep, and near regular disjunction does not guarantee fertility, at least in the spontaneous *Mus* fusion. Under such circumstances there is an obvious need to consider each case in some detail before arriving at any conclusion about the functional role of any given fusion. More specifically, because individual fusions evidently behave in different ways, one needs to exercise caution in deciding whether fixed fusions are involved in a polytypic situation, or whether they distinguish biological species.

Only in one case, that of tandem fusion, is it not possible to produce a balanced polymorphic state under any circumstances. This is because such fusions cannot avoid producing substantial infertility when heterozygous (White et al., 1967). Thus if there is maximum pairing, some 50% of the meiotic products are expected to be unbalanced, whereas if there is not maximal pairing, the univalents that result also lead to infertility.

Chromosome speciation

It is now clear that there are different modes of speciation, some of which do not involve chromosome changes, and others that do. The difficulty of dealing with speciation is simply that there is little opportunity to study the process as such, but only its outcome. The fact that two species are distinguished by fixed chromosome differences need not mean that these differences have been responsible for the speciation event. Some fixed chromosome changes clearly cannot play a role as isolating agents simply because they do not depress fertility in the heterozygous state. This is true of the paracentric inversions that distinguish some of the Hawaiian species of *Drosophila*. These species show pronounced morphological divergence but no obvious difference in mitotic metaphase karyotypes and only moderate reorganization of their polytene elements, involving paracentric inversions that are unquestionably incidental to the speciation process and in no sense causally related to it (Carson et al., 1967).

The case for chromosome speciation has recently been championed by White (1978a) in a book that seems to justify Mark Twain's trite comment that "even the truth can be exaggerated." Here special emphasis has been placed on the stasipatric model of speciation, which was invented to explain the relationship between the so-called coastal complex of morabine grasshoppers in Australia (White, 1968), and which has, subsequently, been extrapolated to a variety of other situations as well. In this model

White envisages a widespread species of limited vagility generating, within its own range, a series of daughter species characterized by chromosome rearrangements that lead to reduced fecundity as heterozygotes, because of meiotic irregularities, but which for usually unspecified reasons (see, however, White 1978b, 1979) are adaptive as structural homozygotes. The daughter species are assumed to gradually extend their range at the expense of the parent species though maintaining a narrow parapatric zone of overlap at the periphery of their distributions. Within such zones, hybridization leads to the production of individuals with reduced fecundity because of irregularities stemming from hybrid meiosis.

If we examine the facts of this case, as opposed to the speculations, they are neither simple nor clear-cut. The genus *Vandiemenella* includes 12 chromosomally distinct taxa, the specific status of which is uncertain. Discussions relating to the complex have centered around the 10 coastal forms. Key (1968) is inclined to regard most of these taxa as chromosome races, but White (1978a) believes the complex is made up of five species with the composition shown in Table 2.4.

Members of the complex certainly have different karyotypes, which are presumed to have arisen from a $2n = 19$ progenitor (a karyotype that is retained in the present-day V_{19}) in the manner summarized in Figure 2.1. In this proposed phylogeny notice that what are termed "races" are distinguished by fusion differences, whereas, with one exception (P45c), what are regarded as "species" are differentiated by X chromosome inversions. In those cases where natural hybrids have been examined (Table 2.5), there is as much, if not more, of a reduction in fecundity in the

Table 2.4. *Chromosome taxa in the coastal complex of* Vandiemenella *(formerly* Moraba*)*

Species	Races
Vandiemenella viatica	V_{19}
	V_{17}
P24	XO
	$XY = \widehat{X1}/1$
	$\widehat{B6}/A$ translocation race
P25	XO
	$XY = \widehat{XB}/B$
P45b	XO
	$XY = \widehat{X6}/6$
P45c	

$V_{17} \times V_{19}$ interracial crosses as there is between either of these "races" and P24 in laboratory crosses. The most complete data refer to laboratory crosses between the more distantly related taxa P24 and P25, where it is difficult to determine the extent to which genotypic differences influence

Figure 2.1. Chromosome phylogeny in the coastal complex of the morabine grasshopper *Vandiemenella*, as envisaged by White (1978a). Compare with Table 2.4.

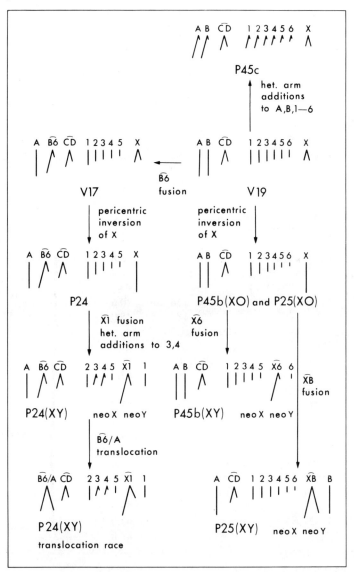

the behavior of those structural differences that exist between the hybridizing forms. Certainly there is a wide range of pairing failure in different hybrid individuals within any given cross, which suggests a not inconsiderable genotypic component. A similar argument may explain why the $\widehat{B6}$ fusion is apparently so effective in distinguishing between P24 and P25

Table 2.5. *Meiotic behavior in field and laboratory bred male hybrids of the coastal taxa of* Vandiemenella *(formerly* Moraba*)*

Cross type	Total cells analyzed	% abnormal cells Uni-valent	Mal-oriented trivalent	Reference
Natural				
(1) $V_{17} \times V_{19}$				
$F_1\male 1$	86	10.4	1.1	White et al.,
$F_1\male 2$	110	8.0	21.0	1964
(2) $V_{17} \times V_{19}$	76	26.3	2.1	Mrongovius, 1979
Laboratory synthesized				
(3) $V_{17} \times$ P24(XO)	100(2\male)	10.0	—	White et al., 1967
(4) $V_{17} \times$ P24(XY)	1171	0.8	—	
P24(XY) $\times V_{17}$	1185	2.4	—	Mrongovius, 1979
(5) $V_{19} \times$ P24(XY)				
$F_1\male 1$	156	20.6	—	
$F_1\male 2$	186	45.0	—	
(6) P24(XO) $\times$ P25(XO)				
$F_1\male 1$	200	69.0	—	
$F_1\male 2$	200	92.0	—	
$F_1\male 3$	200	67.5	—	
P25(XO) $\times$ P24(XO)				
$F_1\male 1$	115	81.7	—	
$F_1\male 2$	115	72.2	—	White et al., 1967
(7) P24(XO) $\times$ P25(XY)	100	29.0	—	
P24(XY) $\times$ P25(XY)	50	38.0	—	
P25(XY) $\times$ P24(XO)	50	82.0	29.0	
P25(XY) $\times$ P24(XY)				
$F_1\male 1$	100	99.0	—	
$F_1\male 2$	80	65.0	—	
$F_1\male 3$	50	48.0	—	

In the laboratory crosses the female parent is given first.

although it is so ineffective in distinguishing between V_{19} and V_{17}. Equivalent criticisms of White's argument were voiced by Key as long ago as 1968, and, so far as I am aware, have never been satisfactorily answered.

There is an additional complication. If the pericentric inversion of the X has indeed played an important role in speciation, then this would require that associated with this inversion there be:

1. Consistent reverse loop pairing in heterozygous individuals at meiosis
2. Regular crossing-over within such reverse loops, coupled with
3. The absence of any compensating mechanisms that might vitiate against the production of unbalanced meiotic products following crossing-over within the reverse loop

It is true that White (1961) has described a single mutant male in *Moraba scurra,* heterozygous for a spontaneous pericentric inversion, in which there was indeed crossing-over within the inverted segment in a small percentage (13%) of cases. But there is no evidence that any of the naturally occurring cases involving pericentric inversions behave in this way in morabines, although some 12% of the known species show polymorphisms for such inversions (White, 1973). In all these cases there is straight, nonhomologous, pairing of the relatively inverted segments at male meiosis and no reverse looping. Because such straight pairing precludes the production of unbalanced gametes in the heterozygotes, it also precludes them from generating hybrid sterility. Indeed there is no case that I am aware of where fixed differences involving genuine pericentric inversions do lead to reproductive isolation. Despite these facts, White (1978a) has written that heterozygotes for rearrangements in *Vandiemenella* "invariably show anomalies of synapsis which lead to the production of a certain number of aneuploid gametes and hence to a reduction in fecundity. In these instances the rearrangements are clearly potentially devisive and we can suspect that they may have played a role in speciation."

Not only is there no substantive evidence that most of the chromosome differences in question are functioning in the way White would wish them to, but the same is true of some of the other cases, interpreted by him, and others, as examples of stasipatric speciation. In some of these cases (e.g., *Diplodactylus vittatus,* King, 1977) it has not been possible to test the fertility of hybrids. In other cases the evidence offered is no less equivocal than that in *Vandiemenella.* For example, the 10 chromosome races of the phasmatid *Didymuria violescens* are distinguished by some four pericentric inversions, at least five and perhaps as many as 10 centric fusions and one tandem fusion (Craddock, 1973). White (1978a), in writing about these, states that "although the data have not been published in full the fecundity of many of the hybrids is rather severely reduced." Craddock herself tells us that "the sterility of chromosome hybrids in *Didymuria* is certainly not complete since backcross and F_2 progenies have been obtained even in the laboratory," and that "the effective reduction in the fertility of chromosome hybrids in *Didymuria* would appear to be small," a fact that is reinforced by her 1975 paper. She also comments on the interesting fact that laboratory hybrids commonly give more infertility than do field hybrids. Thus in the male F_1 laboratory hybrids that she examined the percentage asynapsis in trivalents ranged from 3.3% to 77.8%, whereas the equivalent range in field hybrids was from 0 to 15%.

I do not wish to imply that a stasipatric process might not have operated in these cases, but only that it has not been demonstrated to occur. The difficulty of dealing with situations like this is that although the present-day patterns of geographical distribution are observable and objective facts, they do not, in themselves, necessarily tell us anything at all about the mechanism of speciation or whether the forms in question indeed constitute distinct species. Because present-day patterns of distribution need not even accurately reflect past patterns, and because the events of the past are unavailable to us in detail, stasipatric speciation is a comfortable speculation that, in the absence of the relevant evidence, no one can disprove.

Molecular aspects of chromosome change

There has been a recent trend to seek answers to evolutionary problems in molecular terms; more specifically, in respect to the relationship between chromosome change and evolutionary change, in terms of the satellite DNA component, which, as was mentioned earlier, tends to characterize constitutive heterochromatin. Two aspects of this trend are of particular relevance to the present discussion.

First, there is the argument that there is an important role for satellite DNA in facilitating the evolution of new karyotypes. This is no more than an updated version, another rediscovery, of the long-held claim that heterochromatin similarly facilitates karyotype rearrangement. This was an attractive argument when euchromatin was considered an indispensable, and heterochromatin a dispensable, component of the genome. Both these assumptions are now open to question. As already mentioned, a large proportion of the unique DNA appears to be transcriptively inactive in eukaryotes and, in this sense, is no different from the repetitive DNA component that so often characterizes constitutive heterochromatin. Added to this, heterochromatin may not be as dispensable as it is often assumed to be. Hilliker (1976) has identified several mutants with known activity within the heterochromatic segments of *Drosophila melanogaster.* Loss of heterochromatic regions containing these kinds of loci can, therefore, be expected to lead to deleterious effects. Moreover, in *Clarkia,* interchange exchange points formerly assumed to be within the proximal heterochromatic segments have been more precisely located within the centromere itself (Bloom, 1974). Similarly, whereas interchange exchange points in the cockroach *Blaberus discoidalis* occur within heterochromatic segments (John and Lewis, 1959), those in the American cockroach *Periplaneta americana* do not (John and Lewis, 1958).

The foremost advocates of the argument that claims a relationship between satellite DNA and karyotype change have been Hatch and his colleagues (Hatch et al., 1976; Hatch and Mazrimas, 1977), who, from their studies in *Dipodomys,* conclude that the amount of satellite DNA is positively correlated with a predominance of biarmed chromosomes. This is taken by them to imply that a high satellite content promotes the formation of submetacentric or metacentric chromosomes. It is not entirely clear what these authors are claiming in detail. At least three kinds of change are capable of generating biarmed elements, namely, fusion, pericentric inversion, and growth of additional heterochromatic arms. Although all three types of change may well have played a role in the phylogeny of the genus *Dipodomys,* Hatch and his colleagues are unable to distinguish which of these changes have actually been involved in any particular instance. Nevertheless they argue that, most probably, the heavy satellite DNA promotes the initiation or the fixation of pericentric inversions, and this despite the fact that such inversions cannot of themselves conceivably account for all the karyotypic changes that have gone on within the group.

More telling evidence can be found in the genus *Peromyscus,* which includes some 57 different species, all of which share the same diploid number, $2n = 48$, but in which there is extensive variation in the total number of chromosome arms present (N.A. = 56–96). Using a combination of G- and C-banding, it is possible to show that at least two types of change appear to have been involved in producing this range of biarmed chromosomes, namely, pericentric inversions and probable heterochromatic arm growth (Arrighi et al., 1976; Greenbaum et al., 1978). Moreover, in at least some of the species of *Peromyscus* these heterochromatic arms have been shown to be major sites of satellite DNA (Hazen et al.,

1977). What is abundantly clear in this genus is that the biarmed chromosomes with heterochromatic arms have an unchanged G-band sequence in their euchromatin, whereas those biarmed elements that have an obviously inverted G-band sequence show an unchanged C-band pattern. Here, then, the Hatch hypothesis is patently invalid.

An especially interesting situation obtains in bovids, where autosomal constitutive heterochromatin, as revealed by C-banding, is exclusively centromeric in location. Different species have differing numbers of biarmed elements but the same total number of arms indicating a Robertsonian system, which, as is common in such cases, is presumed to depend on fusion. Whereas the amount of centric heterochromatin is generally reduced in biarmed elements relative to uniarmed members of the same complement, a phenomenon that could be argued to support a hypothesis that the presence of heterochromatin facilitates fusion, there are clear exceptions in which this is not the case. These exceptions imply that the correlations that do exist are unlikely to be causative. Thus in the barbary sheep *Ammotragus lervia* and the eland *Taurotragus oryx* there is virtually no demonstrable centromeric heterochromatin in either uni- or biarmed autosomes (Buckland and Evans, 1978). Other fusion systems known to occur with no loss of C-band material include *Mus* (Comings and Avelino, 1972) and *Ovis* (Bruère et al., 1974).

A majority of the Hawaiian species of *Drosophila* have five pairs of telocentrics and one pair of microchromosomes. The subgenus *Engiscaptomyza,* however, regularly includes a pair of metacentric fusion chromosomes that contain a large concentration of centric heterochromatin. The equivalent telocentrics of related species, by comparison, have relatively little heterochromatin (Yoon and Richardson, 1976). Indeed, considering the entire group of species, it is evident that concentrations of heterochromatin are correlated with the absence of chromosome rearrangements, the exact antithesis of the claim made by Hatch et al. Finally in the *virilis* group of *Drosophila* it is evident that not only can there be overall karyotypic constancy in the presence of large amounts of satellite DNA, but that rearrangements can occur in the virtual absence of satellite DNA (see discussion in John and Miklos, 1979). Thus one can find all kinds of correlations if one considers different groups of species, and it is not clear which, if any, of these are meaningful in a causal sense.

Second, there is the argument that satellite DNA performs some essential function in the mechanism of speciation. In part this stems from the assumed association between satellite DNA and karyotype rearrangement that was criticized in the preceding paragraphs. In part too it stems from the belief that gain of satellite DNA can be expected to create a divergence with respect to chromosome homology that may lead to a pairing failure between otherwise homologous chromosomes in hybrids and so impose a degree of reproductive isolation (Corneo, 1976; Fry and Salser, 1977). Observations of pairing behavior in cases where homologues are known to differ in satellite content make it clear that this argument too is unsound. Thus *Mus musculus molossinus* is a subspecies of mouse native to Japan and Okinawa that has 40% less light satellite DNA than *Mus musculus musculus*. Chromosomes 13, 16, and the X consistently have less C-band material than their homologues in *M. m. musculus,* whereas chromosomes 3, 5, 7, 12, 14, 15, 17, and 19 show reduced C-bands in some cases but not others. Additional C-band material is present in chromosomes 1, 2, 4, and occasionally 12. Individual pairs of homologues in these two subspecies can thus differ strikingly in heterochromatin and hence satellite content. Despite this, these homologues pair regularly in F_1 hybrids, and these hybrids are fully viable and fully fertile (Dev et al., 1975).

An extension of the argument that satellite DNA performs an essential function in speciation has recently been proposed by Jones (1978). This is based on the claim of Singh et al. (1976) that (1) there is a connection between the presence of a specific satellite DNA and the differentiation of sex chromosomes in snakes, and that (2) it is possible to see the whole process of evolution of chromosome sex differentiation represented in living species of snakes. Arguing from these two premises, Jones assumes that the mechanism of the evolution of sex differentiation is likely to contain general clues to the mechanism of evolution of species.

Such an argument rests on two fallacies. The first is that a comparison between a series of living forms can provide an unequivocal basis for constructing a phylogenetic sequence. The inadequacies of such an approach have been repeatedly demonstrated in classical comparative anatomy. They are no less formidable in comparisons involving karyotypic data. Moreover, the wider such comparisons are, the less meaningful they are likely to be. Only when dealing with groups of closely related species are we likely to gain potentially useful comparisons, and, even then, they need to be treated with considerable caution.

The second fallacy is that the isolation of a system of linked genes leading to a chromosome polymorphism is akin to the isolation of entire genomes leading to speciation. It has long been clear that the various kinds of chromosome change are available for two alternative evolutionary roles – for binding species together by balanced polymorphisms or for splitting them apart by intersterility. Because sex chromosomes form the basis of certainly the most common, often the most conspicuous, and probably the most important stable polymorphism known in animals, it is clear that this system cannot be used as a sensible model for speciation.

Conclusion

In dealing with evolutionary situations not only do we lack evidence, but it is obvious that we never will have all the evidence. This makes it more necessary than ever to carefully and objectively evaluate the evidence we do have. Charles Darwin once commented that "without speculation there is no good or original observation." Surprisingly, for such an astute observer, Darwin failed to recognize that his proposition is capable of being reversed. Whereas speculation and fact are both essential components in evaluating any biological situation, they have tended to play nonoverlapping roles insofar as they relate to the relationship between chromosome change and evolutionary change. Indeed there are still too many biologists who, in the name of hypothesis, fail to appreciate the difference between evidence and assumption, or else who are only too ready to ignore the one in favor of the other.

I have benefited from the constructive criticisms and comments of my colleagues George Miklos and John Gibson. The inaccuracies that remain, whether conceptual or factual, are mine. This contribution is dedicated to MJDW in recognition of his singular contribution to cytogenetics and the continued stimulus his work has provided for others like myself.

References

Arrighi, F. E., Stock, A. D., and Pathak, S. 1976. Chromosomes of *Peromyscus* (Rodentia, Cricetidae). V. Evidence of pericentric inversions. In *Chromosomes Today,* Vol. 5 (ed. Pearson, P. L., and Lewis, K. R.), pp. 323–9. N.Y.: Wiley.

Baker, R. J., Bleier, W. J., and Atchley, W. R. 1975. A contact zone between karyologically characterised taxa of *Uroderma bilobatum* (Mammalia: Chiroptera). *Syst. Zool. 24:*133–42.

Bloom, W. L. 1974. Origin of reciprocal translocations and their effects in *Clarkia speciosa. Chromosoma (Berl.) 49:*61–76.

Britten, R. J., and Davidson, E. H. 1971. Repetitive and non-repetitive DNA sequences and a speculation on the origins of evolutionary novelty. *Q. Rev. Biol. 46:*111–38.

Bruère, A. N. 1974. The segregational patterns and fertility of sheep heterozygous and homozygous for three different Robertsonian translocations. *J. Reprod. Fert. 41:*453–64.

Bruère, A. N., Zartman, D. L., and Chapman, H. M. 1974. The significance of the G-bands and C-bands of three different Robertsonian translocations of domestic sheep. *Cytogenet. Cell Genet. 13:*479–88.

Buckland, R. A., and Evans, H. J. 1978. Cytogenetic aspects of phylogeny in Boviidae II. C-banding. *Cytogenet. Cell Genet. 21:*64–71.

Bunch, T. D., Foote, W. C., and Spillett, J. J. 1976. Translocations of acrocentric chromosomes and their implications in the evolution of sheep (*Ovis*). *Cytogenet. Cell Genet. 17:*122–36.

Capanna, E. 1976. Gametic aneuploidy in mouse hybrids. In *Chromosomes Today,* vol. 5 (ed. Pearson, P. L., and Lewis, K. R.), pp. 83–9. New York: Wiley.

Capanna, E., Gropp, A., Winking, H., and Noak, G. 1976. Robertsonian metacentrics in the mouse. *Chromosoma (Berl.)* 58:341–53.

Capanna, E., Gristaldi, M. V., and Noak, G. 1977. New Robertsonian metacentrics in another 22-chromosome mouse population in central Apennines. *Experientia 33:*173–5.

Carson, H. L., Clayton, F. E., and Stalker, H. D. 1967. Karyotype stability and speciation in Hawaiian *Drosophila. Proc. Natl. Acad. Sci. U.S.A. 57:*1280–5.

Cattanach, B. M., and Moseley, H. 1973. Non-disjunction and reduced fertility caused by the tobacco mouse metacentric chromosomes. *Cytogenet. Cell Genet. 12:*264–87.

Chapman, H. M., and Bruère, A. N. 1975. The frequency of aneuploidy in the secondary spermatocytes of normal and Robertsonian translocation carrying rams. *J. Reprod. Fert. 45:*333–42.

Comings, D. E., and Avelino, E. 1972. DNA loss during Robertsonian fusion in studies of the tobacco mouse. *Nature New Biol. 237:*199.

Corneo, G. 1976. Do satellite DNAs function as sterility barriers in eukaryotes. *Evol. Theory 1:*261–5.

Craddock, E. M. 1973. Chromosomal evolution and speciation in *Didymuria.* In *Genetic Modes of Speciation in Insects* (ed. White, M. J. D.), pp. 24–42. Sydney: Australia and New Zealand Book Co.

– 1975. Intraspecific karyotypic differentiation in the Australian phasmatid *Didymuria violescens.* (Leach) I. The chromosome races and their structural and evolutionary relationships. *Chromosoma (Berl.)* 53:1–24.

Davidson, E. H., Klein, W. H., and Britten, R. J. 1977. Sequence organisation in animal DNA and a speculation on hnRNA as a coordinate regulatory transcript. *Dev. Biol. 55:*69–84.

Dev, V. G., Miller, D. A., Tantravahi, R., Schreck, R. R., Roderick, T. H., Erlanger, B. F., and Miller, O. J. 1975. Chromosome markers in *Mus musculus:* differences in C-banding between the subspecies *M. m. musculus* and *M. m. molossinus. Chromosoma (Berl.)* 53:335–44.

Dobzhansky, Th. 1961. On the dynamics of chromosomal polymorphism in *Drosophila.* In *Insect Polymorphism* (ed. Kennedy, J. S.), pp. 30–42. London: Roy. Ent. Soc.

Evans, E. P. 1976. Male sterility and double heterozygosity for Robertsonian translocations in mouse. In *Chromosomes Today,* vol.5 (eds. Pearson, P. L., and Lewis, K. R.), pp. 75–81. New York: Wiley.

Evans, E. P., Lyon, M. L., and Daglish, M. 1967. A mouse metacentric translocation giving a metacentric marker chromosome. *Cytogenetics 6:*105–19.

Ford, C. E., and Evans, E. P. 1973. Robertsonian translocations in mice: segregational irregularities in male heterozygotes and zygotic imbalance. In *Chromosomes Today,* vol.4 (ed. Wahrman, J., and Lewis, K. R.), pp. 387–97. New York: Wiley.

Ford, C. E., and Hamerton, J. L. 1970. Chromosome polymorphism in the common shrew *Sorex araneus. Symp. Zool. Soc. Lond. 26:*223–36.

Fry, K., and Salser, W. 1977. Nucleotide sequences of Hs-α satellite DNA from kangaroo rat *Dipodomys ordii* and characterisation of similar sequences in other rodents. *Cell 12:*1069–84.

Goldschmidt, R. 1955. *Theoretical Genetics.* Berkeley: University of California Press.

Greenbaum, I. F., Baker, R. J., and Ramsey, P. R. 1978. Chromosomal evolution and mode of speciation in three species of *Peromyscus. Evolution 32:*646–54.

Hatch, F. T., and Mazrimas, J. A. 1977. Satellite DNA and cytogenetic evolution. Molecular aspects and implications for man. In *Human Molecular Cytogenetics* (ed. Sparkes, R. W., Comings, D. E., and Fox, C. F.), pp. 395–414. New York: Academic Press.

Hatch, F. T., Bodner, A. J., Mazrimas, J. A., and Moore, D. H. 1976. Satellite DNA and cytogenetic evolution. DNA quantity, satellite DNA and karyotypic variation in kangaroo rats (genus *Dipodomys*). *Chromosoma (Berl.) 58:*155–68.

Hazen, M. W., Arrighi, F. E., and Johnston, D. A. 1977. Chromosomes of *Peromyscus* (Rodentia, Cricetidae) VIII. Genome characterisation in four species. In *Chromosomes Today*, vol.6 (ed. de la Chapelle, A., and Sorsa, M.), pp. 167–76. Amsterdam: Elsevier/North Holland Biomedical Press.

Hewitt, G. M., and John, B. 1972. Inter-population sex chromosome polymorphism in the grasshopper *Podisma pedestris* II. Population parameters. *Chromosoma (Berl.) 37:*23–42.

Hilliker, A. J. 1976. Genetic analysis of the centromeric heterochromatin of chromosome 2 of *Drosophila melanogaster:* deficiency mapping of EMS-induced lethal complementation groups. *Genetics 83:*765–82.

Hotta, Y., and Stern, H. 1979. Absence of satellite DNA synthesis during meiotic prophase in mouse and human spermatocytes. *Chromosoma (Berl.) 69:*323–30.

Hsu, T. C., and Mead, R. A. 1969. Mechanisms of chromosomal changes in mammalian speciation. In *Comparative Mammalian Cytogenetics* (ed. Benirschke, K.), pp. 8–17. Berlin: Springer-Verlag.

John, B. 1973. The cytogenetic systems of grasshoppers and locusts II. The origin and evolution of supernumerary segments. *Chromosoma (Berl.) 44:*123–46.

– 1976a. Myths and mechanisms of meiosis. *Chromosoma* (Berl.) *54:*295–325.

– 1976b. *Population Cytogenetics.* Studies in Biology no. 70. London: Edward Arnold.

John, B., and Freeman, M. 1975. Causes and consequences of Robertsonian exchange. *Chromosoma (Berl.) 52:*123–36.

John, B., and Lewis, K. R. 1958. Studies on *Periplaneta americana* III. Selection for heterozygosity. *Heredity 12:*185–97.

– 1959. Selection for interchange heterozygosity in an inbred culture of *Blaberus discoidalis. Genetics 44:*251–67.

– 1965. Genetic speciation in *Eyprepocnemis plorans. Chromosoma (Berl.) 16:*308–44.

John, B., and Miklos, G. L. G. 1979. Functional aspects of satellite DNA and heterochromatin. *Int. Rev. Cytol. 58:*1–114.

John, B., and Weissman, D. B. 1977. Cytogenetic components of reproductive isolation in *Trimerotropis thalassica* and *Trimerotropis occidentalis. Chromosoma (Berl.) 60:*187–203.

Jones, K. 1977. The role of Robertsonian change in karyotype evolution in higher plants. In *Chromosomes Today*, vol. 6 (ed. de la Chapelle, A., and Sorsa, M.), pp. 121–9. Amsterdam: Elsevier/North Holland Biomedical Press.

– 1978. Speculations on the functions of satellite DNA in evolution. *Z. Morphol. Anthropol. 62:*143–71.

Key, K. H. L. 1968. The concept of stasipatric speciation. *Syst. Zool. 17:*14–22.

King, M. 1977. Chromosomal and morphometric variation in the gekko, *Diplodactylus vittatus* (Gray). *Aust. J. Zool. 25:*43–57.

Lewis, K. R., and John, B. 1963. *Chromosome Marker.* London: Churchill.

Long, S. E. 1977. Cytogenetic examination of pre-implantation blastocysts from ewes mated to rams heterozygous for the Massey I (t_1) translocation. *Cytogenet. Cell Genet. 18:*82–9.

Lucchesi, J. C., and Suzuki, D. T. 1968. The interchromosomal control of recombination. *Annu. Rev. Genet. 2:*53–86.

Moss, J. P. 1966. The adaptive significance of B-chromosomes in rye. In *Chromosomes Today*, vol. 1 (ed. Darlington, C. D., and Lewis, K. R.), pp. 15–23. Edinburgh: Oliver and Boyd.

Mrongovius, M. J. 1979. Cytogenetics of the hybrids of three members of the grasshopper genus *Vandiemenella* (Orthoptera: Eumastacidae: Morabinae). *Chromosoma (Berl.) 71:*81–107.

Nur, U. 1977. Maintenance of a "parasitic" B-chromosome in the grasshopper *Melanoplus femur rubrum. Genetics 87:*499–512.

Ohno, S. 1974. Protochordata, Cyclostomata and Pisces. In *Animal Cytogenetics,* vol. 4, *Chordata* (ed. John, B.) p. 36. Stuttgart: Borntraeger.

Singh, L., Purdom, I. F., and Jones, K. W. 1976. Satellite DNA and evolution of sex chromosomes. *Chromosoma (Berl.) 59:*43–62.

Skinner, D. M. 1977. Satellite DNA's. *Bioscience 27:*790–6.

Smith, G. P. 1978. Non-Darwinian evolution and the beard of life. *Stadler Symp. 10:*105–18.

Stolla, R., and Gropp, A. 1974. Variation of the DNA content of morphologically normal and abnormal spermatozoa in mice susceptible to irregular meiotic segregation. *J. Reprod. Fert. 38:*335–46.

Tettenborn, U., and Gropp, A. 1970. Meiotic non-disjunction in mice and mouse hybrids. *Cytogenetics 9:*272–83.

Thoday, J. M. 1953. Components of fitness. *Soc. Exp. Biol. Symp. 7:*90–113.

– 1975. Non-Darwinian "evolution" and biological progress. *Nature (Lond.) 255:*675–7.

Timberlake, W. E. 1978. Low repetitive DNA content in *Aspergillus nidulans. Science 202:*973–5.

Valdez, R., Nadler, C. F., and Bunch, I. D. 1978. Evolution of wild sheep in Iran. *Evolution 32:*56–72.

Wahrman, J., and Goitein, R. 1972. Hybridisation in nature between two chromosome forms of spiny mice. In *Chromosomes Today,* vol. 3 (ed. Darlington, C. D., and Lewis, K. R.), pp. 228–37. London: Longman Group.

Wahrman, J., and Gourevitz, P. 1973. Extreme chromosomal variability in a colonising rodent. In *Chromosomes Today,* vol. 4 (ed. Wahrman, J., and Lewis, K. R.), pp. 399–424. New York: Wiley.

White, M. J. D. 1961. Cytogenetics of the grasshopper *Moraba scurra* VI. A spontaneous pericentric inversion. *Aust. J. Zool. 9:*784–90.

– 1968. Models of speciation. *Science 159:*1065–70.

– 1973. Speciation in the Australian morabine grasshoppers – the cytogenetic evidence. In *Genetic Modes of Speciation in Insects* (ed. White, M. J. D.), pp. 57–68. Sydney: Australia and New Zealand Book Co.

– 1978a. *Modes of Speciation.* San Francisco: W. H. Freeman and Co.

– 1978b. Chain processes in chromosomal speciation. *Syst. Zool. 27:*285–98.

– 1979. Speciation – is it a real problem? *Scientia 114:*455–80.

White, M. J. D., and Morley, F. 1955. Effects of pericentric rearrangements on recombination in grasshopper chromosomes. *Genetics 40:*604–19.

White, M. J. D., Carson, H. L., and Cheney, J. 1964. Chromosome races in the grasshopper *Moraba viatica* in a zone of geographic overlay. *Evolution 18:*417–29.

White, M. J. D., Blackith, R. E., Blackith, R. M., and Cheney, J. 1967. Cytogenetics of the viatica group of morabine grasshoppers I. The coastal species. *Aust. J. Zool. 15:*263–302.

Williams, E. E., and Hall, W. P. 1976. Primitive karyotypes. In Lizard karyotypes from the Galapagos Islands: chromosomes in phylogeny and evolution. *Breviora 441:*6–18.

Wilson, A. C., White, T. J., Carlson, S. S., and Cherry, L. M. 1977. Molecular evolution and cytogenetic evolution. In *Human Molecular Cytogenetics* (ed. Sparkes, R. W., Comings, D. E., and Fox, C. F.), pp. 375–93. New York: Academic Press.

Yoon, J. S., and Richardson, R. H. 1976. Evolution of Hawaiian Drosophilidae II. Patterns and rates of chromosome evolution in an *Antopocerus* phylogeny. *Genetics 83:*827–43.

– 1978. Rates and roles of chromosomal and molecular changes in speciation. In *The Screwworm Problem* (ed. Richardson, R. H., and Rogers, L. L.), pp. 129–43. Austin: University of Texas Press.

3 Components of genetic systems in mammals

DAVID L.HAYMAN

A comparison of the last two editions of M. J. D. White's valuable book *Animal Cytology and Evolution* effectively demonstrates the enormous increase in our knowledge of the chromosomal features of mammals. This is the result of the development of tissue culture techniques, particularly the short-term lymphocyte culture technique and the introduction of various banding and specialized staining techniques. It is now possible to distinguish the individual chromosomes in an otherwise apparently indistinguishable complement by G banding, and further to differentiate chromosome regions. In particular the differentiation of euchromatin and heterochromatin and the relationship of these regions to the type of nucleic acid present has given us a new approach to examining chromosome variation.

These techniques, however, have been almost exclusively applied to mitotic chromosomes so that our understanding of the significance of much of the information obtained from chromosome studies in mammals is consequently limited. This contrasts with the situation in insect and angiosperm cytogenetics where there is a great deal of information about meiotic events. Compared to the situation in these organisms, there is very little comparative information about the distribution and frequency of chiasmata in mammalian species. Consequently there is an absence of a most important piece of biological information which would provide further insight into the significance of the chromosome variation that exists in this group of organisms.

Plant cytogeneticists, in particular, have been able to develop the concept, and explore the implications, of the notion of the genetic system of a group of plants in a manner that is still not possible for mammals. This concept was developed by Darlington (1958) and has been amplified and illustrated by others since then (Grant, 1958; Stebbins, 1958; Lewis and John, 1963). The proposition put forward by Darlington was that the chromosomal phenotype at mitosis and meiosis is the result of the action of natural selection, and that the recombination products engendered by this phenotype provide the gene combination upon which natural selection can act. Various factors increase the release of variability, others reduce it or restrict it, and the combinations of these factors are integrated into a unit system. Some combinations act synergistically, others antagonistically. An increase in variability produced by one factor may be compensated for or added to by another factor. The level reached by the interaction of these factors was considered to be a compromise between evolutionary flexibility and immediate fitness (Mather, 1953).

Grant (1958) listed the components that could regulate the release of variability among plants as length of generation, chromosome number, crossing-over, breeding system, pollination system, dispersal potential, and population size. His survey showed groups of species in which there were combinations of antagonistic or of complementary factors. As a generalization, cross-fertilizing species tend to have lower chiasma frequencies than self-fertilizing species, and perennials tend to have lower chiasma frequencies than annuals. Thus an increase in variation from one source could be balanced by a reduction from another. Additional examples of interactions of a complex kind are found in the work of Rees and his colleagues (Rees and Hutchinson, 1973; Rees and Dale, 1974) with particular emphasis again on the interaction of chiasma frequency with other factors.

It is the purpose of this necessarily brief account to point out the lack of knowledge of genetic systems in mammals, to indicate areas where something is known, and to indicate situations that would be worth exploring. Because chiasmata are the cytological manifestation of genetic exchange at meiosis (Peacock, 1970), they are a sensitive measure of one of the components of the genetic system. They provide one measure by which we may gauge the comparative recombination of a population. It is recognized that it is not always easy to obtain the quality of meiotic preparations necessary to engender confidence in their interpretation. The "standard" techniques of Evans et al. (1964) and Meredith (1969) often have to be modified for individual species. Further, the determination of the number of chiasmata is only part of the necessary information; it is also important to examine the distribution of chiasmata because this may have profound effects upon the release of variability. The most striking technical advance in this area is that of Shaw (1976), who has computerized the determination of the position of chiasmata and so revealed variation at a very fine level of resolution.

An illustration of what can be done is the study of chiasmata in man by Hultén (1974). Using combined orcein Q- and C-banding of meiotic chromosomes in a healthy (but aged) male, she showed, among other features, that there was chiasma interference along the chromosome arm but not across the centromere; chiasmata were usually distal and terminal rather than medial; and there was no competition for chiasmata between the different autosomes in the same cell. With this work as a basis it will be possible to make comparisons between man and the related primates with similar G-banding chromosomes (reviewed by Dutrillaux, 1979).

There are few genetic studies on chiasmata in mammals. Kyslikova and Forejt (1972) have shown that there is a significant difference in chiasma frequencies between the males of three inbred lines of the house mouse, and de Boer and Van den Hoeven (1977) have obtained heritability estimates of the capacity to form chiasmata in mice heterozygous for an interchange. Because genetic control of the features of chiasmata has been well documented, for example, in grasshoppers (Shaw, 1972, 1974) and plants (Rees, 1961; Riley, 1974), there is no reason to suppose that this will not be found in mammals.

There is also a lack of much of the necessary ecological data about mammalian species, particularly of the patterns of reproduction (Asdell, 1964). It is necessary to know the longevity of the species, the age of sexual maturity, the manner in which mates are chosen such that inbreeding can be estimated, the numbers of young born at each conception, the reproductive potential of a species, and a measure of its dispersal, because all of these could interact with chiasmata in the genetic system. Because the compilation of relevant information is necessary before patterns can be detected, it is inevitable that progress will be uneven. The analysis of genetic systems usually poses the same problem as most studies in natural selection; namely, the initial identification of how the variation is brought about presents fewer difficulties than does the subsequent identification of the adaptive role or roles such variation plays.

The most obvious genetic system in mammals is the existence of separate sexes, which Darlington has referred to as a two-track heredity because such separation has the potential for a difference between the sexes with respect to the frequency and location of chiasmata. There is good evidence for this in the house mouse, where Polani (1972) reports that both the relative positions of chiasmata and the total number differ between the sexes, the female having a higher chiasma frequency. Speed (pers. comm.) finds a higher chiasma frequency in the female marmoset (*Callithrix jacchus*) than in the male. Extrapolation from the genetic evidence compiled by Dunn and Bennett (1967) suggests that this is a common feature of mammals. Those situations where recombination between particular genes is higher in the male than in the female are suggested by Lyon (1976) to be the result of a difference in chiasma localization between the sexes. Not only may the features of chiasmata differ between the sexes, but so may their response to the effects of age. Speed (1977) finds a reduction in chiasma frequency in the female but not in the male mouse as age increases. There appears to be no change in chiasma frequency in the male with increasing age in man (Lange et al., 1975). The significance of this latter source of variation is not clear.

The existence of these differences between the sexes means that the assessment of the role of chiasma variation in any species may require information from both sexes. The determination of the features of meiosis in the females is technically more difficult than it is in males (Tarkowski, 1966; Payne and Jones, 1975); even so, it is worth emphasizing that the considerable success in studies with plants referred to earlier comes largely from an examination of pollen mother cells alone.

The sex chromosome system in mammals so far as it determines sex appears to be largely conserved; the Y chromosome is male-determining (reviewed by Short, 1979), and X-linked genes in one species are usually X-linked in others (Ohno, 1967). The usual consequence of this XX/XY sex chromosome system is approximate equality of the sexes at sexual maturity. The evolutionary advantage of such equality has been considered by Fisher (1930) and Bodmer and Edwards (1960). It is of great interest, therefore, that in at least two (and possibly more) species of lemming there exist modifications of the usual sex chromosome system that are related to the excess of females observed in natural populations. Fredga and his colleagues (Fredga et al., 1976, 1977) find that in *Myopus schisticolor* Lilljeborg there are two types of female, those producing female progeny only and those producing progeny of both sexes. Cytologically there are two sorts of female, XX and XY, whereas all males are XY. The X chromosome in XY females was postulated to be a mutant type that inactivates the male-determining capacity of the Y chromosome and accounts for the cells from this type being *H-Y* antigen negative (Wachtel et al., 1976). This mutant X chromosome has been shown to have a different G-banding pattern from the normal X chromosome, and this difference has enabled the demonstration of XX females heterozygous for the mutant X chromosome (Herbst et al., 1978). Such females produce three female progeny to one male. XY females are believed to produce only X-bearing eggs by nondisjunction of the X chromosome and elimination of the Y chromosome, forming an XX germline. A similar system apparently exists in *Dicrostonyx torquatus* Pall. (Gileva and Chebotar, 1979), where again there are two sorts of female, one producing all female offspring and the other progeny of both sexes. Again the aberrant female is cytologically like the normal male (the authors confusingly call these XO but there is no difference in chromosome number between the sexes), and a mutant X with Y suppressive properties is postulated. Although not all the details of these unusual variants of the normal situation have been worked out, this does seem to be a sex chromosome system that has adaptive significance with respect to population numbers.

Both groups of workers relate the occurrence of the mutant X chromosome and the polygamous habits of the lemming to the periodic rapid increase in numbers found in the wild in both species. Whether there are differences in the features of chiasmata associated with the occurrence of two types of X chromosome is not known. There is an additional variable in that supernumerary chromosomes are found commonly in *Dicrostonyx*. As is discussed later, these chromosomes may have a role in regulating features of chiasmata, and the analysis of their effects in this species will be important. It is possible that combined with a system that is related to a rapid increase in numbers, there is another system that affects the variability found in the population.

There are a number of mammalian species in which the X and/or Y chromosomes have been modified by exchanges with autosomes without changes in the sex ratio of the species, and with no lethality in any zygotic combinations. One such relatively common variation is a chromosomal fusion uniting the X chromosome and an acrocentric autosome, leading to an $XX♀$ and an $XY_1Y_2♂$ (where the Y_2 is the free autosome, found only in the male). At meiosis in the male a trivalent is formed, one element of which involves pairing and exchange between the originally autosomal arms, and the other, a nonchiasmate association between the original X and Y chromosomes; the problem is to account for the success of this derived form of sex-chromosome compound over the original complement. One obvious consequence of this X autosome fusion is that pairing conditions of the X-autosomal compound chromosome differ between the sexes because in the female a homomorphic bivalent occurs with chiasmata capable of being formed all along the chromosome. Either as a result of this structural difference or because of some inherent difference between the sexes there may be a difference in the location of chiasmata in the originally autosomal arm. If such occurred, it might provide the differential for the success of this multiple sex chromosome system. The localization of chiasmata away from the centromeric region would lead to sex-limited inheritance of this region. In any case this system allows the exploitation of partial sex linkage of the type first inferred by Koller and Darlington (1934). There are no species known that are polymorphic for this system; so the only test for these arguments would be a comparison of chiasma frequency and distribution in the two sexes. Without belaboring the point, the other forms of X-autosome or Y-autosome interchanges leading to multiple sex chromosome systems such as $X_1X_1X_2X_2♀/X_1X_2Y♂$ would also repay investigation from this viewpoint. The system of this type described by Matthey (1965), in which different but related forms of a complex multiple sex chromosome system occur within the species complex *Mus minutoides,* offers the opportunity to make a comparison between closely related species (or perhaps subspecies). There is evidence for such an effect in the observations of John and Freeman (1975), who identified a metacentric chromosome derived by fusion of two rod chromosomes in the grasshopper *Percassa rugifrons.* In organisms of this sort chiasma frequency is a function of chromosome length. The metacentric chromosome has a different chiasma frequency and distribution from that expected on the simple addition of its individual component arms. In this species the authors produce evidence for the fusion being between centromeres, and enumerate the range of different fusion processes possible. Whether the products of these different fusion processes result in the same consequence upon chiasma pattern is not known.

A most elaborate genetic system has been described in all three extant genera of monotremes (Murtagh, 1977). In the echidnas *Tachyglossus* and apparently *Zaglossus* there is an $X_1X_1X_2X_2♀/X_1X_2Y♂$ sex chromosome system and six morphologically unpaired autosomal elements. A multivalent chain forms at meiosis involving the sex chromosomes and these unusual autosomes, and disjunction is alternate. No homozygote for these autosomes is found; presumably their formation or survival is prevented. Such a complex interchange system maintains a substantial portion of the genome in a state of permanent heterozygosity. In addition, in *Tachyglossus* polymorphism for chromosome size is present in the six largest autosomes, which introduces a further variable degree of heterozygosity. A similar system of permanent heterozygosity is found in the platypus *Ornithorynchus* without apparently the additional variable heterozygosity. Such elaborate systems must have a long evolutionary (i.e., selective) history, but their adaptive significance in these different species is not known.

Many closely related mammalian species differ in chromosome number with consequent alteration in the number of linkage groups. The development of G-banding has enabled the chromosomal changes between these species to be accurately defined in terms of chromosomal regions. The simplest of these changes to analyze are those in which whole arm fusions or fissions have occurred, and there are many groups of species in which these changes have been identified. This diversity of chromosome form found among related species is a constant reminder that we do not have any explanation at present for the adaptive role of this character. White (1975) has drawn attention to the tendency of similar structural changes to establish themselves in one member of the karyotype after another, which he has called "karyotypic orthoselection." He suggests that one of the reasons for this is that similar rearrangements may have similar effects upon the phenotype. It is possible that one form of generalized effect would be that on chiasma frequency and distribution.

G-banding studies have shown that morphologically similar metacentric or submetacentral chromosomes in different but related species may be derived from the fusion of different combinations of acrocentric chromosomes (e.g., Bickham and Baker, 1977; Rofe, 1979). Comparative studies of the features of chiasmata in such species would be most informative. The observations of Cattanach and Moseley (1973) on *Mus* are relevant in this context. *Mus poschiavinus* (2n = 26) has seven metacentric chromosomes, which are represented in *Mus musculus* (2n = 40) by 14 acrocentric chromosomes. Intercrossing between these two species has enabled the breeding of mice individually homozygous and heterozygous for each of the different metacentric chromosomes in an otherwise acrocentric house mouse complement. The effects of nondisjunction were detected at MII in males and inferred from zygote death in females. Cattanach and Moseley attribute the variation that occurred in these metrics between the mice homozygous and heterozygous for the same metacentric and for different metacentrics as due to genetic background rather than to structural heterozygosity per se. In other words, a chromosomal pattern of association and orientation of a trivalent has been introduced into a genetic background that is not adapted to it. The authors report a suppression of crossing-over in the centromeric regions in heterozygotes between the tobacco-mouse metacentrics and house mouse acrocentrics, indicating an alteration in the distribution of chiasmata. The complementary study in which limited numbers of acrocentrics from *Mus musculus* are introduced into a *Mus poschiavinus* complement would be expected on this hypothesis to show similar disharmonies. Certain of the examples of karyotypic orthoselection might be expected to follow from or to be favored by a genetic modification that accommodated the pairing and disjunctional modifications of an initial alteration in chromosome number and morphology. The system could become "preadapted" to accommodate subsequent changes.

The presence of polymorphisms for chromosome fusions is well documented in mammalian populations, but there is little information about their consequences upon chiasmata. White (1958) has pointed out that a trivalent consisting of a metacentric and two acrocentrics would be at a mechanical advantage when disjunction occurred at anaphase I if the chiasmata were distal rather than proximal. Where there are three chromosomal phenotypes possible in a population, there may well be three associated patterns of chiasma localization.

Such a situation has been described by White and Morley (1956) in grasshopper populations polymorphic for a structural change – probably an inversion. In these situations there is a clear parallel to the Schultz-Redfield effect (Schultz and Redfield, 1951), and there are many opportunities in mammalian populations to examine whether this effect is present and to study the role it may play. Two examples in particular illustrate the point. Yonenaga (1972) and Yonenaga-Yassuda (1979) described heteromorphism for pericentric inversions in several autosomes of the chromosomes of animals from a number of populations of *Akodon arviculoides*. The data available on chiasmata are insufficient at present to allow any conclusions to be drawn. The especially favorable cytological features of this species with 2n = 14 and the possibility of interaction between the different heteromorphisms make this a potentially informative situation. Patton (1970) reports a clear example of clinal variation in chromosome morphology in the pocket gopher (*Thomomys bottae grahamensis*), in which acrocentrics replace metacentrics as the altitude increases. This phenomenon is interpreted by Patton as the result of selection. The part played by chiasma frequency and distribution in this variation is not known as yet, and may be a significant feature of the situation.

A similar comment may be made about the changes that occur in chromosome morphology as a result of fluctuations in the amount of heterochromatin present. This material may be localized in certain chromosome regions or make up nearly all of a chromosome arm. Natarajan and Gropp (1971) have demonstrated the limitation of chiasmata to regions away from the large fixed heterochromatic blocks present in the hedgehog. The regulation of the chiasma frequency and distribution by the presence of heterochromatic regions in a variety of grasshopper species (John, 1973; Miklos and Nankivell, 1976) provide further pointers to the sorts of questions that could be asked in mammalian species; for example, in *Rattus rattus,* in which extensive polymorphism for C-banding regions has been reported (Yosida and Sagai, 1975; Raman and Sharma, 1974). Arnason et al. (1978) have favored an explanation of this sort to account for the C-banding variation found in the otherwise conserved complements of species of whales, but unfortunately with no supporting data. The study by Novotna and Forejt (1975) of variation in centric heterochromatin in mouse reported no difference in the chiasma frequency about the centromere between homozygotes and their heterozygotes varying in the amount of centric heterochromatin, but they do not comment on chiasma distribution. There is, of course, no reason to suppose that the effect of heterochromatin is always the same; and there is a real shortage of data to test ideas like this.

Another means of increasing the amount of chromatin in the nucleus without perhaps any restriction of its effect to specific chromosomes is the polymorphism found in some populations of mammals for supernumerary or B chromosomes. In plants and grasshoppers these chromosomes have been shown to modify the frequency of chiasmata, in some cases by increasing and in other cases decreasing it. Patton (1977) has examined in detail the effects of one of the three morphological types of B chromosomes present in the North American pocket mouse *Perognathus baileyi*. He finds that the presence of these chromosomes is associated with an overall increase in the chiasma frequency of the regular set of chromosomes. The effect is not uniformly additive as the number of supernumerary chromosomes increases but is optimized at a particular frequency – a clear case of regulation. Although the increase in chiasmata is small – rather less than one per cell, Patton is perhaps unwise to suggest that it is insufficient for biological purposes. No data are given of the variance in mean chiasma frequency per cell associated with varying doses of supernumerary chromosomes, which could well be significant.

This species shows the interaction of a number of factors because the three types of supernumeraries observed provide variation at the population level. No population contains all three, but usually two types are found in a population. It will be of considerable interest to establish if the other supernumeraries have any effect on chiasma frequency, and whether interaction occurs between combinations of the supernumerary types. The distribution of supernumeraries is nonrandom over the species range, being highest in the southern populations, where records suggest that the species is extending its range. It may well be that the action of B chromosomes in increasing the number of chiasmata is an advantage to this species in such extension. A further potential system in which interactions could be examined in relation to chiasmata appears to exist in *Rattus rattus*. In addition to the C-band polymorphisms referred to earlier, Yosida (1977) has identified supernumerary chromosomes in three geographic variants of this species. Is this variation in the source of this additional, presumably noncoding, DNA an example of different ways of doing the same thing?

We do not know whether there are changes in chiasma frequency or location when a species is extending its range or where species have been subdivided into isolated populations such as those found on islands. Searle et al. (1970) report that wild male mice from the population of Skokholm have lower chiasma frequencies than inbred laboratory mice, suggesting that data from a wider sample of island populations could be of interest.

It is also possible that laboratory mice and other domesticated mammals such as the various breeds of dogs and cats and the hamster have different chiasma features relative to those found in the equivalent feral representatives because of the intense selective history the former have had. The presence of a higher chiasma frequency in highly selected populations of several outbreeding plant species compared to their less intensively selected related populations has been shown by Rees and colleagues (e.g., Rees and Dale, 1974). They interpret this situation as a consequence of the selection for extreme phenotypes generated by individuals with altered chiasma frequency and distribution, and thus by selection for these altered cytological features. While it is also possible that such altered chiasma frequencies are "compensatory" in populations with a depleted variability, Rees and Dale argue that they are basically relics of the cytological processes on which they were founded.

Among the rodents, in particular, many species are characterized by reproductive traits that contribute to high intrinsic rates of natural increase – such as short gestation periods, multiple litters per annum, large litter sizes, and early sexual maturity. These are features of so-called r-selected species which exhibit high-amplitude population fluctuations and are in direct contrast to large mammals in particular or small tropical species, which appear not to have these features. What role do chromosome number and chiasma frequency play in these contrasting patterns? It may be that a high inherent capacity to increase in number is balanced by a reduced capacity of the chromosome complement to increase the genetic variability. Kirkland and Kirkland (1979) report that in the same area of New York State there was a pattern of considerable fluctuation in numbers for the small rodents and nonhibernating animals and of stability in numbers for three species of hibernating animals. Comparisons of the chromosome number and chiasmata in such groups of species may enable a general pattern to be detected.

A further potentially valuable group of species for studies of these problems is that in which there is little variation in chromosome number or morphology, but a substantial variation in reproductive traits. On the hypothesis that such variation would be reflected by variations in the genetic system, the major variable in that system would be the features of chiasmata. The marsupials are one group in which such a comparison may be made because the same chromosome number and very similar morphology are found in representatives of the American and Australian super-families (Hayman and Martin, 1974). The same G-banding pattern has been conserved in Australian species with these morphologically similar chromosomes (Rofe, 1979). A study of the features of chiasmata in this group has begun, and the data presently available on frequency are summarized in Table 3.1. There is an almost twofold variation over the range of species. There is no apparent relationship between chiasma frequency and either DNA value (which is positively correlated with the amount of centromeric C-banding material) or body weight (which is negatively correlated with longevity and litter size). The work is in progress.

Table 3.1. *Number of chiasmata at diplotene in species of marsupials with 2n = 14, based on 20 cells per individual male.*

Species	Sample size	Chiasmata per cell	Relative DNA content	Size of animal (kg)
Perameles gunni	2	12.6 ± 0.44	ca. 129	0.65–0.75
Isoodon macrourus	1	16.6 ± 0.44	130	0.8–1.3
Isoodon obesulus	1	16.6 ± 0.32	129	0.7–1.0
Sminthopsis crassicaudata	1	16.9 ± 0.32	100	0.012–0.018
Dasyuroides byrnei	2	17.5 ± 0.44	100	ca. 0.1
Dasyurus viverinus	1	18.0 ± 0.34	100	ca. 1
Sarcophilis harrisii	1	21.3 ± 0.29	102	8–9
Lasiorhinus latifrons	1	22.8 ± 0.36	115	20–28

Source: Unpublished data of P. J. Sharp.

It should be emphasized that the desirability of knowledge of the diversity of chiasma patterns in mammals as well as of chromosome number, morphology, and banding patterns is not urged in the belief that it will solve all the problems posed by evolutionary phenomena. It is clear, however, that the potential value of many of the studies conducted on the diversity of mammalian species will not be realized unless this additional information is available and collated.

References

Arnason, U., Purdom, I. F., and Jones, K. W. 1978. Conservation and chromosomal localisation of DNA satellites in Balenopterid whales. *Chromosoma (Berl.)* 66:141–59.

Asdell, S. A. 1964. *Patterns of Mammalian Reproduction.* London: Constable.

Bickham, J. W., and Baker, R. J. 1977. Implications of chromosomal variation in Rhogeesa (Chiroptera: Vespertilionidae). *J. Mammal.* 58:448–53.

Bodmer, W. F., and Edwards, A. W. 1960. Natural selection and the sex ratio. *Ann. Eugen.* 24:239–44.

de Boer, P., and van den Hoeven, F. A. 1977. Son sire regression based heritability estimates of chiasma frequency using T70H mouse translocation heterozygotes, and the relation between univalence, chiasma frequency and sperm produced. *Heredity* 39:335–43.

Cattanach, B. M., and Moseley, H. 1973. Non-disjunction and reduced fertility caused by tobacco mouse metacentric chromosomes. *Cytogenet. Cell Genet.* 12:264–87.

Darlington, C. D. 1958. *The Evolution of Genetic Systems*, 2nd ed. Edinburgh: Oliver and Boyd.

Dunn, L. C., and Bennett, D. 1967. Sex differences in recombination of linked genes in animals. *Genet. Res.* 9:211–20.

Dutrillaux, B. 1979. Chromosomal evolution in Primates: tentative phylogeny from *Microcebus murinus* (prosimian) to man. *Hum. Genet.* 48:251–314.

Evans, E. P., Breckon, G., and Ford, C. E. 1964. An air drying method for meiotic preparations from mammalian testes. *Cytogenetics* 3:289–94.

Fisher, R. A. 1930. *The Genetic Theory of Natural Selection.* Oxford: Clarendon.

Fredga, K., Gropp, A., Winking, H., and Frank, F. 1976. Fertile XX- and XY-type females in the wood lemming *Myopus schisticolor. Nature (Lond.)* 261:225–7.

– 1977. A hypothesis explaining the exceptional sex ratio in the wood lemming (*Myopus schisticolor*). *Hereditas* 85:101–4.

Gileva, E. A., and Chebotar, N. A. 1979. Fertile XO males and females in the varying lemming, *Dicrostonyx torquatus* Pall. (1779). *Heredity* 42 (1):67–77.

Grant, V. 1958. Regulation of recombination in plants. *Cold Spring Harbor Symp. Quant. Biol.* 23:337–63.

Hayman, D. L., and Martin, P. G. 1974. Monotremata and Marsupialia. Mammalia 1. In *Animal Cytogenetics,* vol. 4, *Chordata* 4 (ed. B. John). Stuttgart: Borntraeger.

Herbst, E. W., Fredga, K., Frank, F., Winking, G., and Gropp, Q. 1978. Cytological identification of two X-chromosome types in the wood lemming (*Myopus schisticolor*). *Chromosoma (Berl.)* 69:185–91.

Hultén, M. 1974. Chiasma at diakenesis in the normal human male. *Hereditas* 76:55–78.

John, B. 1973. The cytogenetic systems of grasshoppers and locusts. II The origin and evolution of supernumerary segments. *Chromosoma (Berl.)* 44:123–46.

John, B., and Freeman, M. 1975. Causes and consequences of Robertsonian exchange. *Chromosoma (Berl.)* 52:123–36.

Kirkland, G. L., and Kirkland, C. J. 1979. Are small mammal hibernators K-selected? *J. Mammal.* 60:164–8.

Koller, P. C., and Darlington, C. D. 1934. The genetical and mechanical properties of sex chromosomes, 1. *Rattus norvegicus* ♂. *J. Genet. 29:*159–73.

Kyslikova, L., and Forejt, J. 1972. Chiasma frequency in three inbred strains of mice. *Folia Biol. (Prague) 18:*216–18.

Lange, K., Page, B. M., and Elston, R. C. 1975. Age trends in human chiasma frequencies and recombination fractions. 1. Chiasma frequencies. *Am. J. Hum. Genet. 27:*410–18.

Lewis, K., and John, B. 1963. *Chromosome Marker*. London: Churchill.

Lyon, M. F. 1976. Distribution of cross-over in mouse chromosomes. *Genet. Res. 28:*291–9.

Mather, K. 1953. The genetical structure of populations. *Symp. Soc. Exp. Biol. 7:*66–95.

Matthey, R. 1965. Un type nouveau de chromosomes sexuels multiples chez une souris Africaine du groupe *Mus (Leggada) minutoides* (Mammalia – Rodentia). *Chromosoma (Berl.) 16:*351–64.

Meredith, R. 1969. A simple method of preparing meiotic chromosomes from mammalian testis. *Chromosoma (Berl.) 26:*254–8.

Miklos, G. L. G., and Nankivell, R. N. 1976. Telomeric satellite DNA functions in regulating recombination. *Chromosoma (Berl.) 56:*143–67.

Murtagh, C. E. 1977. A unique cytogenetic system in monotremes. *Chromosoma (Berl.) 65:*37–57.

Natarajan, A. T., and Gropp, A. 1971. The meiotic behaviour of autosomal heterochromatic segments in hedgehogs. *Chromosoma (Berl.) 35:*143–52.

Novotna, B. and Forejt, J. 1975. Polymorphism of centromeric heterochromatin and its relation to chiasma formation in the paracentromeric regions of murine chromosomes. *Folia Biol. (Prague) 21:*1–7.

Ohno, S. 1967. *Sex Chromosomes and Sex Linked Genes*. Berlin: Springer-Verlag.

Patton, J. L. 1970. Karyotypic variation following an elevational gradient in the pocket gopher *Thomomys bottae grahamensis* Golman. *Chromosoma (Berl.) 31:*41–50.

– 1977. B chromosome systems in the pocket mouse *Perognathus baileyi* Merriam. *Chromosoma (Berl.) 60:* 1–14.

Payne, H. S., and Jones, K. P. 1975. in an appendix to Brewen, J. G., Payne, N. S., Jones, K. P., and Preston, R. J. Studies on chemically induced dominant lethality. 1. The cytogenetic basis of MMS-induced dominant lethality in post-meiotic male germ cells. *Mutat. Res. 3:*239–50.

Peacock, W. J. 1970. Replication recombination and chiasmata in *Goniaea australasiae* (Orthoptera–Acrididae). *Genetics 65:*593–617.

Polani, P. E. 1972. Centromere localisation at meiosis and the position of chiasmata in the male and female mouse. *Chromosoma (Berl.) 36:*343–74.

Raman, R. and Sharma, T. 1974. DNA replication G and C bands and meiotic behavior of supernumerary chromosomes in *Rattus rattus* (Linn). *Chromosoma (Berl.) 45:*111–19.

Rees, H. 1961. Genotypic control of chromosome behaviour. *Bot. Rev. 27:*288.

Rees, H., and Dale, P. J. 1974. Chiasmata and variability in *Lolium* and *Festuca* populations. *Chromosoma (Berl.) 47:*335–51.

Rees, H., and Hutchinson, J. 1973. Nuclear variation due to B chromosomes. *Cold Spring Harbor Symp. Quant. Biol. 38:*175–82.

Riley, R. 1974. Cytogenetics of chromosome pairing in wheat. *Genetics 78:*193–203.

Rofe, R. 1978. G-banded chromosomes and evolution of Macropodidae. *Aust. J. Mammal. 2:*53–63.

– 1979. G-banding and chromosome evolution in Australian marsupials. Unpublished Ph.D. thesis, University of Adelaide.

Schultz, J. and Redfield, H. 1951. Interchromosomal effects on crossing-over in *Drosophila*. *Cold Spring Harbor Symp. Quant. Biol. 16:*175–98.

Searle, A. G., Berry, R. J., and Beechey, C. V. 1970. Cytological radiosensitivity and chiasma frequency in wild living male mice. *Mutat. Res. 9:*137–40.

Shaw, D. D. 1972. Genetic and environmental components of chiasma control. II The response to selection in *Schistocerca*. *Chromosoma (Berl.) 37:*297–308.

– 1974. Genotypic and environmental components of chiasma control. III Genetic analysis of chiasma frequency variation in two selected lines of *Schistocerca gregaria*. *Chromosoma (Berl.) 46:*365–74.

– 1976. Comparative chiasma analysis using a computerised optical digiter. *Chromosoma (Berl.) 59:*103–27.

Short, R. V. 1979. Sex determination and differentiation. *Br. Med. Bull. 35:*121–8.

Speed, R. M. 1977. The effects of ageing on the meiotic chromosomes of male and female mice. *Chromosoma (Berl.) 64:*241–54.

Stebbins, G. L. 1958. Longevity, habit and release of genetic variability in higher plants. *Cold Spring Harbor Symp. Quant. Biol. 23:*365–77.

Tarkowski, A. K. 1966. An air-drying method of chromosome preparations from mouse eggs. *Cytogenetics 5:*394.

Wachtel, S., Koo, S. G. C., Ohno, S., Gropp, A., Dev, V. G., Tantravahi, R., Miller, D. A., and Miller, O. J. 1976. *H-Y* antigen and the origin of XY female wood lemmings (*Myopus schisticolor*). *Nature (Lond.) 264:*638–9.

White, M. J. D. 1958. Restrictions on recombination in grasshopper populations. *Cold Spring Harbor Symp. Quant. Biol. 23:*307–17.

– 1975. Chromosome repatterning – regularities and restrictions. *Genetics 79:*63–72.

White, M. J. D., and Morley, F. H. W. 1956. Effects of pericentric rearrangements on recombination in grasshopper chromosomes. *Genetics 40:*604–18.

Yonenaga, Y. 1972. Chromosomal polymorphism in the rodent (*Akodon arviculoides* spp.) (2n = 14) resulting from two pericentric inversions. *Cytogenetics 11:*488–99.

Yonenaga-Yassuda, Y. 1979. New karyotypes and somatic and germ cell banding in *Akodon arviculoides* (Rodentia, Cricetidae). *Cytogenet. Cell Genet. 23:*241–9.

Yosida, T. H. 1977. Supernumerary chromosomes in the black rat (*Rattus rattus*) and their distribution in three geographic variants. *Cytogenet. Cell Genet. 18:*149–59.

Yosida, T. and Sagai, T. 1975. Variation of C bands in the chromosomes of several subspecies of *Rattus rattus*. *Chromosoma (Berl.) 50:*283–300.

4 Sex chromosomes, parthenogenesis, and polyploidy in ticks

JAMES H. OLIVER, JR.

Well over 30,000 species and 1700 genera of acarines have been described, and it is believed that more than half a million more species are living today (Krantz, 1978). These species have been grouped into the subclass Acari, which represents a polyphyletic group of organisms obviously sharing many similarities, but also differing a great deal. Mites and ticks are the common names given to these animals, but the reader should not interpret this nomenclature as reflecting a true taxonomic dichotomy (i.e., many families of mites are much more closely related to ticks than they are to other families of mites). Ticks are grouped into one suborder Metastigmata, which is comprised of three families. Approximately 800 species of ticks have been described, only a few of which have been studied cytogenetically. Chromosome data are available on approximately 79 species of hard ticks (Ixodidae) and 25 species of soft ticks (Argasidae), and no cytogenetic data are available for the Nuttalliellidae. Even though most tick species have not been studied cytogenetically, there is a significant amount of information available, which indicates that various sex chromosome systems are operative and that different reproductive mechanisms exist. The data that are discussed in this paper deal with the Argasidae and the two major groups of the Ixodidae, Prostriata (Ixodinae) and Metastriata (Amblyommatinae). Only one genus, *Ixodes,* comprises the Prostriata, whereas the Metastriata contains several genera.

Sex chromosomes

Data on sex chromosomes from species representing the major genera of ticks indicate that males are the heterogametic and females the homogametic sex (Table 4.1). In a few species the males have not yet been demonstrated to be heterogametic, but it seems probable that they are, and that the cytological differences between the sex chromosomes are so slight that they have not been noticed.

The major sex chromosome systems described in other animals seem to be present in ticks. The XX–XO chromosome system is most prevalent, the XX–XY mechanism is common in some taxa, and multiple sex chromosomes are known from two Australian species. The XY system is the only type thus far described among the Argasidae, and is the dominant sex chromosome mechanism among the Prostriata.

Sex chromosomes are known from approximately 69 species of Metastriata, and all are XO except for four cases of XY and two of multiple sex chromosomes. The XY species include *Amblyomma darwini* from the Galapagos Islands, the Brisbane Australian population of *Amblyomma*

Table 4.1. *Chromosomes in ticks*

Genera	No. of species karyotyped	Ranges of chromosome numbers	Male sex chromosomes
Argasidae			
Argas	11	20–26	XY
Ornithodoros	12	12–34	XY
Otobius	2	20	Presumed XY
Ixodidae			
(Prostriata)			
Ixodes	10	23–28	XY in 7 species
Ixodidae			
(Metastriata)			
Amblyomma	19	19–22	XO in 14 species
			XY in 3 species
			X_1X_2Y in 2 species
Aponomma	4	17–21	XO
Boophilus	3	21–22	XO
Dermacentor	10	20–22	XO in 9 species
			XY in 1 species
Haemaphysalis	14	19–22	XO in 13 species
			XY in 1 species
	1	30–35	Parthenogenetic
Hyalomma	12	21–22	XO
Rhipicephalus	5	21–22	XO
	1	Possibly 24	

References to data are too numerous to include in this summary table. Contact author if specific references are needed.

moreliae (Oliver and Bremner, 1968), an undescribed species of *Dermacentor*, near *taiwanensis*, from Japan (Oliver and Tanaka, unpub.), and *Haemaphysalis hystricis* from Japan (Oliver et al., 1974). Multiple sex chromosomes prevail in *Amblyomma limbatum* and a Sydney, Australia population of *A. moreliae*, whose males possess X_1X_2Y chromosomes (Oliver, 1965; Oliver and Bremner, 1968). Among the Prostriata all males are XY except *Ixodes holocyclus*, which is XO (Oliver and Stone, unpub.).

Morphological characters, life cycle data, and observations on reproductive biology all suggest that the soft ticks (Argasidae) are more primitive than the hard ticks (Ixodidae). It is also generally agreed that species of *Ixodes* (Prostriata) represent products of an evolutionary line originating prior to the lines from which other hard ticks (Metastriata) evolved. Chromosome data (karyotypes and sex chromosome mechanisms) of ticks, although neither abundant nor without exceptions, generally support these phylogenetic conclusions.

The argasids have a wider range and greater number of chromosomes than the Prostriata and Metastriata (Table 4.1). Their numbers range from 12 to 34, suggesting that some argasid species have been conservative and retained some of the chromosome characteristics of their ancestors, whereas others have specialized a great deal. Other aspects of the biology of these species also suggest the same thing. *Ixodes* species, in general, also have a high number. The 2n range is from 24 to 28, which agrees with the hypothesis that *Ixodes* branched off from the ancestral line as indicated above. Finally, chromosome numbers of the Metastriata are lower and have a narrower range, in general, than the other two groups. Most species possess a diploid number of 20 + XX in females and 20 + X in males, although *Aponomma hydrosauri* from Australia has 16 + XX and 16 + X (Oliver and Bremner, 1968).

More often than not, chromosome numbers per species have tended to decrease as evolution and specialization within a taxon occurred. Although there are examples supporting chromosome number increase with specialization in some taxa, reduction of chromosome numbers appears to more often accompany evolution and specialization of species. Cytologically and genetically it would seem easier to understand how loss of centromeres and small amounts of heterochromatin surrounding them could be tolerated than to account for production of new centromeres and chromatin as species evolved. If chromosome breaks were near the centromeres, most of the genetic information of the euchromatin would remain available by translocations and centric fusions, but would be distributed into fewer chromosomes or linkage groups. Reduction of chromosome numbers would cause a reduction for potential genetic recombination at meiosis. This might well be correlated with a reduction in genetic adaptability as species evolve into specialized niches.

The taxonomic distribution of sex chromosome systems of ticks further supports the evolution of higher taxa as indicated above. The XY chromosome mechanism is thought to be evolutionarily more primitive in origin than the XO condition (White, 1973), even though it is clear from several examples that sometimes XY mechanisms evolve from XO ancestors. As far as is known, all argasid tick species have only XY males, and all species of *Ixodes* have XY males except *I. holocyclus,* which has XO males (Oliver and Stone, unpublished). As would be expected, *I. holocyclus* has fewer chromosomes than other *Ixodes* except for two other Australian species (*I. tasmani* and *I. cornuatus*), which have the same number, and which will probably be shown to have XO males. Presumably the Y chromosome of *I. holocyclus* broke, and the noncentromeric portion either fused onto an autosome or was lost. Either case would have resulted in one less chromosome. The fact that sex chromosomes are frequently largely heterochromatic means that any loss of the Y may have been tolerated genetically by the species.

The more highly evolved Metastriata have fewer chromosomes than the Prostriata and Argasidae, and almost all species have an XO sex chromosome system in males. These and other facts suggest a monophyletic origin of the Metastriata and an early evolutionary change from an XY to an XO system. The Metastriata, containing many more species than the Prostriata and Argasidae combined, show a much more conservative range in chromosome numbers, as already indicated. Nevertheless, there are more cases of changes in sex chromsome systems in the Metastriata, probably owing to selection acting on a much greater number of species. As already mentioned there are at least four species or geographic "races" that have XY males, which almost certainly evolved from species with XO males. These changes probably occurred in all four species via breakages of the sex chromosome and an autosome, followed by a translocation of most of the sex chromosome onto the broken autosome and subsequent loss of the original sex chromosome centromere and a small portion of the autosome. Of these four species more information exists for *Haemaphysalis hystricis* and *Amblyomma moreliae* than for *A. darwini* and the undescribed species of *Dermacentor*. In *H. hystricis* we believe there were breaks in the sex chromosome and one of the longest autosomes, after which reciprocal translocations occurred. Part of the sex chromosome fused with the large part of the broken autosome and became the neo-X chromosome. The former homologue of the broken autosome would now be designated as the Y chromosome because it pairs with the neo-X. The part of the original sex chromosome containing the centromere fused with the small piece of broken autosome. This newly fused smaller chromosome was lost in some cases but retained in others. Progeny derived from the former line contain individuals with 20 chromosomes. In other cases in which this smaller chromosome was not lost, it was later recognized as a supernumerary chromosome in individuals with 21 chromosomes. In both cases, the sex chromosome system was changed from an XO to an XY condition. Cytological data support the proposed derivation (Oliver et al., 1974). Differential contraction and staining within the supernumerary and X chromosome show certain similarities at diakinesis, and chromosome dynamics at metaphase and anaphase I also conform with expectations. Moreover, the origin of supernumerary chromosomes in several other organisms has been attributed to sex chromosomes.

Another similar situation probably occurred in the Brisbane population of *A. moreliae* as it evolved from a typical population with 20 autosomes plus 1 sex chromosome in males to a situation in which males had 18 autosomes plus 2 sex chromosomes (XY). Most likely a translocation of most of the sex chromosomes onto a heterobrachial autosome with subsequent loss of the original sex chromosome centromere occurred. The original sex chromosome was probably cephalobrachial or nearly so and probably broke near the centromere. The long non-centromere-containing arm probably then fused with a break on one of the heterobrachial autosomes, converting it to a long isobrachial chromosome, half of which was composed of sex chromatin and half of autosomal material. Its autosomal homologue would, of course, remain unaltered and be designated the Y chromosome because of its pairing behavior with the X chromosome (Oliver and Bremner, 1968).

The Sydney population of *A. moreliae* and *A. limbatum*, which have diploid numbers of 22 and 21 and $X_1X_1X_2X_2$:X_1X_2Y sex chromosome condition in females and males, respectively, probably also evolved from a species with 20 + XX females and 20 + X males. A sex trivalent was present that was formed by an unequal reciprocal translocation between the large sex chromosome and one of the autosomes. Because the autosome involved in the translocation now possessed a piece of the sex chromosome, it became a neo–sex chromosome (i.e., X_2). The chromosome previously homologous to the X_2 is now designated as the Y chromosome because it pairs with and segregates from the X chromosomes (Oliver, 1965).

Parthenogenesis and polyploidy

Thelytoky is the only type of parthenogenesis known among ticks (Oliver, 1971). Occasional parthenogenesis or tychoparthenogenesis occurs in a number of normally bisexual species, but when it is present it occurs sporadically and in a small percentage of eggs. Moreover, the resulting larvae are frequently weak and inviable. Obligatory parthenogenesis is reported for *Amblyomma rotundatum* (=*A. agamum*) and for certain geographic races of *Haemaphysalis longicornis*. More will be said about these two species later, particularly the latter.

Although parthenogenesis has often been reported among ticks (Oliver, 1971), it seems likely that genes for parthenogenesis are more common than one might conclude from the literature. Parthenogenesis is certainly more widespread and common if one defines the phenomenon as broadly as is often done by those interested in its occurrence in birds (Olsen, 1956, 1960, 1965; Olsen and Buss, 1967; Olsen and Marsden, 1954; and others) and mammals (Beatty, 1967). If this broader definition were used in conjunction with invertebrates, many more species would be considered to display various degrees of parthenogenesis. Most cases of early parthenogenetic development in invertebrates are never recognized, and invertebrate species are usually not considered parthenogenetic unless larvae are produced. Even though relatively little is known about the genetics of parthenogenesis, Carson (1967) demonstrated in *Drosophila mercatorum* that both sexes transmitted the parthenogenetic trait, and that predisposition for parthenogenesis was induced by genes at a number of independent loci. Although critical genetic experiments have yet to be conducted on ticks, circumstantial evidence suggests a similar situation of a number of independent loci being involved in ticks.

In several species of ticks genes for thelytoky are quite common in some females and rare among others of the same species. The data indicate that distribution of genes for thelytoky is variable among individuals of the same geographic population, and also certain populations obviously have greater frequencies of these genes than others. In the tick *Dermacentor variabilis* parthenogenesis is extremely rare in specimens collected in the southeastern part of the United States around Statesboro, Georgia, but more common in *D. variabilis* from Texas. In a Texas sample of 100 virgin females, 93 engorged on blood and detached from the host (usually mating is required for full engorgement and detachment); 52 of these subsequently produced eggs, of which an average of 1.5% hatched. The bisexually derived eggs in the control sample had 75.9% hatchability. Most of the parthenogenetic larvae appeared weak and died soon after eclosion, but two attached and engorged on a host (Gladney and Dawkins, 1973). Nagar (1967) reported on a different geographic population of *D. variabilis,* which possessed an even greater degree of parthenogenesis. Unfortunately the geographic origin of these ticks was not mentioned, yet almost certainly they were from a different location from the Georgia and Texas populations cited above. He found that hatchability of parthenogenetically produced eggs ranged from 5% to 50%, but failed to report on the viability and activity of the resulting larvae.

Parthenogenesis was reported in *Amblyomma dissimile* (Bodkin, 1918), and was later refuted (Brumpt, 1934) when efforts to demonstrate it failed. It seems almost certain that both authors were correct, and that gene frequency for parthenogenesis in this species varies in different geographic populations. *Amblyomma dissimile* (originally from Central America) in my laboratory contained sufficient genes for parthenogenesis to allow development of quite a few embryos and hatching of some eggs. Development was absent in many eggs, but present to various degrees in others. Partially formed larvae were common, and some eggs contained "fully developed" larvae that exhibited movement if they were helped by breaking open the eggs. Other species of normally bisexual ticks occasionally express parthenogenesis. The soft tick *Ornithodoros moubata* obviously has parthenogenetic tendencies, and one virgin produced 48 larvae that upon feeding and ecdysing several times developed into 38 adult females (Davis, 1951). Two species of hard ticks, *Boophilus microplus* (Stone, 1963) and *Hyalomma anatolicum* (Pervomaisky, 1949), have also been reported to occasionally exhibit parthenogenesis. Unfortunately, no cytogenetic data are available for these parthenogenetic individuals.

At least one instance of artificial production of parthenogenesis is documented (Nuttall, 1915) in the hard tick *Rhipicephalus bursa.* One experiment using egg batches from seven females involved subjecting four of the batches to immersion in a normal salt solution and gentle rubbing with a camel hair brush, while allowing the remaining three batches to serve as controls. Three of the four experimental egg batches and none of the controls produced larvae. A repetition of the experiment with six egg masses and 18 control batches resulted in larvae from three of the six experimental groups and none from the controls. Only 218 larvae emerged from 15,296 eggs (1.4%) in the experimental groups, and none emerged from the controls. The parthenogenetic larvae were weak and died without feeding. There seems little doubt that genes for parthenogenesis are more common in ticks than is generally acknowledged. Surely there are many instances of it occurring that go unnoticed.

The most interesting case of parthenogenesis among ticks and one from which we have much cytogenetic and other data involves *Haemaphysalis longicornis*. Most populations are bisexual in southern Japan, but obligatorily thelytokous over most of its range. Only parthenogenetic races occur in Australia, New Zealand, New Caledonia, New Hebrides, Fiji, and northern Japan (Hokkaido and northern quarter of Honshu) and eastern USSR (Hoogstraal et al., 1968). Parthenogenetic populations also occur in southern Japan including Kyushu and even the small island of Yakushima off the southern tip of Kyushu (Saito, 1972). Bisexual races are sympatric with parthenogenetic races on southern Honshu and Kyushu islands and also in Korea and extreme Primorye (eastern USSR). Probably both types of races occur in northeastern China. At least one race on Cheju Do (island off the southern tip of Korea) reproduces bisexually and parthenogenetically (Oliver et al., 1973). Polyploidy (3n) is invariably associated with the obligatorily parthenogenetic races, and chromosome numbers range from 30 to 35; most have approximately 32. The bisexual races are diploid (20 + XX female: 20 + X male), and the races on Cheju Do are aneuploid (22–28 chromosomes).

Haemaphysalis longicornis is an interesting example of a species actively evolving chromosomally and reproductively and one in which various chromosomal and reproductive strategies can be compared. Hybridization attempts failed in the laboratory between diploid and triploid races, but succeeded between bisexual diploid males and parthenogenetic aneuploid females. Numbers of F_1 and F_2 progeny were produced, and most germinal cells of male and female F_1 were diploid with karyotypes typical of the bisexual race. A few had irregular chromosome numbers, and it is unknown whether they would have developed into functional eggs and sperm. The F_2 progeny had the normal diploid karyotype in females, and no chromosome data exist for F_2 males. The F_1 and F_2 progeny appeared more like progeny of diploid parents than parthenogenetic ones when tested for parthenogenetic ability and other reproductive parameters. Parthenogenetic ability was almost completely lost in F_1 and F_2 females. Moreover, crossing of F_1 progeny to a bisexual race was successful (Oliver et al., 1973).

The above-cited laboratory data suggest that there is no successful hybridization between naturally occurring diploid bisexual and triploid parthenogenetic *H. longicornis*. This is probably true, but because it is clear that diploid bisexual males can successfully fertilize *aneuploid* parthenogenetic females, and progeny of both sexes be produced, it seems likely that this could occur in nature as well as in the laboratory. The female parents used in the laboratory experiments were from Cheju Do and could reproduce bisexually or parthenogenetically. It seems highly likely that these variously aneuploid (chromosome numbers range from 23 to 28) females with their great reproductive plasticity mate with diploid and aneuploid males when they are available in nature.

The discovery on Cheju Do of aneuploid females that can reproduce parthenogenetically and bisexually is of great significance, and demonstrates the remarkable chromosomal and reproductive plasticity of a naturally occurring population. It also suggests that evolution from a diploid bisexual condition to a triploid parthenogenetic state probably did not occur in one jump (i.e., fusion of haploid sperm with diploid egg, fusion of three haploid cleavage nuclei, or some other such phenomenon causing triploidy). Because even a small sample of Cheju Do females shows aneuploidy ranging from 23 to 28, it seems likely that a greater range of chromosome numbers may be present in some individuals, and may reach numbers usually seen in triploid races (30–33). It is not known when or how this race acquired the ability to become parthenogenetic, but clearly it still possesses the ability to reproduce bisexually. Although it cannot be proved how triploid parthenogenetic individuals evolved, the information now available indicates that it was a gradual process. It is clear from the chromosomal data that complete triploidy is not required prior to display of parthenogenesis. Chromosomal analyses do not allow us to know at this time which chromosomes carry the genes for parthenogenesis or the minimum number of chromosomes needed. It is possible that as few as one or two extra of the particular chromosomes might allow parthenogenesis, and that the various states of aneuploidy arise via nondisjunction.

Acquisition of parthenogenetic ability and triploidy appear to be beneficial steps in the evolution of *H. longicornis,* particularly because the diploid bisexual race has not been eliminated. At present this species has the best of two worlds. One race has retained the advantages of sexual reproduction, allowing for a more flexible and presumably adaptable gene pool, whereas another acquired the evolutionary short-term advantages of parthenogenesis. One of the advantages of parthenogenesis is that it allows species to colonize new areas more readily than bisexual ones do. Presumably this has been a factor in the extensive increase in geographic distribution of the parthenogenetic race of *H. longicornis* when compared to the bisexual one. Of course, *H. longicornis* on Cheju Do has the advantages of both types of reproduction.

Theoretically, parthenogenetic species are considered evolutionary dead ends, and this view is supported by the presence of relatively few thelytokous species among many bisexual ones. One reason given for this is the lack of genetic variability in parthenogenetic species. In considering this presumed lack of genetic variability (or the reduced potential for variability) among parthenogenetic species, the question arises of whether this presumed reduction in variability is due to reproductive method (parthenogenesis) alone, or whether ploidy level and sex might have an influence. Oliver and Herrin (1974, 1976) attempted with partial success to shed light on certain aspects of these questions by obtaining estimates of heterogeneity and generalized variance in *H. longicornis* and the tropical rat mite, *Ornithonyssus bacoti*. The latter species has haploid males and diploid females, and, as already indicated, *H. longicornis* has bisexual diploid males and females, parthenogenetic triploid females, and aneuploid females capable of parthenogenesis and bisexuality.

Univariate and multivariate statistical analyses of large numbers of morphological characters on many mite specimens indicate that variation is greater in the diploid female than in the haploid male *O. bacoti* (Oliver and Herrin, 1974). It is uncertain, however, whether the greater variability of these females is due to diploidy or to femaleness per se. Results obtained from similar investigations of *H. longicornis* suggest that variance is not correlated with sex per se because variance of diploid females is not significantly different from that of diploid males (Oliver and Herrin, 1976). If these results from *H. longicornis* can be extrapolated to the haploid male and diploid female system of *O. bacoti,* then the greater variance noted in female *O. bacoti* can be attributed to effects of ploidy level and/or sexual reproduction. Most likely the greater variability is due to ploidy level because even though the haploid males are produced parthenogenetically, they originate from sexually produced females, and the usual advantages attributed to sexual reproduction have not really been denied the gene pools from which males originated.

Assuming that the above conclusions are correct and increased variability in *O. bacoti* females is due to diploidy, one might suspect that additional ploidy levels would allow even greater variance (i.e., genomes might act in an additive manner on variability). Indeed, preliminary data on *H. longicornis* seem to support this suspicion. In one analysis field-collected triploid parthenogenetic females had greater morphological variability than field-collected bisexual females. This suggested that not only did triploidy allow greater variability than diploidy, but it outweighed the effects of bisexual reproduction. One might have assumed that the presumed greater genetic variability of the bisexual females should have been expressed by greater morphological variability. Caution is recommended before definitive conclusions are made from these data, however, for two reasons. First, only one field-collected presumed triploid parthenogenetic population was analyzed in this comparison, and, second, this population was *presumed* triploid parthenogenetic because of its geographic location. Whereas the presumption was almost certainly correct, parthenogenesis was not examined in the laboratory. Although it appears that diploidy is responsible for greater variability than haploidy, and there is evidence suggesting triploidy might be responsible for greater variability than diploidy, the relative roles or effects of ploidy level and method of reproduction (parthenogenesis vs. bisexuality) on morphological variability still remain unanswered. Some conflicting data from other analyses indicate more variability among *laboratory-reared* diploid bisexual females than among *laboratory-reared* triploid parthenogenetic females. The question arises as to why triploidy was associated with greater variability in field-collected females and diploidy associated with greater variability in laboratory-reared specimens. Samples were larger in the comparisons that indicated more variability in the diploid females so there is a tendency to place more confidence in those results. Unfortunately, however, the roles of bisexuality and parthenogenesis confuse the issue. The studies on *H. longicornis* do show that greater variability exists in field-collected specimens than among laboratory-reared individuals in both parthenogenetic (triploid) and bisexual (diploid) populations. Presumably this is due to the availability of a larger gene pool from which to sample as well as the influence of more diverse selection pressures in nature.

The work reported here was supported in part by U.S. Public Health Service Research Grant AI-09556 from the National Institute of Allergy and Infectious Diseases.

References

Beatty, R. A. 1967. Parthenogenesis in vertebrates. In *Fertilization, Comparative Morphology, Biochemistry, and Immunology,* vol. 1 (ed. Metz, C. B., and Monroy, A.), pp. 413–40. New York: Academic Press.

Bodkin, G. E. 1918. The biology of *Amblyomma dissimili* (Koch), with an account of its power of reproducing parthenogenetically. *Parasitology 11:*10–17.

Brumpt, E. 1934. L'ixodine, *Amblyomma dissimile* du Venezuela ne présente pas de parthénogenèse facultative. *Ann. Parasitol. 12:*116–20.

Carson, H. L. 1967. Selection for parthenogenesis in *Drosophila mercatorum. Genetics 55:*157–71.

Davis, G. E. 1951. Parthenogenesis in the argasid tick *Ornithodoros moubata* (Murray, 1877). *J. Parasitol. 37:*99–101.

Gladney, W. J., and Dawkins, C. C. 1973. Experimental interspecific mating of *Amblyomma maculatum* and *A. americanum. Ann. Entomol. Soc. Am. 66:*1093–7.

Hoogstraal, H., Roberts, F. S. H., Kohls, G. M. and Tipton, V. J. 1968. Review of *Haemaphysalis (Kaiseriana) longicornis* Neumann (resurrected) of Australia, New Zealand, New Caledonia, Fiji, Japan, Korea and northeastern China and USSR, and its parthenogenetic and bisexual populations (Ixodoidea, Ixodidae). *J. Parasitol. 54:*1197–213.

Krantz, G. W. 1978. *A Manual of Acarology,* 2nd ed. Corvallis: Oregon State University Book Stores.

Nagar, S. K. 1967. Parthenogenesis in *Dermacentor variabilis* (Say). *Acarologia 9:*819–20.

Nuttall, G. H. F. 1915. Artificial parthenogenesis in ticks. *Parasitology 7:*457–61.

Oliver, J. H., Jr. 1965. Cytogenetics of ticks. 2. Multiple sex chromosomes. *Chromosoma (Berl.) 17:*323–7.

– 1971. Parthenogenesis in mites and ticks. *Am. Zool. 11:*283–99.

Oliver, J. H., Jr. and Bremner, K. C. 1968. Cytogenetics of ticks. 3. Chromosomes and sex determination in some Australian hard ticks. *Ann. Entomol. Soc. Am. 61:*837–44.

Oliver, J. H., Jr. and Herrin, C. S. 1974. Morphometrics of sexual dimorphism in an arrhenotokous mite, *Ornithonyssus bacoti* (Acari: Mesostigmata). *J. Exp. Zool. 189:*291–302.

– 1976. Differential variation of parthenogenetic and bisexual *Haemaphysalis longicornis* (Acari: Ixodiae). *J. Parasitol. 62:*475–84.

Oliver, J. H., Jr., Tanaka, K., and Sawada, M. 1973. Cytogenetics of ticks. 12. Chromosomes and hybridization studies of bisexual and parthenogenetic *Haemaphysalis longicornis* races from Japan and Korea. *Chromosoma (Berl.) 42:*269–88.

– 1974. Cytogenetics of ticks. 14. Chromosomes of nine species of Asian haemaphysalines. *Chromosoma (Berl.) 45:*445–56.

Olsen, M. W. 1956. Fowl pox vaccine associated with parthenogenesis in chicken and turkey eggs. *Science 124:*1078–9.

– 1960. Nine year summary of parthenogenesis in turkeys. *Proc. Soc. Exp. Biol. Med. 105:*279–81.

– 1965. Twelve year summary of selection for parthenogenesis in Beltsville Small White turkeys. *Br. Poult. Sci. 6:*1–6.

Olsen, M. W., and Buss, E. G. 1967. Role of genetic factors and fowl pox virus in parthenogenesis in turkey eggs. *Genetics 56:*727–32.

Olsen, M. W., and Marsden, S. J. 1954. Natural parthenogenesis in turkey eggs. *Science 120:*545–6.

Pervomaisky, G. S. 1949. Parthenogenetic development of ticks in the family Ixodidae. *Zool. Zh. 28:*523–6.

Saito, Y. 1972. Investigation of ticks in Kyushu district (in Japanese, abstract). *Jpn. J. Sanit. Zool. 22:*255.

Stone, B. F. 1963. Parthenogenesis in the cattle tick, *Boophilus microplus. Nature (Lond.) 200:*1233.

White, M. J. D. 1973. *Animal Cytology and Evolution.* Cambridge: Cambridge University Press.

5   Differentiation of heterochromatin

W.J.PEACOCK, E.S.DENNIS, A.J.HILLIKER, AND
A.J.PRYOR

There seems little doubt that each eukaryote chromosome is one
DNA molecule. This has been shown in yeast by direct isolation of
chromosomal DNA (Petes et al., 1973), and in *Drosophila melanogaster*
by visco-elastic measurement (Kavenoff and Zimm, 1973). Direct studies
are not available in other higher organisms, but many observations support
the generality of one DNA molecule per chromosome (Peacock, 1979).

Another feature of eukaryote chromosomes is that they have regions
differing in their staining properties through the cell cycle; some regions,
the euchromatin, are condensed and dark staining only in late prophase
through to telophase, whereas other segments remain condensed and dark
staining throughout the whole cell cycle including interphase. These heterochromatic segments (Heitz, 1929) must be associated with long domains of
chromosomal DNA molecules that have a different sequence composition
from those corresponding to the euchromatic segments.

Owing to its uniform appearance and apparent lack of associated
phenotype, heterochromatin has often been assumed to be a homogeneous
entity lacking genetic effects. This is not the case. In some species,
heterochromatin is differentiated cytologically and cytogenetically, and in
D. melanogaster the presence of functional loci in heterochromatin has
been clearly demonstrated. Studies of the DNA sequences present in
heterochromatin have also revealed considerable heterogeneity. This
review stresses the differentiation of heterochromatin at genetic, cytogenetic, and DNA sequence levels. Discussion is restricted mainly to *Drosophila* and maize, two organisms that we have found useful in probing the
basic properties of heterochromatin.

Genetic elements in heterochromatin

In early genetic experiments with *D. melanogaster* it was found that large deletions of the basal heterochromatin of the X chromosome did not lead to inviability. These observations (Muller and Painter, 1932) led to the concept that heterochromatin is genetically inert, but there is now abundant evidence that this is not so. Heterochromatic deficiencies in the X do not lead to inviability, but they do affect a diverse array of characters, including developmental time, sex chromosome pairing, and spermatid differentiation (Peacock and Miklos, 1973). On the other hand, the density of genes in heterochromatin is much lower than in euchromatin.

In *D. melanogaster* the entire Y chromosome, the proximal half of the X chromosome, the proximal regions of the arms of chromosomes 2 and 3, and a major proportion of chromosome 4 are heterochromatic (Kaufmann, 1934; Figure 5.1). Analysis of second chromosome heterochromatin has defined 13 gene loci (Hilliker and Holm, 1975; Hilliker, 1976). These loci appear to be "ordinary" vital genes in the sense that they are associated with well-defined complementation groups and can be mapped unambiguously (Figure 5.2). Furthermore, two of the loci have recessive-viable, visible alleles in addition to lethal alleles. Mutations of all of the loci have been induced with EMS (ethyl methanesulfonate), which is thought to induce point mutations, largely as single nucleotide changes. The mutation characteristics of the loci are the same as those for euchromatic genes. Their only distinguishing feature is that they are located in the pericentric heterochromatic regions, and that their density per unit length of DNA (one locus per 500–1000 kb) is substantially lower than that found for genes in the euchromatic regions, where the average density is one locus per 25 kb. Schultz (1947) argued that particular sets of functional genes could be expected to be located in heterochromatin, but at present the primary activities of the heterochromatic genes are not known.

In the X heterochromatin, two loci, mapping to a region between the nucleolar organizer and the heterochromatin–euchromatin junction, have been defined by their effects on the expression of other gene loci. Xh^{cr+} (compensation response) is associated with the regulation of the compensatory somatic increase in ribosomal genes in flies hemizygous for the *bobbed* locus (*bb*) (Procunier and Tartof, 1978), and the Xh^{abo} locus affects the expression of the maternal-effect lethal mutation *abnormal oocyte* (*abo*), located in the second chromosome euchromatin (Parry and Sandler, 1974). Indirect evidence suggests that Xh^{abo} may be correlated with the level of redundancy of ribosomal RNA genes in the nucleolus organizer

Figure 5.1. Mitotic metaphase in a ganglion cell from a male larva of *Drosophila melanogaster.* In this photograph of fluorescence following quinacrine staining, the heterochromatin shows segmental differentiation of bright and dull regions. The Y chromosome is the B^sYy^+. (Reproduced by permission of T. Kaufman, Indiana University)

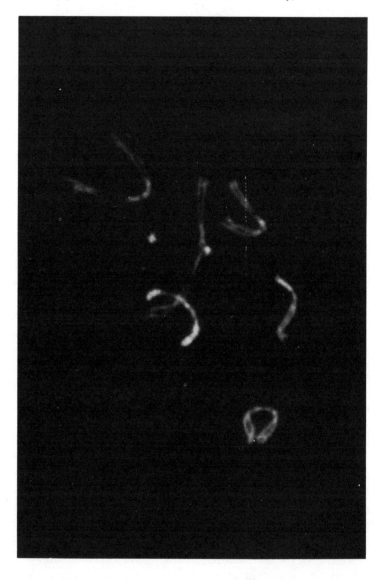

(Krider and Levine, 1975; Krider et al., 1979). Xh interacts with several maternal-effect loci located in the second chromosome euchromatin near *abo* but it has not been shown whether Xh^{abo} is the element involved (Sandler, 1977). These two heterochromatic loci have "allelic" regions in the Y chromosome, as does the *bb* locus, which is associated with the nucleolus organizer region in both the X heterochromatin and the short arm of the Y. X and Y heterochromatin share other genetic properties such as the pairing sites concerned with meiotic conjugation and segregation during spermatogenesis (Cooper, 1964). These sites map to several regions of Xh and to both arms of the Y chromosome.

It has long been known that the Y chromosome is essential for male fertility (Bridges, 1916), and that a number of fertility genes must be present on both arms of the Y chromosome (Stern, 1929; Neuhaus, 1939; Brosseau, 1960). Two recent analyses have mapped the genes and have found two on the short arm and four or five on the long arm (Kennison, 1979; Kaufman, pers. comm.). These genes, like the autosomal heterochromatic loci, have similar mutational properties to euchromatic genes; for example, they have EMS-induced temperature-sensitive alleles and presumably represent unique sequence elements (Ayles et al., 1973). The fertility loci may be associated with secondary constrictions of the chromosome. This may reflect a functional arrangement of different types of sequence within heterochromatin, with alternating regions of condensed DNA and less tightly packaged DNA. The *bobbed* locus is associated with a prominent secondary constriction on both the X and the short arm of the Y chromosome (the NOR). In euchromatin, alternating regions of differential condensation occur, but on a much finer scale as evidenced by the band/interband regions of polytene chromosomes. There are a number of

Figure 5.2. Map of the pericentromeric heterochromatin of chromosome 2 of *Drosophila melanogaster* showing the genetic loci demonstrated by the complementation analysis of Hilliker (1976). The map also shows the position of genes with visible phenotypes. The heterochromatic loci, *lt* and *rl*, also have lethal alleles. The bracket around three 2L heterochromatic loci signifies that their relative order has not been established. Details of the lethal and visible gene loci are given in Hilliker (1976).

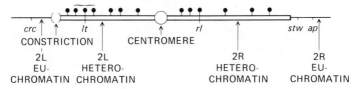

sets of data correlating one essential gene function to each band (e.g., Judd et al., 1972); however, other data exist that map more than one coding segment to a length of DNA of single band proportions (Lis et al., 1978).

Sequence elements within heterochromatin

The existence of gene loci and of differential packaging of the DNA molecules within heterochromatin suggest a complexity of sequence organization. This complexity may be superimposed on general sequence properties that distinguish heterochromatin from euchromatin. The initial in situ hybridization experiments (Pardue and Gall, 1970) showed that the satellite DNA of mouse was located in the pericentric heterochromatic blocks of all of the chromosomes other than the Y. This was the first indication of a different class of sequence organization in heterochromatin relative to that found in euchromatin.

In *D. melanogaster* we fractionated total nuclear DNA using antibiotics

Figure 5.3. Buoyant density profile of the nuclear DNA of *Drosophila melanogaster*. The satellite DNA species that have been isolated in antibiotic-containing gradients are shown. (After Peacock et al., 1973; Brutlag et al., 1977)

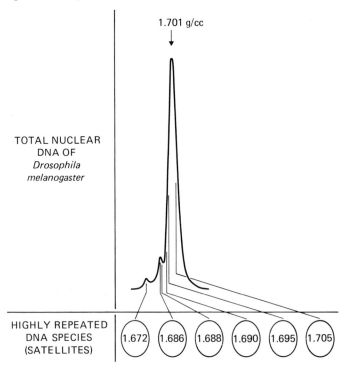

and metal ions known to have base-specific binding preferences. This enabled the isolation of a number of satellites as distinct DNA populations (Peacock et al., 1973; Figure 5.3). Some of these satellite populations have been extensively characterized, three having short simple repeat sequences (1.672, 1.686, and 1.705 g/cc) and one (1.688 g/cc) having a more complex repeat unit (Table 5.1). The simple repeat sequences are very similar to each other and may well have been derived from a common ancestral sequence; they show no obvious relationship to the complex 1.688 satellite. Each satellite comprises about 3%–4% of the genome, and together they account for almost all of the highly repeated DNA of the genome. The alternate repeat forms of the 1.672 and 1.705 satellites are not interspersed, but occur in separate repeating arrays. By measuring the yield of a given satellite over a range of molecular weights of DNA, we were able to show that the repeat arrays of the different satellites were of the order of 500 kb or more (Brutlag et al., 1977). Because each satellite accounts for approximately 8000 kb of DNA, each must occur in relatively few locations in the chromosomal set. For some of the satellites the few locations were concluded to be contiguous because we detected molecules that contained the termini of two of the satellite arrays (e.g., 1.672 and 1.705; Brutlag et al., 1977). These results suggest that the five different DNA molecules that constitute the *Drosophila* genome consist of relatively few domains in which highly repeated sequences are either concentrated or virtually absent.

This conclusion was confirmed at a direct chromosomal level by in situ hybridization (Peacock et al., 1977a,b). The physical properties of different satellites made it possible to conduct well-controlled hybridiza-

Table 5.1. *Nucleotide sequences of* Drosophila melanogaster *major satellites*

Satellite	Repeat length		
	5 bp	7 bp	10 bp
1.672	A T A A T T A T T A	A T A T A A T T A T A T T A	
1.686			A T A G A A T A A C T A T C T T A T T G
1.705	A G A A G T C T T C	A G A G A A G T C T C T T C	
1.688		376 bp repeat	

tion reactions in these experiments and to be certain that there was not any cross-hybridization clouding the results. We were able to do detailed chromosomal mapping of each of the different highly repeated sequences. Each repeat sequence is restricted to heterochromatic regions and most occur on all chromosomes. Nevertheless each chromosome has a distinctive sequence arrangement of these satellite DNAs, the length of arrays and their nearest neighbor associations varying in each major heterochromatic region.

Perhaps the major point that arises from these data is that heterochromatin should be considered a generic term; it encompasses, in any one genome, a number of regions that can each be described and individualized by its sequence composition.

Other sequences in heterochromatin

As well as being segmentally differentiated by highly repeated sequence blocks, heterochromatin must be spatially differentiated with respect to other classes of sequence. Rudkin (1964) has measured the DNA content of the basal heterochromatin of the X, and knowing the overall quantity of each repeat sequence in the genome, we have been able to quantitate the relative amounts of a given repeat in its different chromosomal locations. From these data it appears that the various blocks of satellite DNA, together with the ribosomal DNA, account for over 90% of the heterochromatin in the X chromosome (Peacock et al., 1977a). Our data further suggest that the highly repeated sequences account for the major portions of other heterochromatic regions in *D. melanogaster,* including the Y chromosome, but in every segment there could be substantial numbers of other sequences present.

We have already discussed the genetic demonstration of essential loci in heterochromatin. In situ hybridization has provided knowledge of two other classes of sequence sequestered among the long tandem arrays of simple sequences. Approximately half of the ribosomal genes in the X chromosome of *D. melanogaster* have an insert within the 28S RNA gene (White and Hogness, 1977). The insert, as well as occuring within the ribosomal genes, has other genome locations (Dawid and Botchan, 1977). We have defined some of these other locations by in situ hybridization to polytene chromosomes; of the 50 or more copies that are not in the ribosomal genes, most are in the pericentric heterochromatin of each of the chromosomes including the X, and there is one location in a banded region on chromosome 4 (Peacock et al., in press). By using combinations of X chromosome inversions having one break point in heterochromatin and one near the left end of the chromosome, it has been shown that some of the extra copies occur in the heterochromatic region defined by the break points of the sc^4 and w^{m4} rearrangements (Appels, pers. comm.). The second class of sequence known to have heterochromatic sites encompasses the moderately repeated gene families; these include the *copia*-like families, which in euchromatin are known to change their locations within a relatively short evolutionary time – geographic races having different chromosomal sites (Strobel et al., 1979). There is no information about the stability of the heterochromatic sites, nor is the functional significance of either the heterochromatic or the euchromatic sites of these sequences known.

Heterochromatin, like euchromatin, consists of single-copy, moderately repeated, and highly repeated DNA sequences. The distinction between the two types of chromatin presumably lies in the relative proportions of the sequence classes. The large amounts of highly repeated sequences in heterochromatin may be responsible for the differential coiling and consequent differential staining properties of these chromosome segments (see Appels and Peacock, 1978).

Sequence elements within a satellite DNA

Although in *D. melanogaster* heterochromatic blocks are sequence-differentiated, many organisms have a single satellite DNA that is located on all chromosomes and accounts for the bulk of the heterochromatin. Therefore, in asking whether segmental sequence differentiation of heterochromatic blocks is a general property, we looked at such an organism. The red-necked wallaby has a major satellite amounting to 20% of the genome and approximating the amount of DNA in its set of heterochromatic blocks (Dunsmuir, 1976). It is located in the pericentric heterochromatin of all autosomes. This satellite DNA appears to be homogeneous in buoyant density gradients and by thermal denaturation, but some properties suggested internal sequence heterogeneity. Reannealed molecules had an increased buoyant density and a lowered melting temperature, characteristics indicating differences in sequence between repeat units within the satellite population.

When the satellite DNA was restricted with either of the restriction enzymes, *Bam*HI or *Pst*I, it was totally cleaved into a set of segments of unit length 2500 bp. In addition to the unit length segments there were others corresponding to dimer, trimer, and tetramer segments. These occurred in low frequency and their proportions were consistent with random mutational changes in the recognition sequences for those enzymes. Furthermore, it was established that the different segment lengths were interspersed (Dunsmuir, 1976). However, restriction of the satellite with other enzymes revealed a different pattern of sequence changes, and showed that as well as random base changes, much of the heterogeneity seen in the reannealing experiments could be explained by highly ordered sequence changes. For example, *Hind*III, *Xma*I, and *Eco*RI cleave only some of the satellite DNA, leaving the remainder as high-molecular-weight molecules (Dennis et al., 1979a).

Experiments involving simultaneous and successive digests showed that, in each of these cases, repeats carrying the recognition sequence for particular enzymes were arranged in long tandem arrays. By the use of a number of other restriction enzymes it was possible to show that the satellite DNA was composed of a number of subpopulations of related but different sequences. The molecular arrangement of these subpopulations provides evidence that a number of amplification events must have occurred in the evolution of the satellite. More important in terms of our present discussion, the possibility is raised that the heterochromatic blocks of the red-necked wallaby, despite being largely constituted of this one satellite DNA, have a segmental arrangement of different repeating units. Dunsmuir (1976) isolated another highly repeated DNA species and showed that it also was found on each of the chromosomes, and that particular chromosomes differed in their relative contents of the major and minor satellites.

The magnitude of different subpopulations within the major satellite suggests that some must be represented on several chromosomes, whereas others are potentially restricted to a single chromosome. Restriction enzyme analysis of the mouse satellite, also homogeneous by buoyant density characteristics and occurring on all chromosomes except the Y, has disclosed a subpopulation organization similar to that in the red-necked wallaby (Horz and Zachau, 1977). Recently Beauchamp et al. (1979) have used mouse–human hybrid cells to isolate individual human chromosomes and have been able to show that particular sequence variants of some of the highly repeated DNA satellites of man are restricted to single chromosomes. Because it is known that several different sequences are present on many of the chromosomes, it seems that a segmental organization of the pericentromeric heterochromatin blocks will be established. In many other animals there is enough evidence from either restriction enzyme analyses or in situ analyses to suggest that the segmental sequence differentiation of heterochromatin is of wide occurrence.

Very little information is available for plants, but from analyses we have done in the cereal crops it seems that the same general principle applies. For example, in rye in situ hybridization with the sequences of the total rapidly annealing fraction of DNA shows that the majority are located within the major heterochromatic blocks situated near ends of the chromosomes (Appels et al., 1978). Buoyant density and thermal denaturation analyses show that there are a number of different repeat-sequence families in this fraction. We have purified some of them by cloning individual repeats in bacterial plasmids and have examined their chromosomal distribution (Appels et al., in prep.). The repeating unit present in pSc 11790 occurs in the heterochromatic blocks of all seven chromosomes of rye, with markedly different amounts on different chromosomes, and in some heterochromatic blocks there is more than one discrete location of the repeat sequence (Figure 5.4). The repeat in pSc 1374 is located in only three chromosome pairs (Figure 5.5). Even these two examples of repeat-sequence distribution among the population of heterochromatic regions indicate that, in this plant too, each heterochromatic block will almost certainly contain a number of different sequence arrays arranged in

Figure 5.4. In situ hybridization of a cloned major highly repeated DNA sequence (pSc 11790) to the chromosomes of rye. The sequence is located in all of the telomeric heterochromatic blocks of the complement.

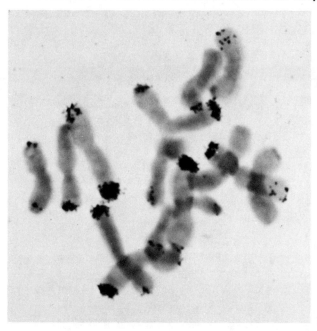

segmental fashion. The segmental disposition of highly repeated sequence units in heterochromatin must be a general feature of eukaryote chromosomes.

Sequence conservation within heterochromatin

Another emerging general feature of heterochromatin is the sequence stability over evolutionary time of its component simple repeats (Salser et al., 1976). An example of the relative stability of a heterochromatic sequence is a satellite that occurs in wheat and in the related cereal crop, barley (Dennis et al., 1980b). When silver ions are bound to the DNA of these species, a satellite can be isolated that has a major sequence component, the trinucleotide segment GAA/CTT (Figure 5.6). In wheat this satellite is found on many of the 21 chromosomes, with major sites on the seven chromosomes of the B genome. The distribution of the arrays of the sequence includes interstitial segments as well as tracts around centromeres and near the ends of the chromosomes. The complexity of the distribution has enabled us to identify, using ditelocentric tester stocks

Figure 5.5. In situ localization of a cloned highly repeated DNA sequence (pSc 1374), which occurs on only three chromosomes of the rye complement.

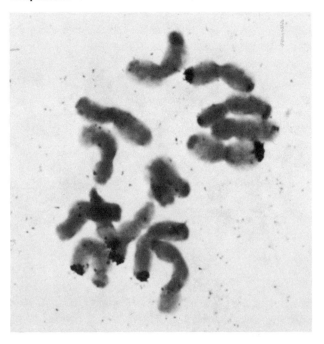

(Sears, 1963), an individual pattern for most of the chromosomes. The distribution pattern of the sequence is much simpler in barley where the heterochromatic blocks are largely pericentromeric. Despite this major difference in the chromosomal distribution of the sequence arrays the satellites have indistinguishable properties. Comparison of moderately repeated and unique sequence elements of wheat and barley has shown that a substantial amount of sequence divergence has occurred during the separate evolutionary histories of these two plant species (Bendich and McCarthy, 1970).

Other examples of the relative evolutionary stability of highly repeated DNA sequences are known between sibling and closely related species of *Drosophila* (Gall and Atherton, 1974; Peacock et al., 1977b) and over much broader phylogenetic distances in rodents (Fry and Salser, 1977), reptiles (Singh et al., 1976), and primates (Gosden et al., 1977).

Lability of sequence repeat representation

In contrast to the conservation of nucleotide sequence of highly repeated DNAs, the number of repeat units present is a property that is

Figure 5.6. Autoradiograph showing the trinucleotide organization of the polypyrimidine–polypurine satellite of barley (*a*) and wheat (*b*) following digestion with the restriction enzyme *Mbo* II. (After Dennis et al., 1980b)

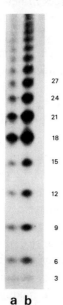

a b

labile over evolutionary time. In the sibling species *D. melanogaster* and *D. simulans*, the representation of a particular repeat can vary by two orders of magnitude despite the maintenance of sequence identity in the two species (Peacock et al., 1977a,b). In the macropod kangaroo group the major satellite sequence of the wallaroo is present in all other kangaroo species we have examined; in the red kangaroo and red-necked wallaby the sequence totals 30% and 15% respectively of the wallaroo value, but in the other species it occurs in very much smaller quantities (Venolia, 1977). The wallaroo satellite also demonstrates that variation in the number of copies of a repeat in a particular heterochromatic segment can occur within a species. In some animals one of the homologues of chromosome pair 5 contains no detectable copies of the sequence. This same polymorphism is found in another subspecies, the euro, and in the related species the antilopine wallaby. Comparable polymorphisms have been documented for a number of chromosomes in the human complement; for example, the pericentromeric heterochromatic block on chromosome 16 is known to be polymorphic in some families, in which its inheritance pattern is according to normal Mendelian expectations (Carnevale et al., 1976).

Such intraspecific polymorphisms suggest that the repeat representation of a satellite may be a highly labile character, but two factors need to be taken into account in considering this point. First of all, the polymorphisms appear to be stable and to be inherited in a normal manner. Despite the fact that a number of polymorphisms are known, changes in heterochromatin block size are not commonly seen in human studies (Gosden, pers. comm.). The knob alleles in maize are another example of stable heteromorphisms in the extent of heterochromatic sequences. The frequency of repeat modulation events must be low, or at least only a few such changes survive successfully. The second factor that should be mentioned is that many polymorphisms in heterochromatic blocks may actually reflect structural rearrangements of the chromosome complement. For example, an apparent polymorphism in humans for the heterochromatic block near the centromere of chromosome 22 has been shown to be a translocation between Y chromosome heterochromatin and chromosome 22 (Schmitt and Engel, pers. comm.). In wheat-rye hybrids reductions in the size of the terminal blocks of heterochromatin of rye chromosomes have been linked with improved fertility (Jagannath and Bhatia, 1972); in at least one case the chromosomal change has been shown to be a translocation of a homologous wheat arm and does not represent a modulation in the amount of rye highly repeated sequences (May and Appels, 1978). Quantitative sequence information should be in hand before sequence modulation is inferred. Nevertheless, in some animals such as rodent species there is evidence for extensive polymorphism of heterochromatic blocks, and it may be that in some genomes modulation of repeat representation is a significant component of the genetic system.

Another example of a highly repeated sequence that shows frequency modulation is found in *Warramaba virgo,* the parthenogenetic morabine grasshopper extensively analyzed by Michael White and his colleagues. *W. virgo* is considered to have been derived by hybridization between the ancestors to two sexual Western Australian species, P169 and P196 (Hewitt, 1975; White et al., 1977). We have cloned a repeated sequence from *W. virgo* and located it on the heterochromatic block on the X + A chromosome, which is thought to have originated from P169. The sequence is present in the P169 genome but in only one tenth the number of copies that are found in *W. virgo.* Restriction enzyme analysis has shown that the sequence has similar molecular organization in the two species, but *W. virgo* shows an additional pattern that is not present in the parental sexual species (Figure 5.7). The new polymeric segment array may be a consequence of an amplification event that has occurred in the parthenogenetic species; its molecular characterization could provide evidence and information about time of origin and geographic migration of *W. virgo.*

In an earlier section we presented evidence for segmental amplification in the generation of a satellite DNA. Amplification of a particular unit repeat appears to be a regular feature of the evolution of heterochromatin. In some instances the amplification involves a modification of the sequence repeat, coamplifying some or all of the basic repeat unit along with intercalated or adjacent sequences. This can result in a significant change in sequence composition of heterochromatin at a given chromosomal site. Possible examples are seen in which the length of the long order repeat of a particular satellite sequence, common in the red-necked wallaby, changes in two other species of kangaroo (Dennis et al., 1980a), and an analysis of the telomeric repeating sequences in rye (Bedbrook et al., 1980) has provided some particularly striking evidence of this form of genome change.

Modulation of repeat unit representation may be a property that is not shared by euchromatin. But in euchromatin too there is recent evidence showing that at least certain sequence components of the genome are far more variable than we have supposed. For example, recent data in *Drosophila* show that a high proportion of the units of families of moderately repeated sequences change their positions in the genome over a short evolutionary time scale (Young, 1979).

Chromosomal sites of specific highly repeated sequences

In the comparison of highly repeated sequences in *D. melanogaster* and *D. simulans* we noted that, despite large differences in the number of repeats, chromosomal locations of each satellite were similar, this applying particularly to the sequence blocks on the Y chromosome (Peacock et al., 1977a). This comparison could only be described in general cytological terms because of the lack of specific chromosomal markers, but within *D. melanogaster* it is clear that there is a striking association of three highly repeated sequences (1.672, 1.686, and 1.688) with the nucleolus organizer regions on both the X and Y chromosome. In polytene nuclei of the salivary gland these satellite sequences are found to be distributed throughout the nucleolus at particular stages of development. It is possible that they have an integral role in nucleolus activity. A conserved association of highly repeated sequences and ribosomal DNA

Figure 5.7. Autoradiograph showing the organization of the cloned highly repeated sequence pWvl in P169 (*a*) and P196 (*b*) and *Warramaba virgo* (*c*) following digestion with the restriction enzyme *Eco*RI. The solid arrow indicates high-molecular-weight DNA without *Eco*RI sites. The open arrow indicates a segment length common to all three species. The length markers are derived from λDNA digested with *Eco*RI.

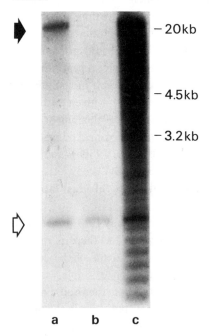

−20kb

−4.5kb

−3.2kb

a   b   c

has also been observed in the primates by Gosden et al. (1978), but these authors noted that there were autosomal sites containing the highly repeated sequences without the rDNA sequences being present. In *Drosophila* too each of the three satellites associated with the nucleolus maps to other regions of the chromosome complement.

An association of a particular repeated sequence and a specific chromosome region is obvious in the telomeric sequences in *D. melanogaster* (Rubin, 1977). Component elements of this complex repeat unit are common to the telomeric regions of all chromosome arms, suggesting a possible functional role of the sequence. However, in *D. simulans* these particular sequences are not present in the telomeric regions (Dunsmuir, pers. comm.). It is not known whether different but comparable repeating elements are present at these sites in *D. simulans*.

Sequence differentiation in relation to heterochromatin differentiation

Because repeated sequences are associated with regions of chromosomes with specialized functions, such as telomeres and the nucleolus organizer regions, and because there is a low frequency interspersion of essential genes in heterochromatin, the possibility of an association between a particular highly repeated sequence and control of particular gene function(s) exists. Control need not necessarily be considered in terms of direct effects on transcription, but might, for example, be mediated through the relative positioning or configurational changes of chromosome regions in the nucleus.

In maize there are separate chromosomal sites for different types of heterochromatin, and there is knowledge of their cytogenetic effects. Maize has heterochromatic elements around the centromeres in each of the 10 chromosomes, and in addition there is a large block of heterochromatin on the short arm of chromsome 6 associated with the nucleolus organizing region. There are also heterochromatic segments (knobs) which can be present in up to 22 specific locations on the 10 chromosomes, most of the locations being in the distal third of the chromosome arms (McClintock, 1978). Finally, supernumerary B chromosomes are characteristic of many populations of maize, and they contain large segments of heterochromatin. Cytologists have noted that knob heterochromatin is different in appearance from other classes of heterochromatin, having, in the light microscope, a uniform dense appearance with smooth contours. This structural feature of knob heterochromatin may be indicative of a specific sequence composition. An observation suggesting such sequence differentiation among the categories of heterochromatin in maize is that each has a characteristic time of replication in the cell cycle (Pryor et al., 1980). The DNA sequences of knobs are the last regions of the chromosomal DNA to complete replication.

The largest knob (K10) occurs in a segment on the long arm of chromosome 10. This segment, when present, defines abnormal chromosome 10, which is known to have a rearrangement of the terminal gene loci (Rhoades, pers. comm.) and contains three prominent chromomeres as well as the large knob. Abnormal 10 produces a spectacular change in the course of meiosis. When it is present in either heterozygous or homozygous condition, all other heterochromatic knobs in the complement act as precocious centromeres (neocentromeres) in the two meiotic divisions. In megasporogenesis, in which only a single meiotic product is functional, the action of the neocentromeres results in preferential delivery of linked alleles into the egg cell. All knobs can be induced by abnormal chromosome 10 to function as centromeres, and no other regions of the chromosome can be so induced. The genetics and cytogenetics of both neocentric activity and preferential segregation have been well characterized by Rhoades and his colleagues (Rhoades, 1978).

DNA from maize stocks having low, intermediate, or high frequencies of knobs in their genome displays a satellite DNA peak whose size directly correlates with knob content (Peacock et al., in prep.). A DNA sequence of ~ 185 bp is the major component of this satellite and is a major sequence component of knob heterochromatin (Figure 5.8); for example, where knob size alleles are known, an increase in size is associated with a corresponding increase in the amount of the ~ 185-bp sequence in in situ hybridization experiments. We cannot exclude the possibility that other sequences are contained within knob heterochromatin; but in all 14 of the different knobs we have examined, the ~ 185-bp sequence is certainly the predominant constituent.

The existence of this sequence in only one class of heterochromatin with specific cytogenetic properties, provides a direct correlation between a highly repeated sequence and an intranuclear function. It is not a function that is essential to maize because stocks can be constructed that lack all but the terminal knob on the short arm of chromosome 6 – and even this knob could probably be excluded if an appropriate chromosome rearrangement could be selected. However, different geographic races of maize have characteristic knob constitutions so it is possible that knobs have a positive selective value in some situations (McClintock, 1978).

The K10 knob must contain sequences other than the ~ 185-bp repeat unit because K10 alone has the capacity to induce neocentric activity in knob heterochromatin. This capacity must itself be dependent on a repeated sequence because dissection of the K10 knob by chromosome breakage has shown that the capacity for preferential segregation and presumably for neocentromere induction is contained in both the proximal and distal segments of the knob (Miles, 1970, as quoted in Rhoades, 1978).

Rhoades and his students have delimited a number of other cytogenetic properties of heterochromatin in maize, and it should be possible to determine whether there are further correlations between specific repeated sequences and specific functions.

Figure 5.8. In situ hybridization of the ~ 185-bp DNA repeat in maize showing that it is localized in the heterochromatic knobs and that it does not occur in the pericentromeric heterochromatin or in the heterochromatin of the nucleolus organizer regions. It also occurs on the small proximal heterochromatic knob of the B chromosome.

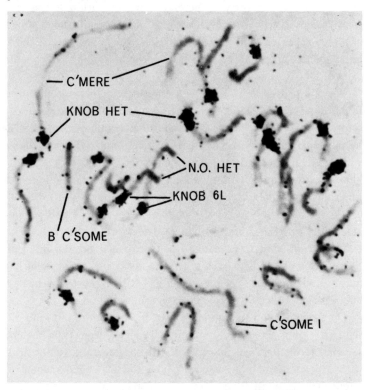

Discussion

Heterochromatin is not a genetically inert component of the cellular chromatin devoid of differentiation. Instead there are defined classes of heterochromatin distinguished on the bases of chromosomal location, differential staining properties, time of replication, and DNA sequence. In maize particular cytogenetic or genetic effects can be associated with each heterochromatic class together with a specific sequence organization.

Even within heterochromatic blocks with similar chromosomal locations and staining properties there is sequence differentiation; for example, the pericentromeric heterochromatin of each of the *D. melanogaster* chromosomes has a different array of satellite sequences. The segmental differentiation of heterochromatin is widespread and extends even to organisms that possess only one major satellite DNA located on all chromosomes. This differentiation within heterochromatic blocks may reflect some functional role or else be a consequence of amplification steps occurring during the generation of the repeated DNA arrays. The observations of constancy of amount and location of heterochromatin within a species and the relative conservation over evolutionary time of specific nucleotide sequence in highly repeated DNAs may also reflect selective pressure for a particular sequence, or it may be the consequence of an inherent property of this class of DNA.

Are the properties of heterochromatin sequence-specific or not? The functioning of the particular loci within the heterochromatin of chromosome 2 of *D. melanogaster* may require a certain environment generated by particular adjacent repeating sequences; equally, it may be associated with general properties of heterochromatin. The fact that there is a knob-specific sequence in maize may indicate that some heterochromatic functions are sequence-specific. Hsieh and Brutlag (1978) showed that one of the satellite DNAs of *D. melanogaster* (the 1.688) is specifically associated with particular protein(s). The neocentric activity of knobs may result from a similar repeated DNA sequence–protein interaction. The aggregation of knobs to each other and not to other classes of heterochromatin, and the fact that they have a distinctive cytological appearance, may also be properties dependent upon such proteins.

The variability in the amount of heterochromatin and kind of repeated sequences seen between closely related species argues that heterochromatin may have functions associated with its general structure rather than with a particular DNA sequence. The classical phenomenon of position effect variegation in *D. melanogaster* is suggestive of this type of function.

Position effect variegation is the phenomenon of the partial suppression of euchromatic loci when placed in juxtaposition to heterochromatin by chromosomal rearrangement (reviewed in Spofford, 1976). Virtually all euchromatic loci appear to be susceptible to this effect of heterochromatin when placed in such juxtaposition. Heterochromatic loci, however, function well amid the highly repeated DNA sequences constituting the bulk of heterochromatin, suggesting that the highly repeated sequence DNA may provide a molecular environment facilitating the function of heterochromatic loci yet partially repressing euchromatic loci brought into close proximity. Explanation for roles of heterochromatin in positioning of chromosomes, or in the neocentromere behavior, could also be interpreted in terms of general structure of heterochromatin rather than the presence of particular DNA sequences.

The functions of heterochromatin may be multiple, some being dependent upon specific DNA sequence and some merely requiring a highly repeated DNA sequence with the correct coiling properties. These different functions may be reflected in the segmental differentiation of heterochromatin seen at the cytological, genetic and sequence levels.

We have benefited from discussions with Professor M. M. Rhoades and Dr. R. Appels. We thank Dr. Appels for permission to include Figures 5.4 and 5.5.

References

Appels, R., and Peacock, W. J. 1978. The arrangement and evolution of highly repeated (satellite) DNA sequences with special reference to *Drosophila*. *Int. Rev. Cytol. Suppl.* 8:69–126.

Appels, R., Driscoll, C. J., and Peacock, W. J. 1978. Heterochromatin and highly repeated DNA sequences in rye (*Secale cereale*). *Chromosoma (Berl.)* 70:67–79.

Appels, R., Dennis, E. S., and Peacock, W. J. Analysis of the heterochromatic blocks of rye chromosomes using specific repeated DNA sequences (in preparation).

Ayles, G. B., Sanders, T. G., Kiefer, B. I., and Suzuki, D. T. 1973. Temperature sensitive mutations in *Drosophila melanogaster*. XI. Male sterile mutants of the Y chromosome. *Dev. Biol. 32:*239–57.

Beauchamp, R. S., Mitchell, A. R., Buckland, R. A., and Bostock, C. J. 1979. Specific arrangements of human satellite. III. DNA sequences in human chromosomes. *Chromosoma (Berl.) 71:*153–66.

Bedbrook, J. R., Jones, J., O'Dell, M., Thompson, R. D., and Flavell, R. B. 1980. A molecular description of telomeric heterochromatin in *Secale* species. *Cell 19:*545–60.

Bendich, A. J., and McCarthy, B. J. 1970. DNA comparisons among barley, oats, rye and wheat. *Genetics 65:*545–65.

Bridges, C. B. 1916. Non-disjunction as proof of the chromosome theory of heredity. *Genetics 1:*1–52, 107–63.

Brosseau, G. E., Jr. 1960. Genetic analysis of the male fertility factors on the Y-chromosome of *Drosophila melanogaster*. *Genetics 45:*257–74.

Brutlag, D., Appels, R., Dennis, E. S., and Peacock, W. J. 1977. Highly repeated DNA in *Drosophila melanogaster*. *J. Mol. Biol. 112:*31–47.

Carnevale, A., Ibanez, B. B., and Del Castillo, V. 1976. The segregation of C band polymorphisms on chromosomes 1, 9 and 16. *Am. J. Hum. Genet. 28:*412–16.

Cooper, K. W. 1964. Meiotic conjunctive elements not involving chiasmata. *Proc. Natl. Acad. Sci. U.S.A. 52:*1248–55.

Dawid, I. B. and Botchan, P. 1977. Sequences homologous to ribosomal insertions occur in the *Drosophila* genome outside the nucleolus organizer. *Proc. Natl. Acad. Sci. U.S.A. 74:*4233–7.

Dennis, E. S., Dunsmuir, P., and Peacock, W. J. 1980a. Segmental amplification in a satellite DNA: restriction enzyme analysis of the major satellite of *Macropus rufogriseus. Chromosoma (Berl.)* (in press).

Dennis, E. S., Gerlach, W. L. and Peacock, W. J. 1980b. Identical polypyrimidine– polypurine satellite DNAs in wheat and barley. *Heredity* (in press).

Dunsmuir, P. 1976. Satellite DNA in the kangaroo *Macropus rufogriseus. Chromosoma (Berl.) 56:*111–25.

Fry, K., and Salser, W. 1977. Nucleotide sequences of HSα satellite from kangaroo rat *Dipodomys ordii* and characterization of similar sequences in other rodents. *Cell 12:*1069–74.

Gall, J. G., and Atherton, D. D. 1974. Satellite DNA sequences in *Drosophila virilis. J. Mol. Biol. 85:*633–64.

Gosden, J. R., Mitchell, A. R., Sueanez, H. N., and Gosden, C. M. 1977. The distribution of sequences complementary to human satellite DNAs I, II and IV in the chromosomes of chimpanzee (*Pan trogolodytes*), gorilla (*Gorilla gorilla*) and orangutan (*Pongo pygmaeus*). *Chromosoma (Berl.) 63:*253–71.

Gosden, J., Lawrie, S., and Sueanez, H. 1978. Ribosomal and human-homologous repeated DNA distribution in the orangutan (*Pongo pygmaeus*). *Cytogenet. Cell Genet. 21:*1–10.

Heitz, E. 1929. Heterochromatin, chromocentren, chromomenen. *Ber. Dtsch. Bot. Ges. 47:*274–84.

Hewitt, G. M. 1975. A new hypothesis for the origin of the parthenogenetic grasshopper *Moraba virgo. Heredity 34:*117–23.

Hilliker, A. J. 1976. Genetic analysis of the centromeric heterochromatin of chromosome 2 of *Drosophila melanogaster:* deficiency mapping of EMS-induced lethal complementation groups. *Genetics 83:*765–82.

Hilliker, A. J., and Holm, D. G. 1975. Genetic analysis of the proximal region of chromosome 2 of *Drosophila melanogaster.* 1. Detachment products of compound autosomes. *Genetics 81:*705–21.

Horz, W., and Zachau, H. G. 1977. Characterisation of distinct segments in mouse satellite DNA by restriction nucleases. *Eur. J. Biochem. 73:*383–92.

Hsieh, T. S., and Brutlag, D. L. 1978. A protein which preferentially binds *Drosophila* satellite DNA. *Proc. Natl. Acad. Sci. U.S.A. 76:*726–30.

Jagannath, D. R., and Bhatia, C. R. 1972. Effect of rye chromosome 2 substitution on kernel protein content of wheat. *Theor. Appl. Genet. 42:*89–92.

Judd, B. H., Shen, M. W., and Kaufman, T. C. 1972. The anatomy and function of a segment of the X chromosome of *Drosophila melanogaster. Genetics 71:*139–56.

Kaufmann, B. P. 1934. Somatic mitoses of *Drosophila melanogaster. J. Morphol. 56:*125– 55.

Kavenoff, R., and Zimm, B. H. 1973. Chromosome sized DNA molecules from *Drosophila. Chromosoma (Berl.) 41:*1–27.

Kennison, J. A. 1979. Studies on the organization of the Y chromosome of *Drosophila melanogaster.* Ph.D. thesis, University of Califonia, San Diego.

Krider, H. M., and Levine, B. I. 1975. Studies on the mutation abnormal oocyte and its interaction with the ribosomal DNA of *Drosophila melanogaster. Genetics 81:*501–13.

Krider, H. M., Yedvobnick, B., and Levine, B. I. 1979. The effect of abo phenotypic expression on ribosomal DNA instabilities in *Drosophila melanogaster. Genetics 92:*879–89.

Lis, J. T., Prestidge, L., and Hogness, D. S. 1978. A novel arrangement of tandemly repeated genes at a major heat shock site in *D. melanogaster. Cell 14:*901–19.

McClintock, B. 1978. Significance of chromosome constitutions in tracing the origin and migration of races of maize in the Americas. In *Maize Breeding and Genetics* (ed. Walden, D. B.) pp. 159–84. New York: Wiley.

May, C. E., and Appels, R. 1978. Rye chromosome 2R substitution and translocation lines in hexaploid wheat. *Cereal Res. Commun. 6:*231–4.

Muller, H. J., and Painter, T. 1932. The differentiation of the sex chromosomes of *Drosophila* into genetically active and inert regions. *Z. Indukt. Abstamm. Vererbungsl. 62:*316–65.

Neuhaus, M. J. 1939. A cytogenetic study of the Y chromosome of *Drosophila melanogaster. J. Genet. 37:*229–54.

Pardue, M. L., and Gall, J. G. 1970. Chromosomal localization of mouse satellite DNA. *Science 168:*1356–8.

Parry, D. M., and Sandler, L. 1974. The genetic identification of a heterochromatic segment on the X chromosome of *Drosophila melanogaster. Genetics 77:*535–9.

Peacock, W. J. 1979. Strandedness of chromosomes and segregation of replication products. *Cell Biology: A Comprehensive Treatise,* vol. 2 (ed. Goldstein, L., and Prescot, D. M.), pp. 363–88. New York: Academic Press.

Peacock, W. J., and Miklos, G. L. G. 1973. Meiotic drive in *Drosophila:* new interpretations of the segregation distortion and sex chromosome systems. *Adv. Genet. 17:*361–409.

Peacock, W. J., Brutlag, D., Goldring, E., Appels, R., Hinton, C. W., and Lindsley, D. L. 1973. The organization of highly repeated DNA sequences in *Drosophila melanogaster* chromosomes. *Cold Spring Harbor Symp. Quant. Biol. 38:*405–16.

Peacock, W. J., Appels, R., Dunsmuir, P., Lohe, A. R., and Gerlach, W. L. 1977a. Highly repeated DNA sequences, chromosomal localization and evolutionary conservatism. In *International Cell Biology* (ed. Brinkley, B. R., and Porter, K. R.), pp. 494–506. New York: Rockefeller University Press.

Peacock, W. J., Lohe, A. R., Gerlach, W. L., Dunsmuir, P., Dennis, E. S., and Appels, R. 1977b. Fine structure and evolution of DNA in heterochromatin. *Cold Spring Harbor Symp. Quant. Biol. 42:*1121–35.

Peacock, W. J., Appels, R., Endow, S., and Glover, D. Chromosomal distribution of the major insert in *Drosophila melanogaster* 28S rRNA genes. *Genet. Res.* (in press).

Peacock, W. J., Dennis, E. S., Pryor, A. J., and Rhoades, M. M. Repeated DNA limited to knob heterochromatin in maize (in preparation).

Petes, T. D., Newlong, C. S., Byers, B., and Fangman, W. L. 1973. Yeast chromosomal DNA: size, structure and replication. *Cold Spring Harbor Symp. Quant. Biol. 38:*9–16.

Procunier, J. D., and Tartof, K. D. 1978. A genetic locus having *trans* and contiguous *cis* functions that control the disproportionate replication of ribosomal RNA genes in *Drosophila melanogaster. Genetics 88:*67–79.

Pryor, A. J., Faulkner, K., Rhoades, M. M., and Peacock, W. J. 1980. Asynchronous replication of heterochromatin in maize *Proc. Natl. Acad. Sci. U.S.A.* (in press).

Rhoades, M. M. 1978. Genetic effects of heterochromatin in maize. In *Maize Breeding and Genetics* (ed. Walden, D. B.), pp. 641–72. New York: Wiley.

Rubin, G. 1977. Isolation of a telomeric DNA sequence from *Drosophila melanogaster. Cold Spring Harbor Symp. Quant. Biol. 42:*1041–6.

Rudkin, G. T. 1964. The structure and function of heterochromatin. In *Genetics Today,* p. 239. Oxford: Pergamon Press.

Salser, W., Bowen, S., Browne, D., Adli, F. E., Federoff, N., Fry, K., Heindell, H., Paddock, G., Poon, R., Wallace, B., and Whitcome, P. 1976. Investigation of the organization of mammalian chromosomes at the DNA sequence level. *Fed. Proc. 35:*23–35.

Sandler, L. 1977. Evidence for a set of closely linked autosomal genes that interact with sex-chromosome heterochromatin in *Drosophila melanogaster. Genetics 86:*567–82.

Schultz, J. 1947. The nature of heterochromatin. *Cold Spring Harbor Symp. Quant. Biol. 12:*179–91.

Sears, E. R. 1963. Chromosome mapping with the aid of telocentrics. In *Proc. 2nd Int. Wheat Genet. Symp.* Published as *Hereditas Suppl. 2* (1966).

Singh, L., Purdom, I. F., and Jones, K. W. 1976. Satellite DNA and evolution of sex chromosomes. *Chromosoma (Berl.) 59:*43–62.

Spofford, J. 1976. Position-effect variegation in *Drosophila.* In *The Genetics and Biology of Drosophila,* vol. 1c, pp. 955–1018. London: Academic Press.

Sten, C. 1929. Untersuchungen uber Aberrationen des Y-Chromosoms von *Drosophila melanogaster. Z. Indukt. Abstamm. Verebungsl. 51:*253–353.

Strobel, E., Dunsmuir, P., and Rubin, G. M. 1979. Polymorphisms in the chromosomal location of elements of the *412, copia, 297* dispersed repeated gene families in *Drosophila. Cell 17:*429–39.

Venolia, L. 1977. Highly repeated DNA and kangaroo phylogeny. M.Sc. thesis, Australian National University, Canberra.

White, M. J. D., Contreras, N., Cheney, J., and Webb G. C. 1977. Cytogenetics of the parthenogenetic grasshopper *Warramaba* (formerly *Moraba*) *virgo* and its bisexual relatives. II. Hybridization studies. *Chromosoma (Berl.) 61:*127–48.

White, R. L., and Hogness, D. S. 1977. R loop mapping of the 18S and 28S sequences in the long and short repeating units of *Drosophila melanogaster* rDNA. *Cell 10:*177–92.

Young, M. W. 1979. Middle repetitive DNA; a fluid component of the *Drosophila* genome. *Proc. Natl. Acad. Sci. U.S.A. 76:*6274–8.

6   A glimpse of the tridactyloid karyotype

ROBERT BLACKITH

The tridactyloids (Orthoptera) are small, usually dark, fossorial insects that graze algal mud in tropical and subtropical rivers. The mud passes through the gut in a manner reminiscent of earthworm digestion. Although it was clearly demonstrated by Dufour as long ago as 1838 that tridactyloids are related to the short-horn grasshoppers (Caelifera), their superficial resemblance to mole-crickets led taxonomists to classify them with the Gryllidae until recently. This fact, together with the paucity of external characters of taxonomic value, has led to an unsatisfactory classification such that, in several major museum collections, as many as six species distinguishable on the basis of the male genitalia stand over a single specific label.

Recently, following Günther's (1974) example, my wife and I have completed a partial revision of the tridactyloids of the western Old World, based on the male genitalia. We have occasionally had the opportunity to examine the karyotypes of some species collected for other purposes, and find that the tridactyloids offer promising cytogenetic material, having a small number of large chromosomes and abundant meiotic divisions visible from testis squashes stained, for instance, with aceto-orcein.

Little is known about tridactyloid karyotypes, and the species to which the few earlier papers refer must be regarded as provisionally determined. Ohmachi (1935) studied *Tridactylus japonicus* de Haan and found six bivalents and the X chromosome at first mitosis, giving 2n = 13 for males. His statement that the testes cannot be taken out seems unduly pessimistic, though the dissection is delicate. Ohmachi found two acrocentric and 10 metacentric autosomes and noted that the X was metacentric.

Helwig (1958) asserts that 2n = 13 for an unspecified tridactyloid of the genus *Tridactylus* and that 2n = 15 for another belonging to the New World genus *Rhipipterix*. He claims that the chromosomes are very small, and considers their size evidence of the primitive nature of the group within the orthopteroid insects. However, chromosomes of *T. variegatus* compare favorably in size with the chromosomes of the eumastacid grasshoppers of the tropics, as drawn by White (1973:423).

Pericentric inversions in *T. variegatus*

T. variegatus is a species that ranges from being the sole representative of the genus in Western Europe to being one member of a richer tridactyloid fauna in the Middle East. The eastern limits of the range of *variegatus* just extend into the northwestern corner of Iran. Reports of this species from farther east should be credited only if the male genitalia have been examined, as the name has been a "dustbin" for phenotypically similar material.

Males of *variegatus* from the Rio Cinca, in northeastern Spain, were

Figure 6.1. First metaphase (side view) of an individual of *T. variegatus* from the Rio Cinca, Spain, showing some banding; 2n (♂) = 11.

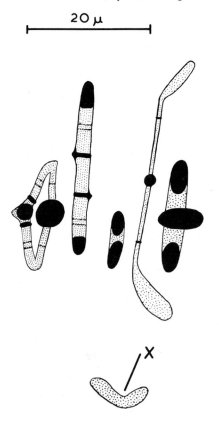

kindly collected by Dr. M. C. D. Speight, and injected with acetic alcohol
on capture. They proved to have the karyotype illustrated in Figure 6.1. In
the specimen illustrated, evidence of banding was obtained even though no
special treatment to reveal such banding was employed. For most of these
individuals from Rio Cinca, 2n (δ) = 11, and there was no strong evidence
of heteromorphism or asymmetry in the arms of the larger metacentric
autosomes. One such individual, however, had the second largest metacentric frankly asymmetric.

Material from Mallemort, on the Durance River in S. France, shows
strong heteromorphism in the three metacentrics, as shown in Figure 6.2.
"Broken" chromosomes were observed in six individuals from Mallemort,
a locality where the tridactyloid habitat has since been almost entirely
destroyed by river works. Here also, 2n (δ) = 11, as it is in material from
Monasterace, on the Neto River, Calabria, Italy, where at least one of the
metacentric autosomes may be heteromorphic. At prophase, the X chromosome is condensed, whereas the autosomes are thin and diffuse, as
occurs regularly in the Caelifera (White, 1973:26). The sex-determining
mechanism throughout the species that we have investigated seems to be
XO.

Figure 6.2. First metaphase (side view) of an individual of *T. variegatus*
from Mallemort, Durance River, France. In this individual the three
long metacentrics are all heteromorphic, and "broken" chromosomes are
visible.

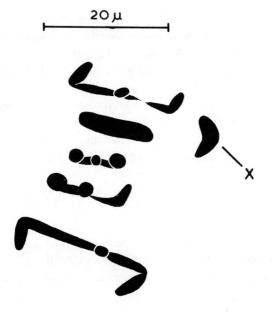

Other species

We have been able to examine testis squashes from material collected from several tropical regions. This material was collected for other purposes and could not be examined until long periods had passed; thus little more than chromosome counts was possible.

T. tartarus Saussure (Chalus, N. Iran)	2n (δ) = 11
T. pfaendleri Harz (Kösk, Aydin, Turkey)	2n (δ) = 11
T. frontomaculatus Günther (Palghat, Kerala, India)	2n (δ) = 11
T. hieroglyphicus Bei-Bienko (Musakaf, Bolan Gorge, Pakistan and Jalalabad, Afghanistan)	2n (δ) = 11
T. nigripennis Chopard (Kizhake Chalakudi River, Kerala, India)	2n (δ) = 13
T. tithonus (Blackith and Blackith, 1979 (Tacazzé River, Ethiopia, under bridge south of Enda Selassie at junction of Tigré and Begemdir provinces)	2n (δ) = 13

The karyotype of the Indian species is shown in Figure 6.3, and the karyotype of the Ethiopian species is shown in Figure 6.4.

Discussion

The tridactyloids offer promising material for cytogenetic studies, having few, large, chromosomes. Although in the past their taxonomy has been unsatisfactory, they can now be uniquely identified from the male genitalia, and many species have already been so characterized (Blackith and Blackith, 1979). Giles and Webb (1972) have drawn attention to the value of cytology in clarifying the confusion arising when the workers concerned with a group of organisms become convinced that there are species in the group that are very variable and cosmopolitan; and such confusion has certainly reigned as far as *T. variegatus* is concerned.

However, there is a particular reason for expecting the cytology of the tridactyloids to reward investigators. In most temperate species, and to varying extents in tropical ones, only a few individuals are capable of flight, and nonflying individuals are very much less likely to found new colonies, given that many tridactyloids live in mountain valleys. Even those individuals that possess long wings may have no adequate flight muscles to power them.

For instance, only 3.0% of males and 4.5% of females were capable of flight in Eurasian *T. pfaendleri,* yet many of the flightless females were gravid. We have never found a specimen of *T. variegatus* capable of flight, though Dufour (1838), who lived in France when that species was evidently far more abundant there than it is now, found about 0.1% long-winged individuals, not all of which were necessarily capable of flight.

Figure 6.3. Presumed spermatogonial metaphase of *T. nigripennis* from the Kizhake Chalakudi River, Kerala State, India. In this individual 2n (♂) = 13.

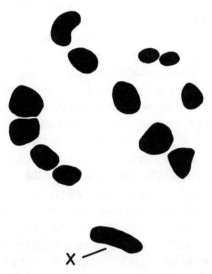

Figure 6.4. Karyotype of *T. tithonus* from the Tacazzé River, Ethiopia. In this individual 2n (♂) = 13.

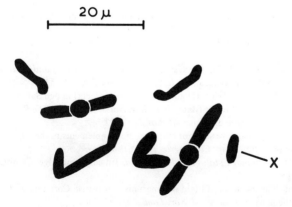

Thus new colonies of *Tridactylus* are likely to depend on a very few individuals alighting in new territory, and the founder principle may well be of enhanced importance. Indeed, unusual karyotypic rearrangements, occurring in a winged individual, may well have exceptional chances of isolation and survival. We do not know whether alary polymorphism is associated with particular karyotypes; but because *T. variegatus,* at least, is polymorphic for pericentric inversions, it would be gratuitous to assume that there was no differential dispersal of the different karyotypes.

Young (1966) has shown that corixid bugs on the surface of unstable water masses develop functional wings at times when the water mass is likely to dry up during the lifetime of the insect; there is no evidence that such a mechanism operates for tridactyloids. There is a general tendency for tropical forms to have more individuals capable of flight, and in the tropics water bodies are likely to be less stable than in temperate climates. Indeed, tropical and subtropical rivers offer a continually renewed series of mudbanks and recently inundated areas ready for colonization by the occasional tridactyloid capable of flight. Although they are occasionally taken at lights in the tropics (*T. berlandi* was taken by us in this way at Siliguri, NE India), none was collected in this way in southern Italy even when light-traps were set up within a mile or so of a riverbed (the Lao) containing numerous *T. variegatus*. The opportunities for the bearer of a karyotypic rearrangement to find an uncolonized habitat free from much competition are good; in addition to dispersal by flight, individuals are no doubt flushed downstream by flood waters. An investigation of possible stasipatric speciation in tridactyloids might well be rewarding, although hybridization experiments are impractical at present. No one has as yet succeeded in breeding them in captivity.

References

Blackith, R. E., and Blackith, R. M. 1979. Tridactyloids of the western Old World. *Acrida (Paris)* 8:189–217.
Dufour, L. 1838. Recherches sur l'histoire du Tridactyle panaché. *Ann. Sci. Nat., Zool.* 9:321–34.
Giles, E. T., and Webb, G. C. 1972. The systematics and karyotype of *Labidura truncata* Kirby 1903 (Dermaptera: Labiduridae). *J. Aust. Entomol. Soc. 11:*253–6.
Günther, K. K. 1974. Uber die Tridactyloidea (Saltatoria, Insecta) in den Sammlungen des Museums für Naturgeschichte der Stadt Genf. *Rev. Suisse Zool. 81:*1027–74.
Helwig, E. R. 1958. Cytology and taxonomy. *Bios 29:*58–72.
Ohmachi, F. 1935. A comparative study of chromosome complements in the Gryllodea in relation to taxonomy. *Bull. Imp. Coll. Agric. For. 5:*1–48.
White, M. J. D. 1973. *Animal Cytology and Evolution,* 3rd ed. Cambridge: Cambridge University Press, 961 pp.
Young, E. C. 1966. The incidence of flight polymorphism in British Corixidae and description of the morphs. *Proc. Zool. Soc. Lond. 146:*567–76.

PART III

HYBRID ZONES

7   Hybrid zones and speciation

N.H.BARTON AND G.M.HEWITT

One of the most central, and yet most difficult tasks of population genetics is to find out how new species are formed. Many models of speciation have been put forward (see Mayr, 1970; Bush, 1975; White, 1978a); they range from those in which two geographically isolated populations diverge so that they can no longer mate successfully with each other when they meet again (allopatric speciation), to those in which two species crystallize out of a single panmictic population (sympatric speciation). Between these two poles lies a multitude of other possibilities, in which at some stage two partially isolated populations meet and hybridize in a narrow zone. In some models, such hybrid zones are irrelevant to the further evolution of the nascent species, but in others, they are crucial sites for the development of ever stronger genetic isolation (Dobzhansky, 1970; White, 1978b). To assess the plausibility of the many theories on the subject, and to suggest ways of testing them, we need to understand the nature and behavior of hybrid zones between incompatible races.

The nature of hybrid zones

Throughout this discussion, a hybrid zone will be defined as a narrow cline, maintained by some sort of hybrid unfitness. Other definitions, based on the history of the cline, have been proposed, but it seems more practical to use qualities that are directly observable (see Woodruff, 1973). Endler (1977) has defined hybrid zones as "narrow belts within which there is greatly increased variability in fitness." This is a similar definition, though there will be greater variability in fitness at the center of any cline maintained by a conflict between natural selection and dispersal; also, hybrid unfitness is somewhat easier to detect than an increased variance in fitness.

When two expanding populations meet, their dispersal will tend to mix them together, producing a gradually broadening cline between the two types. However, if they have evolved along different paths, hybrids between them may be unfit in some way. Organisms that cross into foreign territory will usually mate with the native race, and so will produce a greater proportion of unfit hybrids than will the native type. These immigrants will therefore contribute less to the next generation, and so the parental population will remain pure despite the effects of random dispersal. A stable hybrid zone will be set up, its width determined by the balance between dispersal and selection against hybrids (Bazykin 1969, 1972a,b, 1973; Key, 1974). The zone might involve a variety of genetic characters; morphological, chromosomal, physiological, or behavioral. There might well be assortative mating for some of these characters; this will not prevent the formation of the stable zone, but will make it somewhat wider because there will be fewer hybrids for selection to act on. Indeed, mating preferences could maintain a hybrid zone if hybrids make unattractive mates. The precise nature and details of the differences between the two abutting races are irrelevant here; the crucial factor is the presence of some hybrid unfitness.

Most complex selective systems allow several stable equilibria because there may be several ways of satisfying the same selective constraints (e.g., Bodmer and Felsenstein, 1967; Feldman and Balkau, 1972; Feldman, et al., 1974). Steep clines can form between races fixed at different equilibria, at least when the organism is distributed in demes (Karlin and McGregor, 1972a,b). Stable clines will also form in a continuously distributed organism, unless, for example, a new and advantageous recombinant was produced when the parental types met. The crucial parameters are the strength of the selection against hybrids, the number of loci involved, the degree of assortative mating, and the dispersal rate of the organism.

The movement of hybrid zones

A particularly interesting feature of hybrid zones is that they can move. Ignoring for the moment the true complexity of the environment, there is no intrinsic reason for them to be in any particular place, because they are maintained by internal, genetic factors. Consequently, a variety of forces can push them from place to place. If one race has a selective advantage over the other, the zone will move forward in favor of the superior type. It is most unlikely that the two pure types are selectively identical, and so this effect would seem to make hybrid zones positionally unstable (Key, 1968; Barton, 1979a). However, other forces are also important. If there are more organisms on one side of the zone than on the other, they will push it forward by weight of numbers; asymmetric dispersal will have a similar effect. Populations at the center of the zone will receive more immigrants from one side than from the other, and so the frequency of the more abundant or more mobile type will rise. The zone will therefore move toward regions of low population density and low dispersal, and so be trapped at natural barriers. For example, suppose that a zone, involving a single locus, is maintained by 10% selection against heterozygotes. If the dispersal rate of the organism is 100 meters in a generation, then the zone will be 1 km wide. (The dispersal rate is measured by the root mean square distance between parent and offspring, and the width is defined as the inverse of the maximum slope.) If one form has a selective advantage of 1% in the homozygote, then the zone will advance in its favor by 2.5 meters per generation; eventually, the inferior race will be eliminated. However, if the zone reaches a region where the population density doubles in 3.5 km or less, it will be halted. The critical density gradient will be roughly proportional to the selective difference driving the zone. In nature, much steeper density gradients are likely, and so stronger forces can be resisted. It is hard to imagine a hybrid zone moving far in the face of such obstacles; only alleles that are advantageous when rare can advance because a few long-distance migrants can carry them past local barriers.

It might be objected that population density varies greatly from year to year, as well as from place to place, and so a hybrid zone could be thrown past a barrier by random forces. A shallow density trough might occasionally be inverted, allowing the zone to move on. Such fluctuations in population structure can certainly move a zone considerable distances, but even so, slight barriers will still impede them. For example, suppose that the coefficient of variation ($\sigma/\bar{x}$) of population density is about 50%, and that fluctuations are positively correlated over distances of a kilometer and periods of 10 generations. These are very favorable conditions for zone movement, and a 1-km-wide zone would move in a random walk of about 20 meters in a generation. However, if the average abundance were not perfectly even, but varied by as little as 5% over distances of a kilometer or so, the zone would only move 1 meter in a generation. Stronger, more realistic barriers would virtually halt it; in addition, a long hybrid zone would move less than the one-dimensional zones considered so far because bulges in one place would be pulled back by the rest of the zone (Barton, 1979a).

Effects on gene flow

We have seen that hybrid unfitness can maintain a sharp, stable, cline. Such a cline will act as a barrier to gene flow at linked loci, and so might encourage further divergence between the two races involved (Key, 1974; White, 1978a; Barton, 1979b). Consider a neutral allele, fixed on one side of the zone, but absent on the other. It can only spread into the other gene pool through recombination within the zone. If linkage between the neutral and selected loci is tight compared with the selection maintaining the zone, the spread of the neutral allele will be considerably delayed. When two populations that have diverged at many loci abut, some alleles may prove incompatible, so that sharp clines form for these genes. Other differences may not matter selectively, and will gradually decay. This decay will be impeded by the selected clines, and so as time goes on a gradually decreasing segment of chromosome will remain differentiated (Figure 7.1).

Of course, selectively neutral differences are not important in the formation of new species, although their behavior must be known in order to interpret observations on natural clines. We are interested in knowing how strong a barrier to adaptive alleles a hybrid zone is likely to be, and in

Figure 7.1. Gene flow past a cline. If selection maintains a cline at a locus, gene flow at nearby loci will be impeded. The decay of geographic variation in neutral allele frequency will be delayed. This figure shows the length of chromosome, $\sim 2\,(s/t)^{1/2}$, at which neutral differentiation remains after various times.

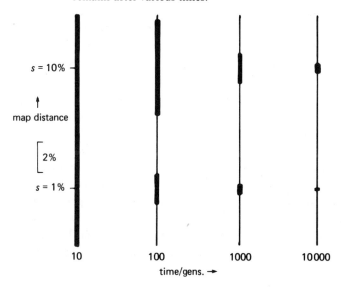

knowing which types of allele are likely to strengthen the isolation between two races. Though neutral alleles can be considerably impeded by a strong zone, advantageous alleles will not be much affected by barriers, whether produced by clines or by regions of low population density (Skellam, 1973). For example, consider a locus that recombines at 0.1% with a cline that is maintained by 10% selection. A neutral difference at such a locus will take 100,000 generations to decay, but an allele with an advantage of only 0.1% will get past the barrier in 2500 generations (Barton, 1979b). A hybrid zone must therefore involve many closely linked loci, and strong selection, in order to have much effect on gene flow. Assortative mating will reduce gene flow, but it must be quite strong in order to help much. The important factor is the number of hybrids present; there must be far fewer than expected with random mating for much of the genome to be effectively isolated.

The origins of incompatibilities

The preceding theoretical discussion suggests that only very strong hybrid zones are of much importance in preventing gene flow and allowing divergence. Such strong clines do exist; indeed, most hybrid zones seem to involve a large fraction of the genome, and seem to be effective barriers to gene flow (see below, section on "The evidence"). It is important to know how these severe incompatibilities arose, and how changes at different loci were brought together, so that one can consider whether the races involved are likely to evolve into full, sympatric species.

It is hard to see how an allele, or a set of alleles, which at present are at a disadvantage when rare, could have become fixed in the first place. Either they were fixed by some stochastic process or they were originally advantageous. The simplest explanation, and the one usually invoked (e.g., White, 1968; Carson, 1975; Hewitt, 1975), is that fixation occurred in a small, isolated population by sampling drift. The chances of fixation of the new type in such a deme can be calculated using the diffusion equation approximation (see, e.g., Crow and Kimura, 1970). Suppose an allele, A, has a heterozygote disadvantage, s, superimposed on some advantage S; the fitnesses of the AA, Aa, and aa genotypes are thus $(1 + 2S) : (1 - s + S) : 1$. The allele will tend to decrease when present at a frequency less than $p^* = (s - S)/2s$. The probability of fixation of a new A mutant, u, is given by the formula:

$$u^{-1} = \frac{2N}{\gamma} e^{-\gamma^2 p^{*2}/2} \int_{-\gamma p^*}^{\gamma(1-p^*)} e^{-t^2/2} \, dt, \, \gamma = (8N_e s)^{1/2}$$

(see Nei, 1975a). The graph (Figure 7.2) shows the expected time before fixation begins $(2Nu\mu)^{-1}$ in a population of size $N = N_e$, with mutation rate $\mu = 10^{-5}$, and $s = 10\%$, for various p^*. It turns out that the initial isolate must be very small (an effective size of 10–100 individuals) for there to be any reasonable chance of fixation in the face of moderate selection, at least for the single-locus case (Bengtsson and Bodmer, 1976). Once it has been fixed locally, the new genotype must somehow spread out to establish itself over a wide area. We have seen that the interface between two incompatible genotypes is likely to be very stable, especially when selection is strong, and when the population is split up into a series of demes. The spread of the new form must therefore be by some sequence of extinction and recolonization, in which diffusion between demes is of negligible importance. Lande (1979) has shown that, provided demes exchange at most a few individuals per generation, and that the selection against heterozygotes is strong ($s > 8m$, where m is the fraction of the population exchanged between demes per generation), the fixation rate in the whole species, R, is equal to that in a single deme. The spread of a new genotype, which is strongly incompatible with the old, through a set of demes, is analogous to the spread of a neutral allele through a panmictic population. Lande shows, using this result, that sampling drift can account for the observed rate of fixation of chromosomal rearrangements, provided that the effective deme size is from a few tens to a few hundreds. This

Figure 7.2. Expected time to fixation of an allele with a heterozygote disadvantage of 10%. See text for explanation.

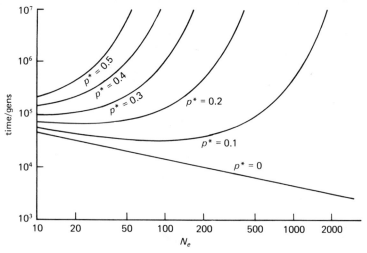

long-term estimate of the deme size is reasonably close to the few present-day measurements of deme size that are available. It should be noted that this analysis only applies to strongly selected changes; the effective deme size for weak changes will be larger, and so they may, paradoxically, evolve more slowly.

If this explanation for the spread of new incompatibilities is valid, each successful change would take roughly $2N_D T$ generations to reach fixation, where N_D is the effective number of demes (possibly much lower than the actual number of demes, if single demes often expand to cover large areas), and T is the average time scale of extinction and recolonization. One would therefore expect to observe a "heterozygosity" of $\sim 2N_D TR/(1 + 2N_D TR)$; this figure is the probability of finding at least one hybrid zone for a given character in a species. For insects, at least, N_D is likely to be very large. Let us take *Podisma pedestris* as an example; each deme (or patch of a few hundred insects) is roughly 10^3–10^4 m^2 in area, whereas the total distribution of this grasshopper in the European Alps is roughly 3×10^{10} m^2. Extinction and recolonization occurs, as a rough guess, over periods of $\sim 10^2$ generations, and so, using Lande's figure of 10^{-7}–10^{-8} for the rate of fixation of chromosomal rearrangements, we reach an estimate of $2N_D TR \sim 6$–600. In this species, one Robertsonian fusion has been fixed in one part of the well-studied Alpine region [it is in fact a rather weakly selected change ($\sim 0.5\%$; see section below on "The evidence")], and so R might be taken to be somewhat larger than Lande's figure. In general, it is very hard to know just how many chromosomal hybrid zones there are in a typical species, partly because species are sometimes described solely on the basis of their karyotype, and partly because information on the geographic distributions of the various karyotypes is scanty. The estimate is rather high, but, in view of the likely discrepancy between the actual and effective numbers of demes, simple drift would seem to be able to account for the observed numbers of chromosomal hybrid zones, as well as for the rate of evolution of the karyotype. The argument applies equally well to the establishment of other types of incompatibilities.

There are a number of variations on this simple, single-locus stochastic model. It has been suggested by White (1968, 1978a) that if the new allele had an advantage superimposed on its intrinsic heterozygote disadvantage (perhaps due to meiotic drive, in the case of a chromosomal mutation), then fixation by drift would be more likely. This is certainly true; Figure 7.2 shows the expected time to fixation of an allele with a heterozygote unfitness of 10%, for various additional advantages. Each allele has a critical frequency below which it is selected against, but above which it rises to fixation. With a critical frequency of 50% (pure heterozygote unfitness), fixation is only likely in a population of 10–20 individuals, whereas with a critical frequency of 20%, it becomes plausible in a population of a few hundred. However, clines with low critical frequencies would be rather unstable, and so one must still postulate a fairly small deme size. An imbalance between the two homozygotes would not affect the rate of spread of the new mutant if this spread were by pure extinction–recolonization, but it might speed the advance of one type if competition for recolonization at sites on the interface between the two types were important; even a slight bias in recolonization rate could greatly speed up the rate of fixation of the new karyotype in the whole species. This interdemic effect could well be more important than the intrademic effect shown in Figure 7.2. It would, however, be very hard to detect any such directional effect because hybrid zones will usually be very stable, and can only move when there are catastrophic changes in population structure.

Carson (1975) has proposed that a population may move from one genetic equilibrium to another more readily during a "flush," as it expands rapidly after colonizing a new site. The reason for this would be that the chance of fixation of a new mutant is greater in an expanding population, and also selection might be relaxed during such an expansion. However, the chances of a mutant arising during the necessarily brief periods involved are slim; the transient increase in the chance of fixation could only aid in the transition between polymorphic equilibria. In addition, though selective pressures may be drastically different during a flush, they are not necessarily any weaker. In the case of a chromosomal rearrangement, for example, a female that produces a certain proportion of aneuploid offspring will be at just as much of a disadvantage as in stabler times. Selection against chromosomal heterozygotes might even increase if sib competition were reduced during a population expansion.

This discussion has so far dealt only with simple single-locus models involving fixed differences. One must also consider the stability of poly-morphic equilibria, and of multilocus systems, to sampling drift. It will be harder to move a complex genetic system from one equilibrium to another because the effects of drift at different loci will tend to cancel each other out (see Barton, 1979a:353). Ohta (1968) has shown that the joint fixation of two coadapted alleles is most unlikely. In general, it is hard to see how "genetic revolutions," involving changes at many loci, can occur by drift unless there is a good deal of polymorphic variation in the system. Transitions between polymorphic equilibria are certainly much easier than when new mutants are needed (Avery and Hill, 1979), but an interface between such equilibria (as envisaged by Carson, 1975) is hardly likely to be strong enough to be of much importance.

Finally, hitchhiking might be important; a slightly disadvantageous mutant might be raised to a high frequency by chance association with a closely linked advantageous mutant (Maynard Smith and Haigh, 1973; Thomson, 1977). However, very high substitution rates would be needed for this effect to be important, even for alleles with quite mild heterozygote unfitness.

So, the simple effect of sampling drift on a highly subdivided population seems to be able to account for the fixation of new incompatibilities, even without the elaborations mentioned above. However, there are a number of difficulties with this model. First, the extreme subdivision envisaged by Lande would have drastic consequences for population genetics in general; new variation would take a very long time to spread through the species (Slatkin, 1976). Second, it is rather hard to account for the karyotypic orthoselection described by White (1978b), and discussed below with reference to *Caledia captiva*. Finally, the action of drift is not the only possible explanation; complex interactions between loci may well send populations along mutually exclusive evolutionary pathways.

If evolution is limited by the availability of the right mutants, then different equilibria may be reached depending on the order in which new mutants become established. For example, suppose that there are three alleles at a locus, and that one of these is initially superior to the other two. If conditions (genetic or environmental) change, so that this allele becomes inferior, then either of the other two may replace it. If the two alleles are incompatible, their heterozygote being less fit, then a patchwork pattern will arise of areas fixed for different alleles, and separated by hybrid zones. This may be an unlikely scenario for one locus, but similar, and perhaps more plausible, situations can be envisaged for multilocus systems. For example, two mutants at different loci that are neutral or advantageous when alone but give sterility in the double homozygote, would give a similar pattern (Wills, 1970, Christiansen and Frydenberg, 1977; see also Dobzhansky, 1937, Müller, 1939, 1942, and Nei, 1975b for similar models of the evolution of reproductive isolation).

The previous possibilities do not rely on environmental variations. In fact, selective pressures may vary widely from place to place, and from time to time. If this variability is taken into account, any number of mechanisms for forming narrow hybrid zones can be thought up. For example, Clarke (1966) has proposed that a cline, maintained at first by an environmental gradient, can be strengthened by the evolution of modifiers. Endler (1977) has also dealt with this model at some length. If there are alleles that are favored only in the genetic background on one side of the cline, these will accentuate the difference between the two types. More and more loci will become involved, and the cline will become narrower and narrower. Eventually the original environmental gradient, which triggered the formation of a geographic pattern, becomes irrelevant to the maintenance of the cline, and a hybrid zone is formed. The plausibility of this model depends on the prevalence of coadaptation, that is, on whether the selective forces on a locus are largely determined by the frequencies of alleles at other loci, or by direct environmental effects. There is rather little evidence on this important question. There seems to be little disequilibrium between allozyme variants in *Drosophila melanogaster* (e.g., Langley et al., 1978). On the other hand, some polymorphisms are known that involve strong epistasis – for example, the segregation distortion complex, again in *D. melanogaster* (Hartl, 1977). If the genome is highly coadapted, then one would expect slight initial differences in genotype to be amplified into large incompatibilities.

In summary, there are many ways in which hybrid zones could arise. Stochastic forces may well be important in many organisms, but are not the only possibility. Differences in environment, and in the availability of new mutations, could allow a uniform population to evolve down a number of different and exclusive pathways. Thus, incompatibilities could develop separately at many loci in two geographically separated populations. However, if evolution is essentially deterministic, one does not need to postulate complete allopatry. A single continuously distributed population can differentiate just as easily as can two large, isolated populations, provided that its range is large compared with the dispersal distance of the organism. This implies that it will be virtually impossible to draw inferences about the origin of incompatibilities from observations of present-day clines, as Endler (1977) has pointed out. The existence of many coincident clines does point to an allopatric origin, but other explanations are possible. The clines may all be part of the same coadapted complex, or a splitting and reunion of the population may have brought many separate hybrid zones together. It would therefore seem profitable to make a thorough analysis of the mechanisms whereby a zone is maintained; its origin must remain a matter of speculation.

The development of a strong barrier

If incompatibilities do arise within continuous populations, as well as in complete allopatry, then there is no particular reason why clines at different loci should coincide. If the clines remained staggered, one would observe a gradual increase in hybrid inviability in crosses between increasingly distant populations (e.g., Oliver, 1972). Distant populations would be regarded as belonging to the same species, even though gene flow might be blocked over the whole genome (see Figure 7.3). Strict application of the biological species concept might lead to different results for different loci; perhaps one can only define "groups of actually or potentially interbreeding natural populations" (Mayr, 1940) at the gene level. However, it is likely that interactions between loci will make a series of coincident clines a much stronger barrier than a set of staggered clines. It is hard to judge the relative importance of the two sorts of barrier, but because strong hybrid zones, involving many coincident changes, do exist, it is important to consider how incompatibilities can be brought together to give a strong barrier.

Several clines could accumulate along the same boundary through changes in population structure, through coadaptation, or through the delaying effect of a barrier. The first possibility is straightforward; the position of a hybrid zone is largely determined by the abundance of the

Figure 7.3. The effect on gene flow of a set of clines at different loci. Each cline blocks gene flow at nearby loci. A set of coincident clines is likely to form a stronger barrier than a set of staggered clines, but the latter could still block gene flow over much of the genome.

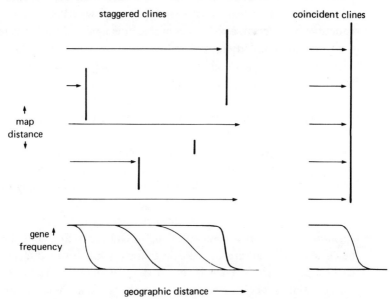

organism, and so all zones will tend to be attracted to density troughs. This demographic restructuring will concentrate differences at different loci; in the extreme case, a splitting and reunion of the population will bring all the clines to the boundary between the reuniting races (Key, 1974).

Alternatively, the evolution of new differences might be encouraged by preexisting incompatibilities, as was discussed in the last section. If the genome is sufficiently coadapted. or very sensitive to a single environmental feature, clines will tend to be formed together in the first place. Clines at linked loci will be attracted to each other, though only over very short ranges (Slatkin, 1975). A rather different cause of divergence is the effect of a cline as a barrier; new advantageous alleles will be delayed in their advance, and thus if conditions (genetic or environmental) change so as to remove their advantage, they will be trapped near the cline. It is hard to judge the relative importance of the three mechanisms, or to distinguish their effects, but it is likely that the effects of population structure will be dominant in the early stages of divergence, when two few loci are involved for interactions to be important.

The modification of hybrid zones

All of the preceding possibilities for the formation of hybrid zones are in essence allopatric. New genotypes arise away from the hybrid zone, and evolution would take much the same course if the zone were replaced by a natural barrier of equivalent strength. However, it is also possible that modification of the zone from within, rather than the accumulation of differences from without, could reduce gene flow. Alleles that lead to a reduction in dispersal, a reduction in selection against hybrids, a reduction in interracial competition, or a reduction in the frequency of heterotypic matings, will all be favored; so, in some multilocus systems, will modifiers reducing recombination. Coyne (1974) has shown that if there is strong sib-competition, early hybrid inviability may be selected for because it may be best to kill hybrids before they can interfere with healthy offspring. White (1978b) has recently argued that chromosomal rearrangements may be selected for near hybrid zones because they will "protect" locally adapted genomes from introgression. However, making maladapted hybrids still less fit, through the nondisjunction of chromosomal heterozygotes, can only be favored under the conditions discussed by Coyne and White if the infertility produced acts *before* other forms of hybrid unfitness, and if there is competition between hybrids and nonhybrids *after* the chromosomal difference has taken effect. Chromosomal incompatibilities produce infertility, which acts at the beginning of the F_2 generation, and so White's (1978b) scheme can only work if there is competition between grandchildren, and if genic incompatibilities primarily produce F_2 breakdown.

The evolution of assortative mating (i.e., the adaptive reinforcement of premating isolation) is especially important because it could lead to speciation; it has been suggested that many of the differences in mating behavior between species evolved as adaptive modifications to increase genetic isolation (e.g., Dobzhansky, 1970). However, there are a number of difficulties in accepting this thesis, some of which have been discussed by White (1978a:328–32) and Paterson (1978). First, it is not clear how often suitable modifiers arise; in particular, it is hard to see how assortative mating could develop in the absence of preexisting behavioral differences between the races involved. (It is important to remember that assortative mating is only likely to develop initially for small sections of chromosome, and will not isolate the whole genome until it is very strong.) Paterson has argued that mating behavior will be under strong stabilizing selection, and so is hard to change. However, this applies equally to allopatric divergence, and in any case, plausible models that circumvent this problem can be devised (see Nei, 1975a,b, and above, section on "The origins of incompatibilities"). There is also experimental evidence that premating isolation can sometimes be produced by strong disruptive selection in an artificial population (Thoday and Gibson, 1962; Thoday, 1972) and also by catastrophic "founder-flush" events (Powell, 1978). The problem might not appear so severe if more were known about the nature of sexual selection – much variance in fitness may be associated with mate choice, which could thus maintain a good deal of polymorphism.

Second, these modifiers must presumably be at a disadvantage outside the zone because otherwise they would almost always arise first in the large populations on either side. Because hybrid zones are narrow, the swamping effect of gene flow from the surrounding regions may considerably impede the fixation of the few modifiers arising within them. (This point was neglected by, for example, Caisse and Antonovics, 1978, who found in a series of computer simulations that genes for reproductive isolation could spread out from a cline in which they were favored. This could only happen if they were exactly neutral in pure populations.)

Third, hybrid zones are narrow, especially if they involve strong selection, and so few suitable mutations will be available. Preexisting polymorphic variation could be used at once, but would probably not be great enough to have much effect.

Some ideas of the effects of the second and third points may be gained from an examination of a simple model. Consider a cline maintained by selection, s, against heterozygotes at a single locus. In an organism with dispersal rate m, this would have width $(2m/s)^{1/2}$. Suppose a modifier has advantage γsh within the cline, where h is the frequency of heterozygotes. It might, for example, reduce the frequency of heterotypic matings by a fraction γ. If it has, in addition, a disadvantage β everywhere, then only mutants with β less than some critical β^* can be established in this limited pocket (Nagylaki, 1975).The cline can be approximated by a rectangular pocket of width $2\ (2m/s)^{1/2}$, in which heterozygotes are present at a frequency of 50%. Nagylaki's formulae show that, for such a pocket, only mutants with parameters (β, γ) lying below the curves shown in Figure 7.4 can become established. Only fairly major modifiers ($\gamma > 0.2$) stand much chance. If the zone is strengthened (s increased), then a larger class of mutants can succeed; their chances of escaping the effects of random drift also increase. However, the zone also becomes narrower, and so fewer mutants arise within it. It is not clear that a narrow zone is subject to much greater modification than a wide one, as has often been assumed.

This simple model neglects several factors; modification will be much more effective in zones involving incompatibilities at many loci because a single modifier could reduce the genetic load at several, or all loci. Although there are some cases where changes at a single locus have led to

Figure 7.4. Conditions for the establishment of a modifier in a narrow cline. Only modifiers with a disadvantage less than some critical value β^* can be established; these will lie in the region of parameter space below the line shown in the graph:

$$\left[\left(2\gamma - \frac{4\beta}{s}\right) \geq \tan^{-1}\left(\frac{2\beta}{\gamma s - 2\beta}\right)^{1/2}, \text{ from Nagylaki (1975)}\right]$$

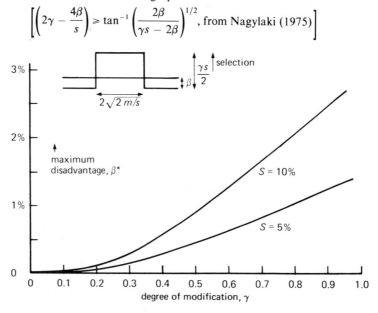

preferential mating (e.g., differences in flowering time in plants), it would often be more realistic to suppose that assortative mating is determined by at least two genes, which determine female preferences and male phenotype separately (O'Donald, 1977). Assortative mating would then result from chance associations between these two loci, and with the genes causing the hybrid unfitness. Reinforcement might well be harder in this more complex model because recombination would continually be breaking down the incipient isolating mechanisms.

A final problem is that any assortative mating (or other changes) that did develop could not spread back from the cline into regions where there is no selective pressure for its presence and where it is at a disadvantage. As premating isolation develops, the zone will become wider, until eventually the two forms become sympatric. The stage at which this happens depends on the ecological competition between the two races, that is, on nongenetic interactions. Interracial competition might prevent full sympatry even in the absence of any cross-mating – the two races presumably occupy similar niches. Conversely, any specialization of the two forms to fill different niches would allow the zone to collapse even without full genetic isolation. Once this happened, there would be selection for the completion of reproductive isolation over a very wide area, and full species would result quite quickly. Taking this possibility to an extreme, sympatric speciation could occur if niche specialization were possible even with no barriers to gene flow.

Three stages can thus be identified in the evolution of a hybrid zone. When the zone is weak, involving few loci, most differences will accumulate from outside. As the zone becomes stronger, and the races more different, modification from within could begin. There will be a conflict between modifiers that widen the zone by reducing the incompatibilities involved, and those that widen it by producing reproductive isolation. If strong assortative mating does develop over most of the genome, the zone will eventually be governed by ecological, effectively interspecific interaction, which will be subject to genetic modification, and may allow the development of either sympatrically or parapatrically distributed species.

The evidence

Some examples of hybrid zones will now be examined in the light of the preceding arguments, in order to assess their relevance and to get some idea of the nature and significance of real zones. Table 7.1, which shows the characteristics of a number of possible examples, is presented; some of the better-studied cases will be discussed in more detail.

Table 7.1. Summary of observations on hybrid zones

Organism	Characters involved	Proposed age (yr)[a]	Width (km)[b]	Field hybrids?	Selection vs. hybrids?	Location	Preferential mating	Enzyme differences[a]	Introgression	Authors
Mus musculus musculus M.m. domesticus (rodent)	Morphology, behavior	~3000	30–80	In Hardy-Weinberg	None noted in F_1	At climatic change	—	41:17:13	Varies between loci; *asymmetric*	Hunt and Selander, 1973; Selander et al., 1969
Spalax ehrenbergi (rodent)	Robertsonian fissions, behavior 2n = 58/60	—	3	In H-W; backcrosses in field	2 in 348 from zone showed mitotic non-disjunction	Partly along river	Demonstrated in laboratory tests. Greater interracial aggression, olfactory discrimination	17:5:1	—	Guttman et al., 1975
	Robertsonian fissions, behavior 2n = 54/58	—	0.7	Fewer than H-W		(Generally follow pptn. contours)		17:6:5	—	Nevo et al, 1975
	Robertsonian fissions, behavior 2n = 54/54	—	(across river)	No hybrids (natural barrier)		(Absolute barrier)		17:5:4	—	Nevo and Shaw, 1972
	Robertsonian fissions, behavior 2n = 52/58	—	0.3	Fewer than H-W; F_2 found in the field		Along cliff		17:3:1 Scattered samples, no mapping	—	Nevo and Bar El, 1973
Sceloporus grammicus (lizard)	Robertsonian fission	~7000	~0.2 (≈ dispersal)	No F_2 seen, few Bx F_1 in H-W	Strong (inferred from lack of F_2 in field)	Oak/pine boundary, suggesting adaptive differences	Probably not	20:12:7	Very limited – a few rare alleles in one population	Hall and Selander, 1973
Warramaba viatica (10 races in all) (grasshopper)	Robertsonian fusion also morphological difference in P24 (male cercus)	Glacial; 17/P24	0.2–0.3	~ In Hardy-Weinberg	Slight	None obvious	Probably not	—	~10 km introgression of cercal hook	White et al. 1967; White et al., 1968
		17/19[d]	0.2	(In limited area)	Sperm transfer failure; abnormal embryos	Transition in soil type	Probably not	—	—	Atchley, 1974
		19/P24	<0.02	One observed			Probably not	—	No introgression of cercal hook	Mrongovius 1975; Mrongovius 1979
Podisma pedestris (grasshopper)	Robertsonian fusion, morphology	5,000–10,000	0.8	~ In Hardy-Weinberg	Strong selection in F_1; cross between distant (~5 km) population	Generally, along mountain range. In detail, along density trough	Probably not	21:6:0	—	Hewitt, 1975; John and Hewitt, 1970; Hewitt and John, 1972; Barton and Hewitt, in prep.

Species	Character		Width	Free crossing	Selection	Cause	Concordant	Allele ratio	Comment	References
Ranidella insignifera / *R. pseudinsignifera* (frog)	Mating call, morphology	—	2.4	Plentiful	None demonstrated (except for mating call discrimination) Some in favor of F₁ hybrid	Broadly parallel to a scarp	Yes	5:5:4 (all esterase)	All changes *sharp*	Bull, 1973 Bull, 1978 Blackwell and Bull, 1978
Lepomis macrochirus (fish)	Morphology	~70,000	~100	In Hardy-Weinberg	—	Across two drainage systems	Probably not	10:5:2	Both enzymes change in parallel – neutral mixing?	Avise and Smith, 1974
Caledia captiva (grasshopper)	Extensive chromosomal differences – all 11 chromosomes involved	~8000	~200 m	Free interbreeding in a narrow region	—	Follows climatic change	—	(Extensive) 30:5	Chromosomes only introgress *one way*	Moran and Shaw, 1977 Moran, 1979
Thomomys bottae / *T. umbrinus* (rodent) (many zones)	Chromosomal	—	—	Only a few, all F₁	—	—	Perhaps	27:13:4	None seen	Patton et al., 1972 Patton and Yang, 1977
Geomys bursarius / *G. arenius* / *G. personatus* (rodent)	Chromosomal	—	—	None located	—	—	—	(Little polymorphism) 24:4:0	—	Selander et al., 1974 Baker et al., 1973 Kim, 1972
Thomomys talpoides (rodent)	Chromosomal (many races)	—	Narrow	Often not, even in (narrow) regions of sympatry	—	---	Yes	Little polymorphism. Only 1 diagnostic locus in 31. 31:25:~1	—	Thaeler, 1968 Nevo et al., 1974
Keyacris scurra	Chromosomal	—	~800 m	Yes	< 10% meiotic nondisjunction from lab. crosses	At edge of range	—	—	—	Key, 1974
Cepaea nemoralis / *C. hortensis* (many "area effects")	Shell color and banding	>5000	<100 m	Yes	—	No obvious cause	—	Some enzyme changes parallel banding changes but there is no strong correlation	—	Jones et al., 1977 Johnson, 1976

125

Table 7.1 (cont.)

Organism	Characters involved	Proposed age (yr)[a]	Width (km)[b]	Field hybrids?	Selection vs. hybrids?	Location	Preferential mating	Enzyme differences[c]	Introgression	Authors
*Macrotus water-housii M. californicus (bat)	Chromosomal	—	Narrow	None even in sympatry	—	—	Yes	21:17:4	—	Greenbaum and Baker, 1976
Partula (snail)	Shell color and banding	—	> 30 m	Intermediates found	One shell color gene recessive lethal – presumably balanced in one genotype by other factors	No obvious ecotone	—	—	—	Johnson et al., 1976 Clarke, 1968 Clarke and Murray, 1969
Clarkia speciosa polyantha C. nitans (plant)	Chromosomal (7 translocations) Morphological (two zones displaced ~ 150 km)	—	~ 40 ~ 80	Chromosomal changes occur at different points, so quite new arrangements exist in zone	Low viability	—	—	—	—	Bloom, 1976
Heliconius (zones within several species) (butterfly)	Müllerian mimicry (~ 6 loci or less)	Quaternary	< 50	Many	Free hybridization in lab.	Along rivers, or ecotones Zones from different species parallel	—	—	—	Brown et al., 1974 Turner, 1971a,b
*Peromyscus maniculatus P. ozarkium P. pallescens (rodent)	Chromosomal (extensive) Morphology	—	—	No hybrids even in sympatry	—	—	—	—	—	Caire and Zimmerman, 1975 Bowers et al., 1973
Didymuria violescens (stick insect) (10 races involved)	Chromosomal	—	< 6 km	In field, 0–15% asynapsis in hybrids	F₁ fit, but F₁♂ have ~ 30% asynapsis	Sometimes follow ecotones	—	—	—	Craddock, 1975
Crinia laevis C. victoriana (frog)	Mating call Morphology	Pleistocene	~ 0.5	Mostly are intermediates within the zone. Slightly	Some inviability	None obvious	Yes	—	—	Littlejohn and Watson, 1973, 1976

This table presents a summary of observations on hybrid zones. It is not complete (see Remington, 1968, and Short, 1969, for example), but all the better studied zones are included. It is not certain that all the cases included are in fact clines maintained by hybrid unfitness; particularly doubtful cases are marked by an asterisk. For example, though the "area effects" in *Cepaea* may be due in part to incompatibilities between different coadapted complexes, many other factors are involved.
[a]The age is usually based on circumstantial climatic evidence, and so should be treated with caution. [b]The width of the zone is not defined consistently by different authors; the most usual definition is the distance between "pure" populations. The figure given here is very approximate. [c]The number of loci examined. [d]The number that proved polymorphic, and the number that proved different between races are given. [e]Meiotic fertility loss (10%) in laboratory.

Species	Character studied	Age[a]	Width[b]	Hybrids found – much abnormality in zone	greater abnormality in hybrids / High frequency of eyeless F_1 hybrids	Forest/grass interface in places		7:7:4 (transferrin)	Notes	References
Litoria ewingi / *L. parawingi* (frog)	Mating call / Body size	~12,000	5–11 km	much abnormality in zone	—	Forest/grass interface in places	Some	7:7:4 (transferrin)	One type has a somewhat slower change	Watson, 1972; Watson et al., 1971; Gartside, 1972a,b; Littlejohn and Loftus-Hills, 1968; Littlejohn and Roberts, 1975
Limnodynastes tasmaniensis (frog)	Mating call – note duration changes slowly, composition changes sharply at center	10,000–20,000	~100	Many	Maybe a cline maintained by heterozygote advantage?	No obvious change	—	—	—	—
Acomys cahirinus (rodent)	Robertsonian change	~20,000?	~16	Yes	Some δ nondisjunction	At narrowest point of distribution	—	—	—	Wahrman and Goitein, 1972
Gerbillus pyramidum (rodent)	7 Robertsonian changes	500–5000	Sharp change over 10 km then slower change over 50 km	Yes	—	Sharp change at a town (barrier), then slower change	—	No differences in esterase, *Hb*, haptoglobin	—	Wahrman and Gourevitz, 1973
Papio anubis / *P. hamadryas* (baboon)	Behavior / Morphology	—	20 (3 troops)	Yes	May have disadvantageous herding behavior	No obvious barrier	Yes	—	—	Gabow, 1975
Sceloporus woodi / *S. undulatus* (lizard)	Morphology / Habitat selection	100,000	0.5	Intermediates found	—	On ecotone	Possibly some	—	—	Jackson, 1972
Pseudophryne bibroni / *P. semimarmorata* (frog)	Morphology (coloration)	—	Color ~2, Mortality ~5, Enzyme ~20	Yes	Much greater embryonic mortality in the zone	No obvious change	No difference in call structure	Fixed difference in *Ldh*	Cline in *Ldh* farther north; some evidence for southward movement of zone	McDonnell et al., 1978; Woodruff, 1979

Sceloporus grammicus

The lizard *Sceloporus grammicus* shows great chromosomal diversity; it has several karyotypic races, which often meet in narrow zones of hybridization. Hall and Selander (1973) studied the zone between the races P1 and F6, which differ by the presence of a polymorphic centric fission in P1, and also by a fixed fission in F6. P1 is found in open pine forest above 3200 meters in the Valley of Mexico, whereas F6 is usually found lower down in dense oak forest. Where they meet, hybrids are produced in a narrow strip a few hundred meters wide, about the same as the dispersal distance. Electrophoresis showed that the two races differed at about half their enzyme loci; analysis of a sample taken in the zone indicated that though F_1 and first generation backcrosses were common, few if any second generation backcrosses, or F_2s, were present. There was almost no introgression of foreign alleles between the P1 and F6 races; even though P1 is a very small isolate, so that a substantial fraction of its numbers are near the hybrid zone, only a few rare F6 alleles have found their way into it.

This seems to be a clear example of a zone maintained by very strong selection against hybrids. The two forms have diverged at many loci, and are now quite incompatible. Despite their great differences, however, they still interbreed freely. Lack of reproductive isolation seems to be a general feature of the *Sceloporus* complex; there may be insufficient variability in the mating system for isolation to develop. It is hard to say whether selection acts strongly at a few loci, or whether it is due to the cumulative effect of many differences. All we know is that recombinant gametes almost all produce sterility or inviability. However it is maintained, it is a virtually impenetrable barrier to gene flow. Because the two races are both thought, on cytogenetic evidence, to have arisen from a third (Standard) race, origin by secondary contact seems likely; this is supported by the presence of many coincident differences. Hall and Selander suggest that during the Thermal Optimum (7000 years ago), *Sceloporus* was confined to a few mountaintops, and that the present pattern of races has been formed by the expansion of these isolates to meet each other. It seems unlikely that such a sharp zone could move far, and so it probably formed near its present position, at the boundary of two forest types.

Spalax ehrenbergi

There are four chromosomal races of the mole rat *Spalax ehren-bergi* in Israel, with diploid chromosome numbers of 52, 54, 58, and 60 (Nevo and Bar-El, 1976). Those races with more chromosomes tend to be found in drier regions, so that their distribution roughly follows precipitation contours. However, there is often no obvious reason for their detailed pattern. Where the races meet, a narrow hybrid zone forms. Races 58 and 60 form a zone about 3 km across, whereas the zone between 54 and 58 is 700 meters wide. Races 52 and 54 are separated by a river. Races 52 and 58 form the narrowest zone, only 300 meters wide. In the two sharper zones, the karyotypic intermediates are not in Hardy-Weinberg proportions, although variations in chromosome frequency, the effects of dispersal, and small sample sizes make interpretation difficult. Analysis of litters found in the field indicates that the backcrosses with 57 and 55 chromosomes are fertile.

Much work has been done on behavioral differences between the races (Nevo et al., 1975; Guttmann et al., 1975; Nevo et al., 1976; Nevo and Heth, 1976). These authors argue that aggression is greater in encounters between individuals of different races than between individuals of the same race, and is greater between races 58 and 60 and between 52 and 58, which abut, than between races 52 and 60, which do not meet (race 54 was not tested). There is also assortative mating, partly based on smell, between races 52 and 58. However, reproductive isolation cannot be strong because many hybrids are found within the zones.

Electrophoretic studies (Nevo and Shaw, 1972; Nevo and Cleve, 1978) show little divergence between races at enzyme loci; of 25 loci, three show a major difference in frequency in one race, whereas there is variation in the frequency of rare alleles at the six other polymorphic loci. Heterozygosity is low in this species, however; so enzymes may just evolve slowly.

It is plausible that these zones are maintained by some sort of selection against hybrids; such selection seems to be greater between the races with greater differences in karyotype because their zones are the narrowest. However, this does not necessarily mean that selection acts through the karyotypic difference; the older, more diverged races may simply have had time to evolve more karyotypic differences. Very few abnormal karyotypes were seen in the zones. the behavioral differences might be more plausible candidates for selection, but there may be many other, as yet undiscovered, differences between the races.

Gene flow between these races is likely to be quite high; fertile backcrosses are found even in the sharpest zone. It is hard to judge the extent of introgression, however, without a geographically detailed analysis of variation at enzyme loci, and in behavioral patterns.

Nevo and his co-workers believe these races to be in the process of speciation; the most telling evidence in support of this idea is the presence of assortative mating, and hence of partial reproductive isolation. However, the isolating mechanisms are not strong enough to widen the zones very much, and have not prevented gene flow. Furthermore, it is hard to say whether the behavioral differences evolved in the zone or outside it. Though the extra aggression between contiguous races supports reinforcement, it seems unlikely that selective pressures in a very small fraction of the population could have influenced the behavior of the large central areas. Only time will tell whether new species are forming.

Podisma pedestris

The montane grasshopper *Podisma pedestris* has an XO sex chromosome system over most of its range in the European Alps and Apennines, but in the Alpes Maritimes, a Robertsonian fusion has given it a neo XY system (Hewitt, 1975). The two races meet in the high mountains on the French–Italian border, and form a long narrow hybrid zone, about 800 meters wide. This zone for the most part follows local regions of low population density (e.g., cliffs or open pasture), but there is no obvious ecological difference between the territory of the two types; this is just what would be expected for a zone maintained by hybrid unfitness. The two types interbreed freely, but laboratory crosses show heavy selection against F_1 hybrids. This selection is not related to the karyotypic difference; nondisjunction is very low. Electrophoresis has not yet shown any differentiation at enzyme loci (Halliday, pers. comm.), but morphometric analysis has shown a consistent difference in shape between the two karyotypes, within mixed populations (Barton, 1980; Barton and Hewitt, in prep.).

Judging from the width of the zone, which is several times the dispersal distance, the selection maintaining the cline in karyotype is about 0.5%; however, the heavy selection found against F_1 hybrids indicates that selection may operate at other, independent loci. Weak differences at several loci would accumulate in the F_1 multiple heterozygote, and make it very unfit. If this were the case, there would be a considerable barrier to gene flow, even though the karyotypic zone is quite wide.

The region now occupied by *Podisma* was covered by ice during the last glaciation, about 10,000 years ago. It seems reasonable, therefore, that the hybrid zone formed in roughly its present position, when two diverged isolates met after the retreat of the ice sheets; the broken mountain terrain would make extensive movement of the zone difficult. The chromosomal difference may be much older, however, and we cannot say how it arose.

Ranidella insignifera/pseudinsignifera

Ranidella insignifera and *R. pseudinsignifera* are two sibling species of frog, originally defined by a difference in mating call. *R.*

insignifera is found on the coastal plain around Perth, Western Australia, and meets the more widely distributed *R. pseudinsignifera* along a 480 km hybrid zone (Bull, 1978). The zone is about 2.5 km wide, and follows the low scarp of the Darling and Whicker ranges. Numerous laboratory crosses have failed to show any hybrid inviability or infertility; in fact, the F_1 hatch rate was higher in hybrid crosses (Bull, 1979). Electrophoretic and morphometric studies (Blackwell and Bull, 1978) reveal extensive differentiation; five esterase loci change alleles in parallel with the change in call, and there is also a distinct change in skeletal shape across the zone.

Although no selection against hybrids was detected in the laboratory, maintenance by hybrid unfitness remains plausible. The zone is quite wide; this could be because partial reproductive isolation is allowing some sympatry. Assortative mating of the two call types has been demonstrated (Bull, 1973), but it must be weak because many hybrids are found in the field. A more likely explanation for the width of the zone is that the selection maintaining it is weak. This selection might be acting through environmental differences, but if this were so, one would expect the zone to vary greatly in width; also the distribution of the two types would be rather more broken. The most likely selective force is hybrid unfitness, either because of undetected differences in viability or fertility, or because the hybrid's intermediate call is unattractive to females. The two types may still be adapted to different environments, but this adaptation may just determine the position of the zone.

It is perhaps surprising that the two forms should be so different if selection is weak. Unless this weak selection acts at many loci, the enzymatic and morphological differences would soon be lost. It may be, therefore, that many weak incompatibilities have accumulated to give a strong barrier to gene flow.

Mus musculus musculus/domesticus

Mus musculus has two semispecies, *musculus* and *domesticus,* which were originally identified by a difference in morphology and behavior (Ursin, 1952). *Musculus* is found in eastern Europe and Scandinavia, and meets *domesticus* in a narrow of zone hybridization that runs north from the Alps to Denmark, and then turns west across Jutland. This parallels a number of other hybrid zones (e.g., those in the carrion crow and in the ringed snake), and also a climatic change. Selander et al., (1969) and Hunt and Selander (1973) have studied the zone in Jutland in great detail, looking at electrophoretic variation at many loci. They found large differences between the two semispecies, 13 out of 17 polymorphic loci showing changes in allele frequency across the zone. No excess of homozygotes was seen, an indication that premating isolation is weak or absent. Changes in enzymes closely parallel morphological and behavioral variation, although there is some variation in cline width; the zone is narrowest in the east, where it is about 20 km wide, and where the change in precipitation is most abrupt. In the west, where the climatic change is more gradual, and the two races are thought to have met more recently, the zone is 70 km wide. The detailed position of the zone is not related to any obvious geographical feature. A striking feature of the enzyme clines is an asymmetric introgression of *domesticus* alleles into the northern *musculus* populations.

There are a number of possible explanations for this zone, all of which present difficulties. Selection must be operating because the zone is thought to have formed several thousand generations ago, and yet is still quite sharp. It must be operating on many loci because most enzymes studied have remained differentiated. It cannot act very strongly (at least, at those loci observed) because the zone is several times wider than the likely dispersal distance of a mouse. This makes it likely that coadaptation causing hybrid inferiority is the major force involved; it seems improbable that so many loci would change in the same place if they each reacted separately to an environmental difference.

There remains the problem of the asymmetry in introgression, and hence in gene flow. Though an asymmetry in cline shape can make it an asymmetrical barrier, there seems no reason why this asymmetry should be consistent in direction between loci. Movement of the zone southward would leave a tail of introgression in the north; but the zone has remained stable for 20 years, and, if anything, has moved northward since its formation (islands tend to have a more northern genotype). There might be a large number of long-distance migrants moving north, but there is no evidence for this. This seems a good example of a hybrid zone due to secondary contact between two diverged races, but the directed introgression remains puzzling.

Caledia captiva

This Australian grasshopper has four races, which differ in karyotype. The Moreton race is found near the coast of S.E. Queensland, and meets the primitive Torresian race along a 150 km hybrid zone (Moran and Shaw, 1977). The two have very different chromosomes; the Torresian complement is entirely acrocentric, whereas the Moreton complement is metacentric or submetacentric. In addition, there is considerable variation in karyotype within the Moreton race. The X chromosome, in particular, is metacentric in the south but acrocentric in the north (though it is always distinguishable from the Torresian X). There are also characteristic differences between the centric heterochromatin of the two races. Preliminary electrophoretic studies show differentiation at 5 out of 30 loci studied, but details of the variation are not yet available.

The position of the zone seems to be determined by climate; it runs close to the 30% contour of the coefficient of variations of rainfall. The climate becomes drier and more seasonal to the west, limiting the Torresian race to only one generation per year. The Moreton race is capable of two. It is thought that the zone is due to secondary contact between populations isolated in the dry Pleistocene interpluvials (~ 8000 years ago).

The zone is very narrow. The width has been measured accurately only in one place; there, most of the change in chromosome frequency occurs within 200 meters. All the chromosomes change in the same place, but there is some variation in width. The X changes most abruptly, whereas chromosomes 7, 8, and 11 change rather more slowly. The most striking feature is that there is highly directional introgression of the Torresian chromosomes some way into the Moreton race. Only the position of the centromere introgresses; the centric heterochromatin is always characteristic of the correct race. The extent of introgression varies between chromosomes, but is always in the same direction. The X shows no introgression, whereas chromosomes 1, 2, 10, and 11 introgress most. There is a difference in introgression between the north and south of the zone; the northern Moreton race, which has the acrocentric X, has more foreign alleles.

Analysis of karyotype frequencies along a transect through the zone indicates that interbreeding is free, but that there is severe, and asymmetric, hybrid breakdown. There is strong disequilibrium between most chromosomes on the Torresian side of the zone, favoring parental combinations. No consistent disequilibria are found on the Moreton side, except with the X, 1, and 2 chromosomes. Laboratory crosses show that the F_1 is viable and fertile, but that there is severe hybrid breakdown in the next generation.

The Torresian and Moreton races have diverged considerably, and now have quite incompatible karyotypes. Strong and asymmetric selection on the first few backcrosses maintains a sharp hybrid zone, which allows gene flow only into the Moreton race. This consistent asymmetry is not so hard to interpret as it is in *Mus* because selection here is very strong. In addition, because the zone consists of many chromosomal differences, there may be some common factor (in meiosis, for example) allowing acrocentrics to introgress into a metacentric karyotype, but not vice versa. Such an explanation is not plausible in *Mus,* where many different enzymatic and morphological characters are involved.

The most puzzling aspect of this zone is the question of how the acrocentric Torresian karyotype zone gives rise to an entirely metacentric karyotype. Such changes have been seen in other organisms, and have been described as "karyotypic orthoselection" (White, 1975). Chance fixation of 11 similar inversions is most unlikely, and yet it is hard to see any selective force favoring metacentricity. An answer to this problem might help in understanding the origin of the zone, and its current maintenance.

Vandiemenella viatica (formerly *Moraba viatica)*

The *viatica* group of morabine grasshoppers has several karyotypic races, which are distributed parapatrically in southern Australia (White et al., 1967; White, 1968). They are separated by very narrow hybrid zones, although these zones have often been destroyed by agriculture. Several chromosomal changes are involved; the distribution of any one rearrangement is continuous, but may overlap with the distribution of other rearrangements. Races therefore share some karyotypic features with neighboring races. We will concentrate on the races $viatica_{17}$, $viatica_{19}$, and P24(XY), which all meet on Kangaroo Island; $viatica_{17}$ and $viatica_{19}$ also meet in a long zone on the mainland. These three have been studied in detail by Mrongovius (née Andre) 1975, 1979), and by White et al. (1968).

$Viatica_{19}$ is thought to have the primitive karyotype of the group. $Viatica_{17}$ differs from it by a fusion between two autosomes, whereas P24(XY) has, in addition, a fusion of the X with an autosome. In addition to these chromosomal differences, considerable morphological variation has been detected. Race P24 has a cercal hook in the male, whereas multivariate analysis reveals differences in general shape, and in the female egg guide (Atchley, 1974; Atchley and Hensleigh, 1974; Atchley and Cheney, 1974). However, the differences between races are obscured by large differences between quite close populations, perhaps due to the extremely low vagility of these wingless grasshoppers. For example, there is a difference in shape between $viatica_{17}$ and $viatica_{19}$ on the mainland, with some evidence for introgression, but it is in a different direction from the variation between more distant populations.

The zones on Kangaroo Island vary in width, and this variation correlates inversely with the strength of selection seen in the laboratory. The zone between $viatica_{17}$ and P24(XY) is the widest, being 200–300 meters across; little nondisjunction was seen in chromosomal heterozygotes, and no effects on viability were detected. There is considerable introgression of the cercal hook of P24 into $viatica_{17}$, as would be expected through such a wide, weak zone. The zone between $viatica_{17}$ and $viatica_{19}$ is much narrower, being only 20 meters wide; few hybrids have been found. In both field and laboratory, there is some ($\sim 20\%$) aneuploidy in the male, but selection acts much more strongly through other characters. There is frequent failure of sperm transfer (a premating isolating mechanism if multiple insemination is common), and a high frequency of parthenogenetic embryos. This zone is found at a transition between two soil types; there are indications that $viatica_{19}$ has a rather different habitat from that of the other races.

The contact between $viatica_{19}$ and P24(XY) is the sharpest; only one heterozygous female has been found. Again, there is frequent failure of sperm transfer in laboratory crosses, and hybrids have a very low hatch rate. F_1 males fail to mature. The P24 race develops earlier in the field, a phenomenon that may give a measure of premating isolation. Such isolation is clearly not effective in allowing sympatry, however.

Thus, these races differ not only in karyotype, but also in genes determining viability and morphology. Though the chromosomal differences lead to some selection against hybrids, through aneuploidy, it seems likely that the zones are largely maintained by these other factors.

The overlapping distribution of the chromosomal rearrangements suggest, if one accepts the idea that the overall distribution has remained more or less unchanged, that the various changes spread out separately to cover different regions. White (1968) and Key (1974), though they differ in their views on the origin of these changes, propose that they spread out behind narrow, mobile hybrid zones. The previous discussion makes this seem unlikely, however, especially in an organism as sedentary as this. The rearrangements may have arisen and spread because they were advantageous even when heterozygous; their advantage may then have been lost, leaving a narrow zone. This does not affect the problem of the coincidence of the karyotypic with the genic variation; it may be that clines at different loci have come together, as suggested by Key (1974), but it is not possible to say whether the chromosomal or genic changes came first.

Pseudophryne bibroni/semimarmorata

Pseudophryne bibroni and *P. semimarmorata* are two southeastern Australian frog taxa, differing primarily in color pattern, that meet along a 380 km long hybrid zone. A transect of 10 sites across the zone north of Melbourne has been studied by McDonnell et al. (1978), following earlier work by Woodruff (1972). The color pattern changes sharply over about 3 km; the *bibroni* form has advanced southward by about a cline width during the 10-year study period.

The frogs also differ in lactate dehydrogenase (*Ldh*) mobility. The cline for allozyme change is much wider ($\sim$ 20 km), and is markedly asymmetric, with a *semimarmorata*-like tail in *bibroni* territory. There is no significant association between color pattern and enzyme type in mixed populations.

Embryonic mortality in egg masses collected from the zone is much greater than in pure populations ($\sim$ 50% inside, $\sim$ 15% outside). The region of increased mortality is somewhat wider than the cline in color pattern, and tails off more slowly to the north, in the same way as *Ldh*.

There is no evidence for assortative mating between the two forms; their breeding calls are indistinguishable, and there is a wide range of variability in color pattern within mixed populations. A significant deficiency of *Ldh* heterozygotes was found at one site, but this could well be the result of hybrid inviability.

It is unlikely that the clines in color pattern and *Ldh* type are maintained directly by environmental factors. There are no obvious ecological differences across the zone, and it is odd that the severe embryonic mortality should be wider than the cline in color pattern. Either embryo viability is not as closely related to fitness as one might imagine, or many more loci are involved in determining the embryonic incompatibilities than in determining color pattern.

Because the color cline is moving rapidly, it is unlikely that it will remain near the other two clines for very long. The northerly tails of the *Ldh* cline and the mortality cline support this notion. There is little or no correlation between *Ldh* type and color within populations, and so there will be little or no tendency for the *Ldh* cline to be dragged along with the advancing color pattern. As for the mortality cline, it is likely to remain fairly static because it is presumably maintained by hybrid unfitness. McDonnell et al. (1978) suggest that the two forms may have come into contact as a result of agricultural changes in the last century; this would explain the present near coincidence of the clines. The *bibroni* color pattern may be advantageous and advancing (at least in the region studied), leaving behind other differences. It would be interesting to know what is happening elsewhere in the zone.

Summary of the evidence

What does this survey of hybrid zones tell us? One of the most striking features is that all, or nearly all, the zones involve divergence and selection at many loci. Yet, a hybrid zone involving only a few genes should be stable. Either the genome is highly coadapted, so that slight initial differences in genotype rapidly increase, or hybrid zones are usually very old, so that there has been time for many incompatibilities to accumulate and be brought together by secondary contact. Although most zones are thought to have been formed after the last ice age (10,000 years ago or less), the differences on which they are based may have arisen long before then. This is an important distinction; the way in which the present zone was formed may be quite unrelated to the way in which the differentiation itself arose.

Most of the hybrid zones considered here involve chromosomal differences. Why? This is related to the intriguing question raised by White (1978a); he asks why species should so often differ in karyotype when their chromosomes are so rarely polymorphic. Many authors (e.g., John and Lewis, 1966) have argued that chromosomal mutants have quite different effects from "point mutation" at single loci. They are often disadvantageous in the heterozygote because meiosis is disrupted. In addition, an alteration in chromosome architecture will change linkage relations between many loci. If epistasis is common, it will in turn alter the allele frequencies at many loci. Thus, a change in karyotype could profoundly affect the rest of the genome. (We should be careful in using this simple model of stable epistasis, however, because recent studies suggest that it may not adequately explain the maintenance of recombination; see Maynard Smith, 1978.) However, interactions work both ways; the selective forces on many loci will affect the success of a chromosomal rearrangement. Although the chromosomes may be intimately involved with coadapted gene complexes, it cannot be argued that they *cause* changes at these loci. It may be that the frequency with which chromosomal changes are seen across hybrid zones is a result of their sensitivity to all the other changes that occur across hybrid zones. The problem is to find out how any strongly coadapted system, whether genic or chromosomal, can change.

The position of these hybrid zones often parallels a climatic or ecological change, rather than a natural barrier or density trough, as might have been expected. Those in *Mus, Caledia, Sceloporus, Spalax,* and perhaps in *Ranidella* and the morabines can all be broadly related to an environmental change, whereas only in *Podisma* is there a clear correlation with a barrier. The reason for this could be that they have moved to an equilibrium position, determined by selection on the homozygotes, after their formation. However, the genome would have to be very sensitive to the effects of the environment for the zone to overcome variations in population structure. Another possibility is that the zones have not moved far, but formed near an ecological shift. This could be so either if the expansion of two reuniting genomes were limited by climatic factors, or if the zone were originally due to modification of an ecotonal cline, as proposed by Clarke (1966) and in theories of parapatric speciation (Endler, 1977).

In general, the degree of introgression correlates well with the width of the zone. (It should be noted that completely successful introgression will not be detected.) However, the marked asymmetry seen in *Mus* and *Caledia* is perplexing. More detailed studies of these zones, and of the theoretical behavior of multilocus clines as barriers, are needed.

Reproductive isolation may or may not be present; none of these zones provides evidence for reinforcement of premating isolation (e.g., McDonnell et al., 1978). There is no reason to suppose that the evolution of reproductive isolation is fast enough to make zones unstable, although this supposition depends critically on the amount of variation in mating behavior present in the species. Indeed, partial premating isolation could maintain a zone, if hybrids between two types with a difference in mating behavior were not recognized by either parental type; this may be the case in *Ranidella*, for example.

Hybrid zones and speciation

Many models of speciation imply the formation of hybrid zones, but they are really only by-products of the process unless evolution within them is important in the development of genetic isolation, or unless they can act as strong barriers to gene flow. The two models that rely most heavily on the behavior of hybrid zones are those of stasipatric and of parapatric speciation. In the first, M. J. D. White (1968, 1978a,b) proposes that a karyotypic change occurs and becomes established in a small region in the interior of the organism's range. This initial fixation occurs by random drift, perhaps assisted by meiotic drive. Because such changes usually produce infertility in the heterozygote, a narrow hybrid zone forms. This zone advances to enclose a large area, and, because it acts as a barrier to gene flow, the two forms can diverge genetically, eventually becoming fully isolated species.

There are a number of problems with this scheme, which have been discussed in previous sections, but which will be summarized here. First, there is no particular reason why the change should be karyotypic. It is certainly striking that so many hybrid zones, and species pairs, should involve chromosomal differentiation, but in all the cases studied in detail, other changes are involved as well (see White 1978a:326). The hybrid infertility produced by nondisjunction in chromosomal heterozygotes is often only a small part of the overall loss in fitness.

Second, the chances of fixation of a deleterious mutant whose effects are strong enough to give a significant barrier to gene flow are minute without the aid of a severe and prolonged population bottleneck. There is compelling, though circumstantial, evidence that speciation is associated with a substantial amount of genic and chromosomal change; species-rich genera show greater chromosomal diversity, and (in the tetraploid Centrarchidae) a greater loss of duplicate gene activity (Levin and Wilson 1976; Bush et al., 1977; Ferris et al., 1979; see also Gould and Eldredge 1977). These authors interpret this as an effect of drift both in fixing new karyotypes and in forming new species. However, other possibilities have not been excluded; population restructuring might act to bring together incompatibilities that arose deterministically. Alternatively, speciation might involve considerable genetic changes (which might or might not be initiated by drift) to which the karyotype reacts. Founder effects may well be important in speciation, but it seems unlikely that drift could fix a strongly deleterious chromosome aberration, as White suggests. The frequent occurrence of "karyotypic orthoselection" (as in *Caledia captiva*) also provides telling evidence against random drift; it is more likely that several similar rearrangements should occur through some adaptive effect in meiosis, for example, than by pure chance.

Third, stasipatric speciation demands that hybrid zones should be able to move out to enclose a large area. A long hybrid zone is more likely to be observed, and so we may be biased in assuming that such zones are necessarily the ones involved in speciation. However, unless a new race could expand well outside its birthplace, it would be subject to swamping by the parental type, and would have little variability available for further evolution. Because extensive zone movement is unlikely on theoretical grounds, this is a serious difficulty.

Finally, and perhaps most important, a cline involving a single chromosomal change cannot present a strong barrier to gene flow; even if heterozygotes are only half as fertile as homozygotes, gene flow on other chromosomes will not be seriously hindered. Key (1974) and White (1978b) have suggested that several weak zones, involving different chromosomal changes, could fuse to give a strong barrier. Most of the zones that have been discussed seem to be quite effective barriers to gene flow, although advantageous mutants may still penetrate them quite easily. However, many of the factors that make them good barriers are probably not karyotypic; for stasipatric speciation to be effective, one or a few chromosomal changes must be able to trigger differentiation in the rest of the genome.

Endler (1977), Clarke (1966), and others have proposed models of "parapatric" speciation. Here, a continuously distributed population splits under the influence of geographically varying selection pressures, as discussed above under "the origins of incompatibilities." This model differs from stasipatric speciation in that changes are not necessarily chromosomal, the clines do not need to move, and the strengthening of the isolation depends on coadaptation rather than on simple barrier effects. Parapatric speciation is quite consistent with the properties of hybrid zones, and the frequent association of zones with ecotones could be taken as support for it. However, the main problem would seem to be the bringing together of the clines at different loci; they would only arise together if the genome were highly coadapted. It may be that secondary contact can fulfill this function by drawing together during contraction clines that formed in different places in a continuous population. This is a reversal of the usual scheme.

In view of the many complex possibilities for the formation of species it is very difficult to make inferences from present hybrid zones and species distributions; perhaps the way forward is to analyze in detail the differences between diverging taxa, and to recognize that the various theories of speciation are not necessarily mutually exclusive.

We are most grateful to the Science Research Council for their financial support of part of this work.

References

Atchley, W. R. 1974. Morphometric differentiation in chromosomally characterized parapatric races of morabine grasshoppers (Orthoptera: Eumastacidae). *Aust. J. Zool.* 22:25–37.

Atchley, W. R., and Cheney, J. 1974. Morphometric differentiation in the *viatica* group of morabine grasshoppers (Orthoptera: Eumastacidae). *Syst. Zool.* 23:400–15.

Atchley, W. R., and Hensleigh, D. A. 1974. The congruence of morphometric shape in relation to genetic divergence in four races of morabine grasshoppers (Orthoptera: Eumastacidae). *Evolution* 28:416–27.

Avery, P. J., and Hill, W. G. 1979. Distribution of linkage disequilibrium with selection and finite population size. *Genet. Res.* 33:29–48.

Avise, J. C., and Smith, M. H. 1974. Biochemical genetics of sunfish I: Geographic variation and subspecific intergradation in the blue gill *Lepomis macrochirus*. *Evolution* 28:42–56.

Baker, R. J., Williams, S. L., and Patton, J. C. 1973. Chromosomal variations in the plains pocket gopher, *Geomys bursarius major*. *J. Mammal.* 54:765.

Barton, N. H. 1979a. The dynamics of hybrid zones. *Heredity* 43:341–59.

– 1979b. Gene flow past a cline. *Heredity* 43:333–9.

– 1980. The fitness of hybrids between two chromosomal races of the grasshopper *Podisma pedestris. Heredity 45:*49–61.

Barton, N. H., and Hewitt, G. M. A chromosomal cline in the grasshopper *Podisma pedestris* (in preparation).

Bazykin, A. D. 1969. Hypothetical mechanism of speciation. *Evolution, 23:*685–7.

– 1972a. The disadvantage of heterozygotes in a system of two adjacent populations. *Genetika 8:*155–61.

– 1972b. The disadvantage of heterozygotes in a population within a continuous area. *Genetika 8:*162–7.

– 1973. Population genetic analysis of disruptive and stabilising selection. Part II. Systems of adjacent populations and populations within a continuous area. *Genetika 9:*156–66.

Bengtsson, B. O., and Bodmer, W. F. 1976. On the increase of chromosome mutation under random mating. *Theor. Pop. Biol. 9:*260–81.

Blackwell, J. M., and Bull, C. M. 1978. A narrow hybrid zone between the Western Australian frog species, *Ranidella insignifera* and *R. pseudinsignifera:* the extent of introgression. *Heredity 40:*13–25.

Bloom, W. L. 1976. Multivariate analysis of the introgressive replacement of *Clarkia nitens* by *C. speciosa polyantha* (Onagraceae). *Evolution 30:*412–24.

Bodmer, W. F., and Felsenstein, J. 1967. Linkage and selection: theoretical analysis of the deterministic two locus random mating model. *Genetics 57:*237–65.

Bowers, J. H., Baker, R. J., and Smith, M. H. 1973. Chromosomal, electrophoretic and breeding studies of selected populations of deer mice *Peromyicus maniculatus*) and black-eared mice (*P. melanotis*). *Evolution 27:*378–86.

Brown, K. S., Sheppard, P. M., and Turner, J. R. G. 1974. Quaternary refugia in tropical America: evidence from race formation in *Heliconius* butterflies. *Proc. R. Soc. Lond. B 187:*368–78.

Bull, C. M. 1973. The interaction of two allopatric frog species at their common boundary. Ph.D. thesis, University of Western Australia.

– 1978. The position and stability of a hybrid zone between the Western Australian frogs *Ranidella insignifera* and *R. pseudinsignifera. Aust. J. Zool. 26:*305–22.

– 1979. A narrow hybrid zone between two Western Australian frog species *Ranidella insignifera* and *R. pseudinsignifera:* the fitness of hybrids. *Heredity 42:*381–9.

Bush, G. 1975. Modes of animal speciation. *Annu. Rev. Ecol. Syst. 6:*339–64.

Bush, G., Case, S. M., Wilson, A. C., and Patton, J. L. 1977. Rapid speciation and chromosomal evolution in mammals. *Proc. Natl. Acad. Sci. U.S.A. 74:*3942–46.

Caire, W., and Zimmerman, E. G. 1975. Chromosomal and morphological variation and circular overlap in the deer mouse, *Peromyscus maniculatus,* in Texas and Oklahoma. *Syst. Zool. 24:*89–95.

Caisse, M., and Antonovics, J. 1978. Evolution in closely adjacent plant populations IX. Evolution of reproductive isolation in clinal populations. *Heredity 40:*371–84.

Carson, H. L. 1975. The genetics of speciation at the diploid level. *Am. Nat. 109:*83–92.

Christiansen, F. B., and Frydenberg, O. 1977. Selection–mutation balance for two non-allelic recessives producing an inferior double heterozygote. *Am. J. Hum. Genet. 29:*185–207.

Clarke, B. 1966. The evolution of morph ratio clines. *Am. Nat. 100:*389–402.

– 1968. Balanced polymorphism and regional differentiation in land snails. In *Evolution and Environment.* (ed. Drake, E. T.), pp. 351–68.New Haven: Yale University Press.

Clarke, B., aıd Murray, J. 1969. Ecological genetics and evolution in land snails of the genus *Partula. Biol. J. Linn. Soc. 1:*31–42.

Coyne, J. A. 1974. The evolutionary origin of hybrid inviability. *Evolution 28:*505–6.

Craddock, E. M. 1975. Intraspecific karyotypic differentiation in the Australian phasmatid *Didymuria violescens* (Leach), I: the chromosome races and their structural and evolutionary relationships. *Chromosoma (Berl.) 53:*1–24.

Dobzhansky, Th. 1937. *Genetics and the Origin of Species*. New York: Columbia University Press.

— 1970. *Genetics of the Evolutionary Process*. New York: Columbia University Press.

Endler, J. A. 1977. *Geographic Variation, Speciation and Clines*. Princeton: Princeton University Press.

Feldman, M. W., and Balkau, B. 1972. Some results in the theory of three gene loci. In *Population Dynamics* (ed. Greville, T. R. H.). New York: Academic Press.

Feldman, M. W., Franklin, I., and Thomson, G. J. 1974. Selection in complex genetic systems. I. The symmetric equilibria of the three-locus symmetric viability model. *Genetics 76:*135–62.

Ferris, S. D., Portnoy, S. L., and Whitt, G. S. 1979. The role of speciation and divergence in the loss of duplicate gene expression. *Theor. Pop. Biol. 14:*114–39.

Gabow, S. A. 1975. Behavioural stabilization of a baboon hybrid zone. *Am. Nat. 109:*701–712.

Gartside, D. F. 1972a. The *Litoria ewingi* complex (Anura: Hylidae) in south-eastern Australia II. Genetic incompatibility and delimitation of a narrow hybrid zone between *L. ewingi* and *L. paraewingi*. *Aust. J. Zool. 20:*423–33.

— 1972b. The *Litoria ewingi* complex (Anura: Hylidae) in south-eastern Australia III. Blood protein variation across a narrow hybrid zone between *L. ewingi* and *L. paraewingi*. *Aust. J. Zool. 20:*435–43.

Gould, S. J., and Eldredge, N. 1977. Punctuated equilibria – the tempo and mode of evolution reconsidered. *Paleobiology 3:*115–51.

Greenbaum, I. F., and Baker, R. J. 1976. Evolutionary relationships in *Macrotus* (Mammalia: Chiroptera): biochemical variation and karyology. *Syst. Zool.* 25:15.

Guttman, R., Naftali, G., and Nevo, E. 1975. Aggression patterns in three chromosome forms of the mole rat, *Spalax ehrenbergi*. *Anim. Behav. 23:*485.

Hall, W. P., and Selander, R. K. 1973. Hybridization in karyotypically differentiated populations of the *Sceloporus grammicus* complex (Iguanidae). *Evolution 27:*226–42.

Hartl, D. 1977. How does the genome congeal? In *Measuring Selection in Natural Populations, Lecture Notes in Biomathematics* 19 (ed. Christiansen, F. B., and Fenchel, T. M.), pp. 65–82. Berlin: Springer-Verlag.

Hewitt, G. M. 1975. A sex chromosome hybrid zone in the grasshopper *Podisma pedestris* (Orthoptera: Acrididae). *Heredity 35:*375.

Hewitt, G. M., and John, B. 1972. Interpopulation sex chromosome polymorphism in the grasshopper *Podisma pedestris* II. Population. *Chromosoma (Berl.) 37:*23–42.

Hunt, W. G., and Selander, R. K. 1973. Biochemical genetics of hybridization in European house mice. *Heredity 31:*11–33.

Jackson, J. F. 1972. The phenetics and ecology of a narrow hybrid zone. *Evolution 27:*58–68.

John, B., and Hewitt, G. M. 1970. Inter-population sex chromosome polymorphism in the grasshopper *Podisma pedestris* I. Fundamental facts. *Chromosoma (Berl.) 31:*291–308.

John, B., and Lewis, K. R. 1966. Chromosome variability and geographic distribution in insects. *Science 152:*711–21.

Johnson, H. S., Clarke, B., and Murray, J. 1976. Genetic variation and reproductive isolation in *Partula*. *Evolution 31:*116–26.

Johnson, M. S. 1976. Allozymes and area effects in *Cepaea nemoralis* on the Berkshire downs. *Heredity 36:*105–21.

Jones, J. S., Lieth, B. H., and Rawlings, P. 1977. Polymorphism in *Cepaea:* a problem with too many solutions? *Annu. Rev. Ecol. Syst. 8:*109–43.

Karlin, S., and McGregor, J. 1972a. Application of the method of small parameters to multi-niche population genetic models. *Theor. Pop. Biol. 3:*186–209.

— 1972b. Polymorphisms for genetic and ecological systems with weak coupling. *Theor. Pop. Biol. 3:*210–38.

Key, K. H. L. 1968. The concept of stasipatric speciation. *Syst. Zool. 17:*14–22.

– 1974. Speciation in the Australian morabine grasshoppers: taxonomy and ecology. In *Genetic Mechanisms of Speciation in Insects* (ed. White, M. J. D.), pp. 43–56. Sydney: Australia and New Zealand Book Co.

Kim, Y. J. 1972. Studies of biochemical genetics and karyotypes in pocket gophers. Ph.D. thesis, University of Texas, Austin.

Lande, R. 1979. Effective deme sizes during long-term evolution estimated from rates of chromosomal rearrangement. *Evolution 33:*234–51.

Langley, C. H., Smith, D. B., and Johnson, F. M. 1978. Analysis of linkage disequilibrium between allozyme loci in natural populations of *Drosophila melanogaster. Genet. Res. 32:*215–29.

Levin, D. A., and Wilson, A. C. 1976. Rates of evolution in seed plants: net increase in diversity of chromosome numbers and species numbers through time. *Proc. Natl. Acad. Sci. U.S.A. 73:*2086–90.

Littlejohn, M. J., and Loftus-Hills, J. J. 1968. An experimental evaluation of premating isolation in the *Hyla ewingi* complex (Anura: Hylidae). *Evolution 22:*659–63.

Littlejohn, M. J., and Roberts, J. D. 1975. Acoustic analysis of an intergrade zone between two call races of the *Limnodynastes tasmaniensis* complex (Anura:Leptodactylidae) in S.E. Australia. *Aust. J. Zool. 23:*113–22.

Littlejohn, M. J., and Watson, G. F. 1973. Mating call variation across a narrow hybrid zone between *Crinia laevis* and *C. victoriana* (Anura: Leptodactylidae) *Aust. J. Zool. 21:* 277–84.

– 1976. Mating call structure in a hybrid population of the *Geocrinia laevis* complex (Anura: Leptodactylidae) over a seven-year period. *Evolution 30:*848–50.

McDonnell, L. J., Gartside, D. F., and Littlejohn, M. J. 1978. Analysis of a narrow hybrid zone between two species of *Pseudophryne* (Anura: Leptodactylidae) in S.E. Australia. *Evolution 32:*602–12.

Maynard Smith, J. 1978. *The Evolution of Sex.* Cambridge: Cambridge University Press.

Maynard Smith, J., and Haigh, J. 1973. The hitch-hiking effect of a favourable gene. *Genet. Res. 23:*23–35.

Mayr, E. 1940. Speciation phenomena in birds. *Am. Nat. 74:*249–78.

– 1970. *Populations, Species and Evolution.* Cambridge, Massachusetts: Harvard University Press.

Moran, C. 1979. The structure of the hybrid zone in *Caledia captiva. Heredity 42:*13–32.

Moran, C., and Shaw, D. D. 1977. Population cytogenetics of the genus *Caledia* (Orthoptera: Acridinae) III: Chromosomal polymorphism, racial parapatry and introgression. *Chromosoma (Berl.) 63:*181–204.

Mrongovius, M. J. 1975. Studies on hybrids between members of the *viatica* group of morabine grasshoppers. Ph.D. thesis, University of Melbourne.

– 1979. Cytogenetics of the hybrids of three members of the grasshopper genus *Vandiemenella* (Orthoptera: Eumastacidae: Morabinae). *Chromosoma (Berl.) 71:*81–107.

Müller, H. J. 1939. Reversibility in evolution considered from the standpoint of genetics. *Biol. Rev. Camb. Phil. Soc. 14:*261–80.

– 1942. Isolating mechanisms, evolution and temperature. *Biol. Symp. 6:*71–125.

Nagylaki, T. 1975. Conditions for the existence of clines. *Genetics 80:*585–615.

Nei, M. 1975a. *Molecular Population Genetics and Evolution.* Amsterdam: Elsevier.

– 1975b. Models of speciation and genetic distance. In *Population Genetics and Ecology* (ed. Karlin, S., and Nevo, E.), pp. 723–65. New York: Academic Press.

Nevo, E., and Bar-El, H. 1976. Hybridization and speciation in fossorial mole rats. *Evolution 30:*831–40.

Nevo, E., and Cleve, H. 1978. Genetic differentiation during speciation. *Nature (Lond.) 275:*125–6.

Nevo, E., and Heth, G. 1976. Assortative mating between chromosome forms of the mole rat, *Spalax ehrenbergi. Experientia 32:*1508.

Nevo, E., and Sarich, V. 1973. Immunology and evolution in mole rats, *Spalax. Israel J. Zool. 23:*210.

Nevo, E., and Shaw, C. R. 1972. Genetic variation in a subterranean mammal, *Spalax ehrenbergi. Biochem. Genet. 7:*235–41.

Nevo, E., Kim, Y. J., Shaw, C. R., and Thaeler, C. S. 1974. Genetic variation, selection, and speciation in *Thomomys talpoides* pocket gophers. *Evolution 28:*1–23.

Nevo, E., Naftali, G., and Guttmann, R. 1975. Aggression patterns and speciation. *Proc. Natl. Acad. Sci. U.S.A. 72:*3250–4.

Nevo, E., Bodmer, M., and Heth, G. 1976. Olfactory discrimination as an isolating mechanism in speciating mole rats. *Experientia 32:*1511.

O'Donald, P. 1977. Theoretical aspects of sexual selection. *Theor. Pop. Biol. 12:*298–334.

Ohta, T. 1968. Effect of initial linkage disequilibrium and epistasis on fixation probability in a small population with two segregating loci. *Theor. Appl. Genet. 38:*243–8.

Oliver, C. G. 1972. Genetic and phenotypic differentiation and geographic distance in four species of Lepidoptera. *Evolution 26:*231–41.

Paterson, H. E. H., 1978. More evidence against speciation by reinforcement. *S. Afr. J. Sci. 74:*369–71.

Patton, J. L., and Yang, S. Y. 1977. Genetic variation in the *Thomomys bottae* pocket gophers: macrogeographic patterns. *Evolution 31:*697–720.

Patton, J. L., Selander, R. K., and Smith, M. H. 1972. Genic variation in hybridizing populations of gophers (genus *Thomomys*). *Syst. Zool. 21:*263–70.

Powell, J. R. 1978. The founder-flush speciation theory: an experimental approach. *Evolution 32:*465–74.

Remington, C. L. 1968. Suture zones of hybrid interaction between recently joined biotas. *Evol. Biol. 2:*321–428.

Selander, R. K., Hunt, W. G., and Yang, S. Y. 1969. Protein polymorphism and genic heterozygosity in two European subspecies of house mouse. *Evolution 23:*379–90.

Selander, R. K., Kaufman, D. W., Baker, R. J., and Williams, S. L. 1974. Genic and chromosomal differentiation in pocket gophers of the *Geomys bursarius* group. *Evolution 28:*557–64.

Short, L. L. 1969. Taxonomic aspects of avian hybridization. *Auk 86:*84–105.

Skellam, J. G. 1973. The formulation and interpretation of mathematical models of diffusionary processes in population biology. In *The Mathematical Theory of the Dynamics of Biological Populations* (ed. Bartlett, M. S., and Hiorus, R. W.), pp. 63–85. London: Academic Press.

Slatkin, M. 1975. Gene flow and selection in a two locus system. *Genetics 81:*787–802.

Slatkin, M. 1976. The rate of spread of an advantageous allele in a subdivided population. In *Population Genetics and Ecology* (ed. Karlin, S., and Nevo, E.), pp. 767–80. New York: Academic Press.

Thaeler, C. S. 1968. An analysis of three hybrid populations of pocket gophers (genus *Thomomys*). *Evolution 22:*543–55.

Thoday, J. M. 1972. Disruptive selection. *Proc. R. Soc. Lond. B 182:*109–43.

Thoday, J. M., and Gibson, J. B. 1962. Isolation by disruptive selection. *Nature (Lond.) 193:*1164–6.

Thomson, G. 1977. The effect of a selected locus on linked neutral loci. *Genetics 85:*753–88.

Turner, J. R. G. 1971a. Two thousand generations of hybridisation in a *Heliconius* butterfly. *Evolution 25:*471–82.

– 1971b. Studies of Müllerian mimicry and the evolution in burnet moths and heliconiid butterflies. In *Ecological Genetics and Evolution* (ed. Creed, E. R.), pp. 224–60. Oxford: Blackwell.

Ursin, E. 1952. Occurrence of voles, mice and rats (Muridae) in Denmark, with a special note on a zone of intergradation between two subspecies of the house mouse (*Mus musculus L.*). *Medd. Danik. Naturhist. Forening 114:*217–44.

Wahrman, J., and Goitein, R. 1972. Hybridization in nature between two chromosome forms of spiny mice. *Chromosomes Today 3:*228–37.

Wahrman, J., and Gourevitz, P. 1973. Extreme chromosomal variation in a colonizing rodent. *Chromosomes Today 4:*399–424.

Watson, G. F. 1972. The *Litoria ewingi* complex (Anura: Hylidae) in south-eastern Australia. II. Genetic incompatibility and delimitation of a narrow hybrid zone between *L. ewingi* and *L. paraewingi. Aust. J. Zool. 20:*423–33.

Watson, G. F., Loftus-Hills, J. J., and Littlejohn, M. J. 1971. The *Litoria ewingi* complex (Anura: Hylidae) in S.E. Australia. I: a new species from Victoria. *Aust. J. Zool. 18:*401–16.

White, M. J. D. 1968. Models of speciation. *Science 158:*1065–70.

– 1975. Chromosomal repatterning: regularities and restrictions. *Genetics 79:*63–72.

– 1978a. *Modes of Speciation.* San Francisco: Freeman.

– 1978b. Chain processes in chromosomal evolution. *Syst. Zool. 27:*285–98.

White, M. J. D., Blackith, R. E., Blackith, R. M., and Cheney, J. 1967. Cytogenetics of the *viatica* group of morabine grasshoppers. I: the "coastal" species. *Aust. J. Zool. 15:*263–302.

White, M. J. D., Key, K., Andre, M., and Cheney, J. 1968. Cytogenetics of the *viatica* group of morabine grasshoppers II. Kangaroo Island population. *Aust. J. Zool. 17:*313–28.

Wills, C. J. 1970. A mechanism for rapid allopatric speciation. *Am. Nat. 111:*603–5.

Woodruff, D. S. 1972. The evolutionary significance of hybrid zones in *Pseudophryne* (Anura: Leptodactylidae). Ph.D. thesis, University of Melbourne.

– 1973. Natural hybridization and hybrid zones. *Syst. Zool. 22:*213–18.

– 1979. Postmating reproductive isolation in *Pseudophryne* and the evolutionary significance of hybrid zones. *Science 203:*561–3.

8 Chromosomal hybrid zones in orthopteroid insects

DAVID D.SHAW

Zones of hybridization between closely related taxa have consistently aroused the interest of both taxonomists and evolutionary biologists (Short, 1972; Woodruff, 1973). Taxonomists view them with caution because their very presence creates problems of definition in terms of the species concept. The evolutionary biologist, however, regards them as important, dynamic phenomena potentially capable of providing some insight into the final phase of the speciation process, namely, the development of premating isolating mechanisms. Of course, not all hybrid zones are candidates for such developments because the formation of hybrid individuals does not necessarily imply hybrid inferiority. Indeed, it is equally plausible that the products of hybridization may have a higher fitness than either parent, under certain environmental conditions, and give rise to "hybrid swarms." However, in the context of speciation phenomena, it is those cases in which hybrids are less fit than either parental form that are of major interest.

The classical theory of speciation, and the one clearly accepted by the majority of evolutionists, proposes that speciation follows a regular sequence of events. Thus initially a single taxon becomes disrupted into two or more geographically isolated populations, between which gene flow is nonexistent (Mayr, 1969). These isolates are subjected to different environmental pressures, and the genetic systems diverge accordingly. Secondary expansion follows climatic amelioration and leads to a reunion of the "populations." The consequences here will, albeit simplistically, depend upon three factors.

First, if genetic divergence between isolated populations has been minimal, then the two isolates may intermate with little or no loss in fitness, and fuse into one single taxon.

Second, it is purported that, in isolation, the different populations may have undergone considerable genetic divergence as a consequence of natural selection for different environmental and phenological optima. Associated with this divergence, the isolates develop different mating patterns such that upon secondary contact the isolates fail to interbreed and may overlap widely, with the resultant recognition of two sympatric species.

The third alternative concerns those cases in which the isolates become contiguous and have still retained the ability to hybridize along the region of contact. This leads to the development of a hybrid zone, the width of which will be proportional to the degree of intertaxon mating, the dispersal pattern, and the fitness of the hybrids (Endler, 1977). Dobzhansky (1940) proposed that in these regions of hybridization selection for premating isolating mechanisms will operate at a level dependent upon the reduction in fitness of F_1 hybrid progeny. Thus it is considered that the final phase of speciation may take place at the interface between the two interacting taxa.

This neat system of selection for premating isolation is plausible but has never been verified as operational in any naturally occurring hybrid zone. This is not to say that it is fallacious; rather it has rarely been investigated with the thoroughness needed for such a conclusion to be reached. Thus the genetical structure of hybrid zones still remains a problem worthy of investigation with the potential of clarifying some of the more turbid aspects of the speciation process.

One special category of hybrid zones includes those cases in which contiguous and hybridizing taxa differ by one or more chromosomal rearrangements, which may appear to be the major phenotypic difference between the taxa. The orthopteroid insects exemplify this type of hybrid zone which has been identified in phasmids (Craddock, 1970) and in the morabine, cantatopine, and acridine grasshoppers (White and Chinnick, 1957; White et al., 1964; Hewitt, 1975; Moran and Shaw, 1977; Moran, 1979; Mrongovius, 1979). All these cases are characterized by extremely narrow zones of overlap between morphologically indistinguishable taxa that differ by one or more pericentric inversions, centric fusions/fissions, or sex-autosome translocations.

The narrowness of these zones and the involvement of chromosomal rearrangements, primarily in morabines but subsequently identified within several species of mammals, lizards, and stick insects, led White et al. (1967) to propose the theory of stasipatric speciation, which deals directly with the function of chromosomal rearrangements as isolating mechanisms in areas of reproductive interaction between taxa. Essentially, the theory proposes that the rearrangements arise as unique events, overcome the initial barrier of heterozygosity, and rapidly reach fixation in small semi-isolated demes. The rearrangement confers an advantage to the carrier in the homokaryotype and would never have existed in a balanced polymorphic state. The theory proposes that chromosomal hybrids will suffer from reduced fecundity because of the irregular meiotic behavior of the heterozygous rearrangements, thus generating a restriction to gene flow between the taxa. The genetic effects of the rearrangements upon the phenotype of the carriers are left rather vague and have never been adequately incorporated into the theory.

The essential difference between the allopatric and stasipatric models is that in the former the interruptions to gene flow arise by extrinsic barriers, whereas in the latter the rearrangement itself acts as a mechanical barrier to gene exchange between taxa that are partially sympatric. The stasipatric model does not accommodate any provision for the development of premating isolating mechanisms in the hybrid zone.

The five cases of chromosomal hybrid zones in orthopteroid insects all show the basic ingredients for speciation by stasipatry, at least superficially. It is our purpose to reexamine these cases in terms of the following criteria: Can the presence of the hybrid zones be directly equated with adaptation to environmental or ecological factors? Do the chromosomal rearrangements generate nondisjunctional products at meiosis in F_1 hybrids at a sufficiently high level to act as a primary isolating mechanism? Can a distinction be made between genic and chromosomal divergence in these cases? Do these types of chromosomal rearrangements produce other effects upon the endophenotype that would provide alternative explanations for their involvement in speciation processes? In order to seek answers to these important questions, it is necessary first to present a brief summary of the pertinent data from each of the five cases, and to conclude by assessing the evidence in terms of the functional role of chromosomal rearrangements in the maintenance of narrow hybrid zones.

Case histories
Keyacris (= Moraba) scurra
This species of morabine grasshopper exists as two chromosomal races in S.E. Australia. The two races are distinguished by a chromosomal dissociation of the ÂB metacentric chromosome into two acrocentric elements, A and B (White, 1957). White considers that the fission of the AB was not a simple one but involved a translocation with the CD (Blundell) chromosome.

The two races meet along a front of at least 250 km, but, unfortunately, the advent of sheep grazing has decimated the natural vegetation of the area, and any evidence of widespread interaction between the races is now absent. Even so, White and Chinnick (1957) did find 17- and 15-chromosome types within 820 meters of each other. Only one natural heterozygote (a 16-chromosome male) was ever found in a population of the 15-chromosome individuals ($8\delta + 4\female$) collected at the contact zone. There was no evidence of any sharp ecological discontinuity at the zone of contact. White and Chinnick proposed that the most adequate explanation for the narrowness of the zone of overlap was selection acting against heterozygotes for the A/B fission. However, evidence obtained from both the natural hybrid and several laboratory hybrids clearly indicates that AB/A + B heterozygotes suffer little infertility through trivalent malorientation ("98%–99% of normal in many individuals"). The limited data do not permit inviability assessments to be made, but even so, with only one marker, derived generations cannot be identified.

One important point to which White and Chinnick refer is that the frequency and position of crossing over are quite different between the two homokaryotypes. The chiasma frequency in A + B bivalents is approximately 50% greater than in the fused AB bivalent. Furthermore, the fused AB bivalents contain only a few proximal chiasmata, whereas the same region in the A + B bivalents shows "quite a high chiasma frequency."

The *Vandiemenella* complex (= coastal *viatica*)

The *Vandiemenella* complex (coastal *viatica* group) consists of ten chromosomal "taxa," of which White (1974) considers five or six warrant species status. In contrast, Key (1974) recognizes only a single species composed of ten chromosomal races. All the taxa show parapatric distribution patterns, but zones of contact with hybridization have only been investigated at one locality on the mainland (White et al., 1964) and on Kangaroo Island (Mrongovius, 1979). The chromosomal differences between these taxa are shown in Table 8.1. Quite clearly the major differences between them involve both centric fusions and pericentric inversions, although the X-autosome fusion in P24(XY) is accompanied by the increase in short arm lengths on autosomes 3 and 4 (White et al., 1967), possibly because of small pericentric inversions.

Contact Zones between P17 and P19. Mainland: As in the case of *Keyacris scurra*, the mainland populations of *Vandiemenella* have been drastically affected by agriculture over the past 100 to 150 years. Even so, White et al. (1964) were able to locate hybrid populations between the P19 and P17 chromosome races (Table 8.2). Two hybrid males were examined meiotically, and, as in the *K. scurra* trivalent, the majority (70%–72%) of cells showed disjunctional behavior; 8%–10% of the metaphase I cells contained a small univalent and B/B6 bivalent, and 20% showed malorientation patterns. This behavior, they argue, would tend to diminish fecundity in chromosomal heterozygotes and thus account for the very narrow hybrid zone. White et al. (1964) also suggest that it is likely that the mixed populations are panmictic.

In B6/B + 6 heterozygotes, it is interesting to note that, although chiasmata occupy medial, submedial, or terminal positions in the B/B association, the 6/6 arms invariably show only a single terminal chiasma. A subterminal chiasma in the B/B gives the trivalent a much more balanced appearance. Unfortunately, the pattern of chiasma distribution in free B/B and 6/6 bivalents has not been examined, so that a change in distribution cannot be ascertained. There is a possibility that a difference in chiasma position in the 6/6 association may arise in the heterozygote if compared to the nonfused 6 bivalent.

In this case there is a coarse correlation between the position of the zone of contact of the two taxa and the 50 cm isohyet, the territory of *viatica*$_{17}$ being characterized by a much lower rainfall and higher rate of evaporation.

Table 8.1. *Summary of the recognizable chromosomal differences between those taxa of* Vandiemenella *that form narrow hybrid zones in nature*

Chromosomal taxon[a]	Recognizable chromosomal rearrangements
viatica$_{19}$ ↓	A, B, ĈD, 1, 2, 3, 4, 5, 6, XO/XX
viatica$_{17}$ ↓	A, B̂6, ĈD, 1, 2, 3, 4, 5, XO/XX
P24(XO) ↓	A, B̂6, ĈD, 1, 2, 3, 4, 5, XO/XX (pericentric inversion)
P24(XY)	A, B̂6, ĈD, 2, 3 (acro), 4 (acro), 5, neo XY (X/autosome fusion)

[a]The sequence *viatica*$_{19}$ → P24(XY) indicates the assumed phylogeny of the races.
Source: White et al. (1967).

Contact between P24(XO) and P24(XY). An area to the north of Port Gawler, South Australia is quoted in White et al. (1967) as containing a mixture of chromosomal types. However, no chromosomal studies were carried out.

Kangaroo Island: White et al. (1969) surveyed the island in 1966 and 1967 and examined the distribution pattern of 24(XY), *viatica*$_{17}$, and *viatica*$_{19}$. They showed that again the races were contiguous at several

Table 8.2. *Summary of the available data obtained after sampling contact zones between the chromosomal races of* Vandiemenella *on the Australian mainland and Kangaroo Island*

Contact zone	Chromosomal constitution	Width of zone
Mainland		
viatica$_{17}$/ *viatica*$_{19}$	Locality 1,27(17):2(hybrid) :2(19)	Not more than 0.5 km wide
	Locality 2,25(17):1(hybrid) :0(19)	
	Locality 3,31(17):2(hybrid) :1(19)	
P24(XO)/ P24(XY)	Not studied "Mixed population a short distance north of Pt. Gawler"	Not stated
Kangaroo Island		
viatica$_{17}$/ P24(XY)	11(17):6(hybrid) :3(P24(XY)) Only females exam- ined – male hybrids indistinguishable	White et al. (1969) approx. 275 m.
viatica$_{19}$/ P24(XY)	28(P24(XY)):10(19)	(No hybrids found) Pure races within 1–3 m of each other.
viatica$_{17}$/ *viatica*$_{19}$	3(19):2(hybrid):2(17)	Prob < 40 m
viatica$_{17}$/ *viatica*$_{19}$	25(19):1(F$_2$?):7(17)	Not determinable
viatica$_{19}$/ P24(XY)	56P24(XY):1(hybrid):7(19)	(No hybrids found) Pure races within 1–3 m of each other.
viatica$_{17}$/ P24(XY)	71(17):23(hybrid)♀ :61(P24(XY))	500 m (Mrongovius, 1979)

localities, and in one locality six female hybrids [P24(XY)/*viatica$_{17}$*] were detected cytologically. Male hybrids are not chromosomally distinguishable because of the nature of the X/autosome fusion – they appear as either XO (*viatica$_{17}$*) or neo XY [P24(XY)] types.

A contact zone between P24(XY) and *viatica$_{19}$* was identified at Mt. Thisby in which pure cytological forms of both races were found as close as 1–3 meters from each other, but no cytological hybrids were found.

Mrongovius (1979) made a more detailed survey of the hybrid zone situation on Kangaroo Island and also attempted to rear laboratory hybrids to seek evidence of a meiotic drive mechanism favoring the rearranged chromosome as proposed by White (1968).

A transect was made across the contact zone of *viatica$_{17}$* and *viatica$_{19}$*, and in one sample of seven animals two chromosomal heterozygotes were identified. There was a complete absence of F$_1$ hybrids in another sample of 33 from an area containing a mixture of racial types (Table 8.2). However, the location of the transect coincided with vegetational and soil discontinuities such that *viatica$_{19}$* and *viatica$_{17}$* may occupy areas with different soil types. If the habitat differences are effective, then random mating may not be expected.

One heterozygous male was examined meiotically, and only 20% of the MI cells showed irregular behavior of the B6/B + 6 trivalent. Twenty-eight percent of the MI cells contained univalents of the B and/or 6 chromosomes. Mrongovius estimated that such chromosome behavior would, in total, generate 20%–35% unbalanced sperm, and by extrapolation a similar level during oogenesis in female hybrids.

The overlap zone between P24(XY) and *viatica$_{19}$* is the same geographical locality as that sampled by White et al. (1969). In this instance, one hybrid female was detected cytologically. In addition, six F$_1$ hybrid males derived from a laboratory cross were examined meiotically, and all showed normal testis development and meiotic behavior. Diakinesis and first metaphase were only studied in detail in two males, and here 80% and 55% of the cells appeared to be normal.

The third zone to be studied by Mrongovius involved that between P24(XY) and *viatica$_{17}$*, and it too was at the same location as that sampled by White et al. (1969). Over an eight-year period, a total of 23 female chromosomal heterozygotes have been found in mixed populations, including the data of White et al. (1969).

Meiosis was studied in laboratory-bred males and was entirely normal when compared with the pure strains. In this case, Mrongovius supplemented her data by obtaining embryos from field-collected females for meiotic analysis, and in most cases the embryos contained the normal diploid karyotype. From these data it was possible to show that the recovery rate of the *viatica*$_{17}$ X chromosome was 0.56 in natural heterozygotes and 0.54 in laboratory-reared heterozygotes. Thus, there was some evidence of meiotic drive, but it favored the "ancestral" *viatica*$_{17}$ X chromosome rather than the derived P24(XY) X chromosome as predicted by White's hypothesis.

From these limited data, Mrongovius (1979) concluded: "It seems likely that major chromosomal rearrangements have played a role through the reduced fertility of the chromosomal heterozygotes, in the initial evolutionary divergence of the group."

In addition, hybrid inviability derived from the limited number of laboratory crosses undertaken is considered a potentially important contributing factor.

The question of the involvement of meiotic drive is still not resolved and it cannot be considered directly pertinent to the population genetics of the *Vandiemenella* group.

Didymuria violescens

The stick insect *Didymuria violescens* is a complex of at least 10 chromosomal taxa that inhabit the eucalyptus forests of eastern Australia. The evolution of the group is complex, and several steps in the presumed phylogenetic pathway from a high chromosome number to a low number are not now available for study (Craddock, 1975). White (1978) uses this particular chromosomal complex as a paradigm of stasipatric speciation. The major types of chromosomal rearrangements that differentiate the 10 known races involve centric fusions and pericentric inversions of both autosomes and sex chromosomes. Unfortunately, and contrary to the views of White (1978), the available knowledge on the distribution patterns of these races and zones of contact is incomplete. "Only a few" natural hybrids have been found because of the difficulties involved in adequate sampling in the field (see Table 8.3). However, artificial hybrids between both parapatric and allopatric races have been made, and these, coupled with evidence from the natural hybrids, reveal the behavior of fusion heterozygotes in meiosis (see Table 8.4). Craddock (1975) was able to show that the disjunctional pattern of trivalents was very regular. In only two crosses (3 and 6, Table 8.4) was a high level of asynapsis found. The former cross also showed a greater variety of metaphase I associations (150 types in 844 cells). In the case of the latter cross, the formation of irregular multiples was observed and interpreted as evidence for nonhomologous association [frequently found in species hybrids (Shaw and Wilkinson, 1978)] rather than chromosomal heterozygosity for a rear-

Table 8.3. *Types of natural hybrids found between races of* Didymuria violescens

Contact zone	Chromosomal constitution[a]	Width of zone
28(♀) × 26(XY ring)♂ 26(♀) × 28(XY ring)♂	F_1♂ has 27 chromosomes Parents differ by one autosomal translocation Meiotic configuration of $1_{III} + 11_{II} + XY$ ring.	Not calculable because samples too widely separated No. of hybrid individuals not given
30(♀) × 26(XY ring)♂ 26(♀) × 30(XY ring)♂	F_1♂ has 28 chromosomes Parents differ by one autosomal translocation, an X-autosome translocation and a neo Y fusion	Not calculable because samples too widely separated. No. of natural hybrids not given

[a]Note that the meiotic behavior of F_1 heterozygous males was mostly regular and disjunctional.

Table 8.4. *Summary of the chromosomal behavior in F_1 males obtained by artificial hybridization*

Artificial hybrid	Chromosomal behavior[a]	Natural association
1. 40(m)♀ × 35♂ 36♀ × 39(m)♂	$2_{III} + 15_{II} + X$ – regular disjunction of trivalent	No natural zone of overlap
2. 32♀ × 35♂	$2_{III} + 13_{II} + X$ – regular disjunction	No natural zone of overlap
3. 32♀ × 39(m)♂ 40(m)♀ × 31♂ }*	$4_{III} + 11_{II} + X$ and $3_{III} + 11_{II} + 1$ unequal biv + $1_I + X$	Not known to be contiguous in nature
4. 32♀ × 39 (XY)♂	$15_{II} + X$ – regular disjunction	Contiguous but no natural hybrids found
5. 32♀ × 26 (XY ring)♂ (XO race)	$2_{III} + 11_{II} + X$ – regular disjunction	Contiguous but no natural hybrids found
6. 40(m)♀ × 26 (XY ring)♂*	Meiotic cells showed a considerable degree of asynapsis and formation of irregular multiples due to nonhomologous association.	Not known to be contiguous in nature

[a]In only two cases* was gross meiotic irregularity observed, and these races are not known to meet or hybridize in nature. The remaining hybrids all showed regular meiotic behavior.

rangement. Furthermore, these two cases are not directly relevant to our analysis because they both do not show zones of overlap and hybridization in nature.

Craddock (1975) suggests that the major classes of rearrangements in *Didymuria* (centric fusions and X/A translocations) lead to reduced fertility of the chromosomal heterozygote even though the meiotic evidence she presents would suggest otherwise. The primitive race [39:40(m)] with the metacentric X chromosome has an extensive distribution pattern, and wide ecological latitude. The evolution toward a decreased chromosome number is considered to be correlated with associated reductions in the amount of, and changes in the pattern of, recombination. This, in turn, is considered to provide a reduction in variability, permitting ecological specialization. Thus the chromosomal rearrangements are ascribed adaptive values other than their purported isolating role during meiosis. The rearranged races are also postulated to have displaced (and presumably still are displacing) the ancestral form by the gradual movement of a narrow hybrid zone in the direction of the less adapted form until an equilibrium point is reached, even though natural hybrids have rarely been found.

Podisma pedestris

This wingless cantatopine grasshopper is widely distributed in Russia and Europe with peripheral populations in the French and Italian Alps. John and Hewitt (1970) have shown that the latter populations can be subdivided into two distinct chromosomal forms on the basis of an X/autosome fusion. In some areas the two forms are contiguous and form narrow hybrid zones (Hewitt, 1975). Unfortunately, the lack of markers, other than the neo XY sex chromosome system, does not allow the recognition of hybrid males because their sex chromosome system is morphologically the same as that of either parent. However, a few hybrid females have been identified mitotically within mixed populations (see Table 8.5). These mixed populations occupy very narrow areas between the two parental forms, a phenomenon that led Hewitt to suggest that the hybrid populations cannot exist in equilibrium (i.e., hybrids have a reduced fitness). Regrettably the relative fitnesses and fertility of the two homo-karyotypes and heterokaryotypes are not known.

Hewitt and John (1972) have made a detailed study of the effect of the X/autosome fusion upon recombination patterns in the autosomal part of the fusion. It is the L_3 autosome that has fused with the X chromosome, and these authors have compared the chiasma frequency and distribution in both fused and free L_3 autosomes. The effect of the fusion is quite dramatic. A comparison of chiasma frequency in the L_3 pair when fused and free reveals highly significant differences due to the reduced frequency in the fused L_3. Moreover, the fusion results in a redistribution of chiasmata such that there are more distal and fewer proximally sited chiasmata in the neo XY bivalent. This means, of course, that there is a considerable restriction in the amount of recombination in the fused race, but any adaptive significance is not known.

The evolution of the fusion system is postulated to involve geographical isolation with severe bottlenecks during glacial and interglacial periods. The present-day distribution patterns of the two races do not seem to be associated with ecotones, and the stability of the zones is unknown.

Caledia captiva

This morphologically monotypic acridine grasshopper is composed of a complex of four distinctive and parapatric chromosomal taxa. Investigations of the differences between taxa have involved analysis of chromosome morphology, C-banded heterochromatin distribution, and allozyme variation (Shaw, 1976; Shaw et al., 1976; Moran et al., 1980).

Two of the taxa (the Moreton and Torresian) differ by at least nine

Table 8.5. *Summary of the chromosomal analysis of five populations of* Podisma pedestris *showing evidence of hybridization between the XO and* neo XY *races*

Contact zone	Chromosomal constitution	Width of zone
Population 1	2 XO♂:9 XY♂:5 X̂A/X̂A♀:1 X̂A/X + A hybrid ♀	300 m to E is a pure neo XY population
Population 2	7 XO♂:2 XY♂:3 X̂A/X̂A♀:1 X̂A/X + A hybrid ♀	500 m to SE the pops. are neo XY type
Population 3	10 XO♂:1 neo XY♂:2 X̂A/X̂A♀:1 X̂A/X + A hybrid ♀	Found on XO, 100 m "or so" from the hybrid population
Population 4	10 XO♂:2 neo XY♂:2 X̂A/X̂A♀:1 X̂A/X + A hybrid♀:2 X̂A/X̂A♀	About 800 m to E is an XO population
Population 5	12 XO♀:1 X̂A/X̂A♀:2 X̂A/X̂A♀	1.5 km to NE is an XO population

pericentric rearrangements (acro- vs. metacentric chromosomes), by a multitude of interstitial and terminal C-bands restricted to the Moreton taxon and located outside the limits of the pericentric rearrangements, and by four fixed allele differences for allozyme variants (Moran et al., 1980). The two chromosomal forms meet along a 200-km front in S.E. Queensland where the pure forms are contiguous and exhibit a very narrow hybrid zone less than 1 km wide (Moran and Shaw, 1977). Two detailed transects have been taken across the zone, one in which the Moreton taxon is characterized by a homozygous acrocentric X chromosome and the other in a region where the X chromosome is homozygous for a metacentric form. (See Table 8.6.) Populations of *Caledia* are continuous across both areas with density estimates of 2500 per hectare as determined by

Table 8.6. *The chromosomal constitution of the Moreton and Torresian taxa, which form a very narrow hybrid zone in S.E. Queensland*

Contact zone[a]	Chromosomal constitution[b]	Width of zone
Transect 1. Torre-sian/Moreton (meta-centric X)	2n = 22 autosomes + XX/XO sex chromosome system. Taxa differ by nine pericentric rearrangements including the X chromosome. All combinations found within an 800-m zone. Chromosomal frequency changes of 60% approx. over a 200-m distance. Numerous heterochromatin differences also segregating. Hybrid zone is asymmetrical about the null point in terms of chromosomal characters.	400 m on either side of the null point. Introgression of Torresian chromosomes shows different levels of penetrance.
Transect 2. Torre-sian/Moreton (acro-centric X)	Taxa differ by eight pericentric rearrangements. All combinations of chromosomal types found within an 800-m zone. Good evidence of asymmetrical introgression of Torresian chromosome. However, pattern of allozyme variation across zone is symmetrical. Allozymes not tightly linked to rearrangements.	400 m on either side of null point. Introgression of Torresian chromosomes occurs at a higher frequency than in (1) above. Population density estimates of 2500 per hectare.

[a]Two transects were analyzed in detail across areas with continuous distribution patterns.
[b]Note that the major change in chromosomal frequency occurs within a 200-m distance.

mark/release/recapture experiments (Craft, pers. comm.). Moran (1979) took samples every 200 meters across the metacentric X hybrid zone and found the following:

1. On the basis of centromeric rearrangements, the population samples over a total distance of 1 km contained a very high frequency of backcross derivative karyotypes, and yet on either side of this 1-km zone, the samples represented pure Torresian on one side and Moreton karyotypes on the other.

2. Chromosomal frequencies changed by an average of 60% over a 200-meter distance. Moreover, the X chromosome showed the greatest change in frequency, 80%, over the same distance.

3. The hybrid zone was shown to be asymmetrical with evidence of one-way introgression of Torresian chromosomes into the Moreton taxon. However, the rate of introgression for different nonhomologues was variable, and there was no evidence for the movement of the acrocentric X chromosome from the Torresian into the Moreton taxon.

Subsequently, Shaw et al. (1980) analyzed a second zone where the Moreton acrocentric X taxon meets the Torresian taxon, and found a similar situation. Moran et al. (1980) also examined allozyme variation in the *same individuals* sampled from this zone and identified four fixed differences between the taxa. Here, it was found that, unlike the chromosomal situation, the hybrid zone structure was much more symmetrical, with evidence of movement of alleles across both sides of the zone. This situation clearly demonstrates the inadequacy of analyzing hybrid zones in terms of one character only. It also demonstrates a very clear example of Key's (1974) concept of a "tension zone."

In this particular case, a comprehensive series of laboratory crosses has been made to analyze the nature of any fertility or viability deficiencies in hybrids. (See Table 8.7.) The results show clearly that the F_1 suffers no significant reduction in either fertility or viability in any of the reciprocal hybrids. Meiosis in these F_1 males is not affected by the presence of eight chromosomal heterozygotes. This is hardly surprising because, unlike the case of *Drosophila,* pairing loops at pachytene do not occur, and pairing is nonhomologous and straight. Consequently, when heterozygous, crossing over is completely inhibited in the region of the chromosomal rearrangement, with a resultant marked repatterning of crossing over but with no effect upon synapsis.

The effects of hybridization are not expressed until later segregating generations. Thus the F_2s are completely inviable even though egg production per female approximates that of the parental taxa. In the backcross generations, all crosses involving a (Moreton × Torresian) $F_1\male$ parent are also 100% inviable. The remaining backcrosses show 50% inviability. The inviability occurs during the embryonic development within the confines of the egg. The stage reached is quite variable within the progeny of single females, suggesting some kind of segregation of factors that determine embryogenesis.

From these data, it is easy to see that the abrupt change in chromosomal frequency observed in the hybrid zone is mainly explained by the absolute breakdown in the F_2 progeny. Thus F_1 individuals must backcross to produce viable progeny. As a consequence, there will be a difference in chromosomal rearrangement frequency of 50% at the interface between the two taxa. One can also estimate the expected number of chromosomal

Table 8.7. *Meiotic behavior and viability data obtained from laboratory hybridization experiments between the Moreton and Torresian races of* Caledia captiva

Cross	Meiotic behavior	Viability
Moreton × Moreton	Normal. Xa frequency 14.6.	98% viability
Torresian × Torresian	Normal. Xa frequency 14.5.	99% viability
Moreton × Torresian (F_1)	Heterozygous for 8 peri-	95%–100% viability
Torresian × Moreton	centric rearrangements. Meiotic behavior very regular. Reduction in chiasma frequency (12.8 per cell) and marked changes in chiasma position.	
(M × T) × (M × T) (F_2)	Not available because of inviability.	100% inviable
Backcrosses		
(M × T) × M	Individuals contain a wide	50% inviability
M × (M × T)	range of chromosomal	100% inviable
(M × T) × T	morphs owing to segrega-	50% inviability
T × (M × T)	tion and recombination.	100% inviable
(T × M) × T	Meiotic behavior very	50% inviability
T × (T × M)	regular. Increased variance in cell chiasma frequency. Evidence of highly significant interchromosomal interactions in backcross progeny.	50% inviability

rearrangements per individual from the binomial expansion $(a + b)^6$. Surprisingly, the observed data from surviving adult backcross progeny do not deviate significantly from random. This finding suggests, superficially, that the rearrangements are not directly involved in the generation of the inviability of F_2s and certain of the backcrosses. However, if one considers that 50% of the gametes will carry novel intrachromosomal recombinants owing to the redistribution of crossing over in the F_1 adults, then it may be that these intrachromosomal recombinants generate some of the developmental blockages observed in the F_2 and backcross generations. Furthermore, there is evidence of strong linkage disequilibria between Moreton chromosome 1 and Moreton metacentrics 2, 4, 5, and 6. Thus interactions both within and between chromosomes may be making a significant contribution to the maintenance of this narrow hybrid zone.

One final point concerns the role of environmental factors in the maintenance of this narrow zone. Moran and Shaw (1977) have shown that there is a gross correlation between the position of the zone and the 30% isocline of the coefficient of variation of the mean weekly moisture index. However, the importance of such correlations is not considered significant in the present state of our knowledge of the ecology of *C. captiva,* as is also the case in those other examples of hybrid zones discussed above.

Conclusions

If we return to those questions that were originally posed in the introduction, we can now attempt to assess the evidence from these five examples of chromosomal hybrid zones that favors a functional role for the observed rearrangement differences.

First, the available evidence does not show the presence of any obvious ecological or environmental differences in the form of ecotones that could explain the sharp discontinuities in karyotypes as adaptations to different environmental optima. Gross correlations with such factors as rainfall and vegetation do exist between taxa, however, but the association is generally on a much greater scale than the observed narrowness of these hybrid zones, and cannot be directly attributable as causal phenomena to the maintenance of the abrupt changes in karyotypes. This does not imply that adaptations to more subtle environmental variants are not present or important, but they are certainly not evident in these examples. In fact, from the available evidence, it seems that the two taxa are inhabiting essentially the same habitat in both space and time within the vicinity of the hybrid zone at least, and in most cases over a much wider range than the confines of the zone itself.

The effects of past climatic changes upon the distribution patterns of the chromosomal taxa have been considered to be significant factors by both Hewitt (1975) and Shaw (1976), particularly in the establishment of the chromosomal rearrangements in allopatry with subsequent secondary expansion and contact upon climatic amelioration. It seems also applicable to the case of *Didymuria violescens,* in which the present-day distribution pattern is certainly suggestive of a breaking up of a widely distributed species into small isolated populations, clearly shown by the presence of the 39:40(m) race in South Australia, which is 600 km from other populations of the same chromosomal constitution. In the case of *Vandiemenella,* however, White (1978) considers that past climatic changes have not been important factors in the establishment of the chromosomal races in this species and favors the hypothesis that the new chromosomal rearrangements became established *within* the distributional range of the parental form. Key (1968), however, considers that the distributional data on the *Vandiemenella* complex are compatible with an *allopatric,* or at least peripheral, origin of the races which then expanded to make secondary contact with other racial forms.

Clearly, arguments concerning the role of past climates in the establishment of chromosomal rearrangements as homozygotes must be inferential, and such inferences cannot make a significant contribution to our analysis of the structure and maintenance of present-day hybrid zones.

The analysis must, in the first instance, be directed toward mapping the present-day distributions, particularly in those areas of intertaxon contiguity and overlap. It should be mentioned here that the micro-ecological aspects of chromosomal hybrid zones are virtually unknown, and, similarly, studies of dispersal patterns have never been made. The fact that all cases of narrow hybrid zones involve organisms of limited mobility is obviously one aspect of zone structure that needs to be studied, particularly in terms of the stability of hybrid zones.

If we now turn our attention to the functional role of chromosomal rearrangements as isolating mechanisms via mechanical irregularity during meiosis, then the evidence presented above is far from convincing. In all cases, the level of nondisjunction is quite small, and highly variable between individuals in both natural and artificial hybrids. For instance, in *Didymuria violescens* the fusion heterozygotes behave in a very regular manner with significant breakdown occurring in only two synthetic hybrids, and, moreover, in these cases the parental forms are not known to be parapatric or to form a hybrid zone in nature. In *Keyacris scurra* meiosis in synthetic hybrids is essentially normal with a reduction in fertility of "at most, 10%" (White and Chinnick, 1957). The same also applies to the *Vandiemenella* hybrids studied by both White et al. (1967, 1969) and Mrongovius (1979). On the mainland, *viatica*$_{17}$ × *viatica*$_{19}$ hybrids were again estimated to show only a 10% reduction in fertility. Natural hybrids between *viatica*$_{17/19}$ and *viatica*$_{19}$/(P24(XY)) showed an estimated reduction in fertility ranging from 2% to 35%.

Two important findings on the behavior of heterozygous chromosomal rearrangements during meiosis are very relevant here. The first concerns the behavior of heterozygotes in hybrid mice that differ by one or more chromosomal fusions. Here Cattenach and Moseley (1973) have shown that the heterozygotes show a nondisjunction frequency between 6% and 35%, with considerable variation both between different individuals carrying the same fusion heterozygote and between individuals with different fusion heterozygotes. However, they also show that the observed nondisjunction could well result from other genic or minor chromosomal differences, reflecting the fact that the chromosomes may be derived from separate species. Thus, the asynapsis and malorientation may not be due *directly* to the structural heterozygosity. There are also several cases of centric fusion polymorphisms occurring within one biological species, as in Swedish cattle (Gustavsson, 1969). Here a centric fusion between the largest and smallest autosomes has been found in populations of SRB cattle at relatively high frequencies (14% heterozygotes, 0.34% homozygotes). Meiosis in male translocation heterozygotes was regular and disjunctional with no observed unbalanced MII cells. White and Tjio (1967) have reported on the behavior of an unequal fusion between the largest and smallest chromosomes of the mouse, and here the disjunctional pattern of the asymmetrical trivalent was very high as judged from the predominantly balanced MII cells (91%). Thus, even the involvement of two very dissimilar-sized chromosomes in the fusion does not affect disjunctional behavior. Again, natural populations of the insectivore *Sorex araneus* are found to be polymorphic for centric fusions that behave in a regular manner (Ford et al., 1957).

Other examples of polymorphic fusions are available, but we may conclude from these examples alone that there is no a priori reason to assume that Robertsonian fusions invariably lead to the generation of unbalanced gametes. The second point, and a very important one, which has been overlooked by most investigators, concerns changes in the disjunctional behavior of chromosomal rearrangements by selection. Lawrence (1958) has clearly show that the orientation pattern and segregational properties of translocation heterozygotes are subjected to the same genotypic control measures that are known to affect other aspects of chromosome behavior (Rees, 1961). Lawrence was able to show that when rye plants, heterozygous for two independent translocations, were selected for seed production, increased productivity was directly associated with an increase in the frequency of alternate disjunctional patterns even though the latter was not the selected character. By extrapolation, one can readily visualize such a system operating on populations that consistently hybridize to generate F_1, F_2, and backcross generations, as is probable in all the chromosomal hybrid zones discussed here even if we assume a slight initial reduction in fertility. One can hardly envisage the opposite effect whereby selection would favor those individuals showing the greatest reduction in fecundity!

In the case of heterozygosity for pericentric rearrangements, as found in *Caledia captiva,* those F_1 individuals found in nature and also generated in the laboratory show no significant effects of the presence of seven autosomal heterozygous rearrangements upon meiotic behavior. Here, however, one could argue that these pericentric rearrangements are known to show straight pairing at pachytene in heterozygotes (White and Morley, 1955) and show a complete lack of crossing over within the region of the presumed pericentric inversion. Thus, unlike the *Drosophila* case, we do not expect to generate duplication/deficiency products during meiosis. This would tend to suggest that they represent a separate class of rearrangement, quite distinct from the centric fusion. However, this is not necessarily the case because they both share a common effect when heterozygous by modifying linkage relationships because of changes in the pattern of crossing-over, which will be referred to later.

A third point concerns the relevance of the "background" genome in hybrid individuals because we already know that segregating generations are likely to occur in chromosomal hybrid zones. Unfortunately most of these chromosomal hybrid zones have not been investigated for the presence of variation in systems other than the observed rearrangements themselves. Segregational and recombinational patterns of markers such as heterochromatin and allozyme variants have only been investigated in the case of *Caledia.* With the aid of these additional markers, distributed among all chromosomes within the genome but showing differences between taxa, the genetic structure of these hybrid zones, in terms of recombinant genotypes, would be possible. Without them, it is always assumed that the background genotype is nonrecombinant (Craddock, 1975). This is certainly not a true representation, and Key (1974) has provided clear evidence to show that intertaxon differences in the male cercal morphology can introgress well beyond the chromosomally defined boundaries of the hybrid zone. Thus, within the hybrid zone between the races P24(XY) and *viatica*$_{17}$, many males showed cercal structure intermediate in character, and, furthermore, the variants could be identified up to 10 km from the hybrid zone. Again, in *C. captiva,* Moran and Shaw (1977) have shown that the hybrid zone is asymmetrical when assessed chromosomally, with a one-way movement of chromosomes from Torresian to Moreton, but not vice versa. However, when assessed on the basis of allozyme frequency variation across the same zone and within the same individuals used for chromosome assay, the zone is now symmetrical with respect to the introgression of alleles at four unlinked loci (Moran et al., 1980). These cases clearly support Key's concept of a tension zone in which he postulates that the function of hybrid zones is analogous to a semipermeable membrane that allows some chromosomes, chromosome segments, or individual genes to filter through unimpeded, whereas the penetration of others is completely inhibited. Such a situation has also been demonstrated by Nagle and Mettler (1969), who showed that following hybridization between *Drosophila mojavensis* and *D. arizonensis,* the fitness of the former was actually improved by the process of selective one-way introgression.

The concept of the tension zone is clearly useful because it incorporates the interaction of two entire genomes rather than the role of the chromosome rearrangement isolated from the total genotype. This brings us to our third point concerning the distinction between chromosomal and genically induced isolation in hybrid zones. If we accept the idea that the chromosomal rearrangement differences between these hybridizing taxa offer minimal isolation via mechanical impairment of meiosis, then we must now consider the evidence in terms of their gene content and the ways in which they influence or modify the fitness of both homo- and heterokaryotypes. This aspect of the structure of chromosomal hybrid zones is most important in any analysis, and yet its involvement is the least understood. Allozyme variation is evident in the case of *Caledia,* and, although informative, and valuable as a source of markers, it reveals little concerning the causes of the developmental breakdown seen in the F_2 and backcross generations. Information concerning the genetic aspects of the structure and maintenance of hybrid zones is needed if we are to obtain insight into their relevance to the speciation mechanism.

One factor that is pertinent to an understanding of the genetic structure of the two hybridizing taxa concerns the nature of the rearrangements themselves and their influence upon the spatial relationships of the genes both within the rearranged chromosome and between it and other members of the genome. Thus a pericentric rearrangement, whatever its cause, produces a tightly linked group of genes within the limits of the rearrangement. Robertsonian fusion reduces the number of linkage groups and hence modifies segregational variation. A second feature, to which attention was drawn earlier, concerns the consequences of chromosomal rearrangements upon the pattern of crossing-over.

Hewitt and John (1972) have shown that the pattern of crossing over in the fused autosomes of the neo XY system in *Podisma* is very different from that seen in the same nonfused autosomal bivalent of the XO race. Shaw and Wilkinson (1980) have also shown that in F_1 hybrids between the two races of *Caledia captiva* not only is crossing over suppressed within the limits of the heterozygous pericentric rearrangements in seven of the autosomes, but as a consequence there is also a major repatterning of crossing over in the remainder of the genome. Again in *Keyacris scurra,* fusion heterozygotes show distally localized crossing over, whereas the dissociated A and B chromosomes in the 17-chromosome race show a high proportion of proximal crossovers. Moreover, the chiasma frequency in the free autosomes of the 17-chromosome race is 50% higher than that observed in the fusion homozygote of the 15-chromosome race. The effect of the rearrangements in the other cases can only be assumed to follow repatternings similar to crossing over because they have not been investigated. However, the repatterning of recombination subsequent to pericentric rearrangement, centric fusion or X/autosome fusion seems to be a general principle in eukaryotes (White, 1973). The consequences of the redistribution of crossovers are generally interpreted in terms of the conservation of coadapted gene sequences within the rearrangement itself, but this reasoning stems mainly from an assessment of their role in intra-population polymorphisms (i.e., inversion systems in *Drosophila* species). In the present context of chromosomal hybrid zones, the situation is just the reverse, with adaptive homokaryotypes and presumed disruptive heterokaryotypes. The effect of homozygosity for the kinds of chromosomal rearrangements considered here is not known to be directly associated with any marked morphological divergence between taxa in any example described above. Furthermore, there is no evidence of adaptation to different ecological or environmental optima. Thus, to date, there is no a priori reason to invoke major genetic differences between the parapatric taxa. In this context the theory of genetic divergence during allopatry does not specify the kinds of changes that are hypothesized to occur during the period of isolation other than those related to presumed environmental changes. Even so, if selection favored a different phenotype during past environmental perturbation, then presumably the same selective processes would operate to produce a phenotype suited to the present-day environment to which it is now exposed. Furthermore, the classical model of allopatric speciation prescribes that reproductive isolation develops as a correlated response to genetic divergence (Mayr, 1963). This is clearly a vague concept that has proved difficult to justify. Similarly, the notion of natural selection favoring the development of reproductive isolation mechanisms after secondary contact has been made between the genetically divergent taxa (e.g., Dobzhansky, 1940), also creates problems, none more difficult than the question of why these isolating genes retain their potency in distant and essentially allopatric populations of the two hybridizing taxa. These problems have remained refractory for many years.

Is it possible that the role of the chromosomal rearrangement lies more in its effect in altering the *spatial* relationships of the genetic material both within and between chromosomes? Can we envisage a system of change that would be large enough to generate a new divergent genotypic optimum with minimal genic change? It seems a possibility, and one that may offer an explanation to White's (1973) concept of karyotypic orthoselection.

The essential feature here is that the observed spatial organization of the genes into a certain karyotypic pattern is a fundamental one and is intimately associated with function. That is, the expression of any gene depends to a greater or lesser extent upon its relationships with its neighbors. Such a system of functional activity is highly stable and conservative in its tolerance to change. This may be reflected in the constraints normally placed upon the pattern of crossing over. Thus most species show a restrictive pattern with marked localization of crossing over to particular chromosomal regions. This system allows recombination to generate an array of variants within certain limits not solely because of allelic differences at individual loci but also as a result of the different nonallelic associations.

If, however, the stability of the chromosomal system is suddenly altered by rearranging the chromosomal material, either intra- or interchromosomally, then changes to the gene order and the relationships with other genes may lead to changes in gene expression resulting in new nonallelic interaction patterns. Thus the genotypic changes induced may be quite abrupt compared with those generated by gene frequency changes and the accumulation of mutations. It may apply particularly to those cases where the alleles are associated as "internally balanced" chromosomes (Mather, 1943), and would be most applicable to very small isolates undergoing significant inbreeding. If we take the case of a pericentric rearrangement as an example, then the permanent repositioning of those genes within the rearranged segment, which includes a significant proportion of the genome, may generate changes in cellular processes and functional activity on a large enough scale to promote the need for further, complementary reorganization elsewhere in the genome. Once the inertia of the stable genotype is broken, a chain reaction may be initiated by this single rearrangement until a new equilibrium condition is attained. In extreme cases, the genic repatterning may involve all members of the genome, as seen in *Caledia captiva.* A similar argument applies to the chromosome fusion races of the *Mus musculus* complex (Capanna et al., 1976). Thus, the karyotypic repatterning that takes place in isolation leads to the appearance of chromosomally homozygous taxa that are similar in phenotype but quite different in the ways they adopt to produce it. This kind of karyotypic change is obviously not applicable to intra-population stable polymorphisms but seems to explain the more disruptive aspects of chromosomal rearrangements as exemplified by the presence of hybrid zones. One consequence of the readjustments to chromosomal organization is that if the two cohesive karyotypes are disrupted by hybridization, then the internally balanced chromosomes of each type may function normally within the F_1 heterozygote as seen in *C. captiva,* but will be disrupted by recombination in subsequent generations, leading to variable levels of inviability.

The changes in the biochemical pathways postulated to arise subsequent to a rearrangement are at present unknown because our knowledge of the cellular process of eukaryote transcription, translation, and gene regulation is not adequate.

However, the putative effect of a chromosome rearrangement acting as an initiator to the development of new adaptive, physiological optima does explain some of the anomalies associated with previous hypotheses, and there is some suggestive evidence in its favor. First, the phenomenon of "position effect" provides evidence that a chromosomal rearrangement can induce changes in gene expression (Lewis, 1950). Second, Wilson et al. (1974) have proposed that chromosomal rearrangements may be important evolutionary mechanisms by inducing changes to "regulatory" systems, independent of the accumulation of mutational changes in structural genes. Third, the available evidence suggests that genic differentiation is minimal between taxa that form chromosomal hybrid zones, and yet in one case recombination and segregation in chromosomal hybrids can generate 100% inviable progeny.

It seems unlikely that conventional selective processes of rather vague definition have been instrumental in the disruption of the developmental pathways seen in the hybrids between the two taxa of *Caledia.* We can find no evidence to support this idea, either on the grounds of differences in environmental conditions between natural populations, or in laboratory-reared stocks where the rates of development are identical between the two taxa. Detectable genetic differences between the chromosomal forms are hard to find, and, indeed, one may ask why should the developmental pathways of two taxa that inhabit similar environments show morphological equivalence and identical phenologies and yet degenerate immediately after recombination and segregation. Is it justifiable to think of chromosomal rearrangements as mechanisms that reshape the genetic system by way of physically restructuring interactions between the *existing* genes without the assistance of a "gradual shifting of gene frequencies," or "genetic indeterminancy," "jack of all trades genes," or "good mixers" (Mayr, 1963)? It is certainly a possibility and one that can be put to the test, but only after much more thorough analyses of chromosomal hybrid zones have been performed at both the level of the population and that of the individual developmental processes.

There is little doubt that we have, as yet, no adequate explanations for the function of the observed chromosomal differences between contiguous hybridizing taxa. This is not surprising because the studies performed to date can only be called preliminary. Future analyses of chromosomal hybrid zones must surely be focused toward studies of the hybrids themselves, which, after all, represent the essence of the problem. We should, however, be grateful to M. J. D. White, who first brought them to our attention.

References

Capanna, E., Gropp, A., Winking, H., Noack, G., and Civitelli, M-V. 1976. Robertsonian metacentrics in the mouse. *Chromosoma (Berl.)* 59:341–53.

Cattenach, B. M., and Moseley, H. 1973. Nondisjunction and reduced fertility caused by the tobacco mouse metacentric chromosomes. *Cytogenet. Cell Genet.* 12:264–87.

Craddock, E. M. 1970. Chromosomal number variation in a stick insect *Didymuria violescens. Science 167:*1380–2.

– 1975. Intraspecific karyotypic differentiation in the Australian phasmatid *Didymuria violescens* (Leach). 1. The chromosomal races and their structural and evolutionary relationships. *Chromosoma (Berl.)* 53:1–24.

Dobzhansky, Th. 1940. Speciation as a stage in evolutionary divergence. *Am. Nat. 74:*312–21.

Endler, J. 1977. *Geographic Variation, Clines and Speciation.* Princeton: Princeton University Press.

Ford, C. E., Hamerton, J. L., and Sharmann, G. B. 1957. Chromosomal polymorphism in the common shrew. *Nature (Lond.) 180:*392–3.

Gustavsson, I. (1969). Cytogenetics, distribution and phenotypic effects of a translocation in Swedish cattle. *Hereditas 63:*68–169.

Hewitt, G. M. 1975. A sex chromosome hybrid zone in the grasshopper *Podisma pedestris* (Orthoptera: Acrididae). *Heredity 75:*375–87.

Hewitt, G. M. and John, B. 1972. Inter-population sex chromosome polymorphism in the grasshopper *Podisma pedestris.* II. Population parameters. *Chromosoma (Berl.) 37:* 23–42.

John, B. and Hewitt, G. M. 1970. Inter-population sex chromosome polymorphism in the grasshopper *Podisma pedestris*. I. Fundamental facts. *Chromosoma (Berl.). 31:*291–308.

Key, K. H. L. 1968. The concept of stasipatric speciation. *Syst. Zool. 17:*14–22.

– 1974. Speciation in the Australian morabine grasshoppers: taxonomy and ecology. In *Genetic Mechanisms of Speciation in Insects* (ed. White, M. J. D.), pp. 43–56. Sydney: Australia and New Zealand Book Co.

Lawrence, C. W. 1958. Genotypic control of chromosome behaviour in Rye. VI. Selection for disjunction frequency. *Heredity 12:*127–31.

Lewis, E. B. 1950. The phenomenon of position effect. *Adv. Genet. 3:*73–115.

Mather, K. 1943. Polygenic inheritance and natural selection. *Biol. Rev. 18:*32–64.

Mayr, E. 1963. *Animal Species and Evolution*. Cambridge, Massachusetts: Belknap Press.

Moran, C. 1979. The structure of a narrow hybrid zone in *Caledia captiva*. *Heredity 42:*13–32.

Moran, C., and Shaw, D. D. 1977. Population cytogenetics of the genus *Caledia* (Orthoptera: Acrididae). III. Chromosomal polymorphism, racial parapatry and introgression. *Chromosoma (Berl.) 63:*181–204.

Moran, C., Wilkinson, P., and Shaw, D. D. 1980. Allozyme variation across a narrow hybrid zone in the grasshopper *Caledia captiva*. *Heredity 44:*69–81.

Mrongovius, M. J. 1979. Cytogenetics of the hybrids of three members of the grasshopper genus *Vandiemenella* (Orthoptera: Eumastacidae: Morabinae). *Chromosoma (Berl.). 71:*81–107.

Nagle, J. J. , and Mettler, L. E. 1969. Relative fitness of introgressed and parental populations of *Drosophila mojavensis* and *D. arizonensis*. *Evolution 23:*519–24.

Rees, H. 1961. Genotypic control of chromosome form and behaviour. *Bot. Rev. 27:*288–318.

Shaw, D. D. 1976. Population cytogenetics of the genus *Caledia* (Orthoptera: Acrididae). 1. Inter and intra specific diversity. *Chromosoma (Berl.) 54:*221–43.

Shaw, D. D., and Wilkinson, P. 1980. Chromosome differentiation, hybrid breakdown and the maintenance of a narrow hybrid zone in *Caledia*. *Chromosoma (Berl.)* (in press).

Shaw, D. D., Webb, G. C., and Wilkinson, P. 1976. Population cytogenetics of the genus *Caledia* (Orthoptera: Acrididae). II. Variation in the pattern of C-banding. *Chromosoma (Berl.) 56:*169–90.

Shaw, D. D., Wilkinson, P., and Moran, C. 1980. A comparison of chromosomal and allozymal variation across a narrow hybrid zone in the grasshopper *Caledia captiva*. *Chromosoma (Berl.) 75:*333–51.

Short, L. L. 1972. Hybridization, taxonomy and avian evolution. *Ann. Missouri Bot. Gard. 59:*447–53.

White, B. J., and Tjio, J. H. 1967. A mouse translocation with 38 and 39 chromosomes but normal N.F. *Hereditas, 58:*284–96.

White, M. J. D. 1957. Cytogenetics of the grasshopper *Moraba scurra*. 1. Meiosis of interracial and interpopulation hybrids. *Aust. J. Zool. 5:*285–304.

– 1968. Models of speciation. *Science 159:*1065–70.

– 1973. *Animal Cytology and Evolution,* 3rd. ed. Cambridge: Cambridge University Press.

– 1974. Speciation in the Australian morabine grasshoppers: the cytogenetic evidence. In *Genetic Mechanisms of Speciation in Insects* (ed. White, M. J. D.). Sydney: Australia and New Zealand Book Co.

– 1978. *Modes of Speciation.* San Francisco: W. H. Freeman.

White, M. J. D., and Chinnick, L. 1957. Cytogenetics of the grasshopper *Moraba scurra*. III. Distribution of the 15- and 17-chromosome races. *Aust. J. Zool. 5:*338–47.

White, M. J. D., and Morley, F. H. W. 1955. Effects of pericentric rearrangements on recombination in grasshopper chromosomes. *Genetics 40:*604–19.

White, M. J. D., Carson, H. L., and Cheney, J. 1964. Chromosomal races in the Australian grasshopper *Moraba viatica* in a zone of geographic overlap. *Evolution 18:*417–29.

White, M. J. D. Blackith, R. E., Blackith, R. M., and Cheney, J. 1967. Cytogenetics of the *viatica* group of morabine grasshoppers. I. The "coastal" species. *Aust. J. Zool. 15:*263–302.

White, M. J. D., Key, K. H. L., André, M., and Cheney, J. 1969. Cytogenetics of the *viatica* group of morabine grasshoppers. II. Kangaroo Island populations. *Aust. J. Zool. 17:*313–28.

Wilson, A. C., Sarich, V. M., and Maxon, L. R. 1974. The importance of gene rearrangement in evolution: Evidence from studies on rates of chromosomal protein and anatomical evolution. *Proc. Natl. Acad. Sci. U.S.A. 71:*3028–30.

Woodruff, D. S. 1973. Natural hybridization and hybrid zones. *Syst. Zool. 22:*213–18.

9 Toward a genodynamics of hybrid zones: studies of Australian frogs and West Indian land snails

DAVID S. WOODRUFF

The biological species concept is a central tenet of evolutionary biology. Hybrid zones of various kinds threaten this concept by revealing that species may lack some of the properties (integrity, cohesion, and reproductive isolation) that we have ascribed to them. Whereas hybrid zones have traditionally been treated as taxonomically bothersome (but not unexpected) phenomena involving borderline cases, they are increasingly being regarded as evolutionists' delights that allow us to monitor the processes of geographic differentiation and ultimately speciation. In my view the elucidation of the roles of gene flow and natural selection in a natural hybrid zone is a prerequisite to the solution of the problems presented by that zone. On a more general level, the need to develop empirical techniques to measure the genodynamics of hybrid zones is critical if we are to understand the processes by which speciation occurs and the entity of the species itself. Until we learn more about what is going on in nature, it strikes me that assertions that the species is not the keystone of evolution are simply premature. Whether species are "mental abstractions which order clusters of diversity in multidimensional character space" (Levin, 1979:381), or whether they have reality, "a reproductive community, a gene pool, and a genetic system" (White, 1978), remains to be seen. Obtaining the answers will certainly require a more rigorous application of the techniques of population biology and genetics; we can no longer entrust these important phenomena exclusively to Darwin's (1859) "naturalists having sound judgement and wide experience."

In this chapter, I will illustrate the extent to which our present theory of hybrid zones is inadequate by describing two groups of animals in which hybridization is a prominent feature. I will point out the nature of the problems encountered in each group and note the extent to which these problems might be alleviated if we understood the genodynamics of the interactions. I will suggest ways in which existing techniques might be applied to these problems. Although no single hybrid zone seems amenable to the full range of investigations that should be conducted, it seems likely that we can make substantial progress by conducting more detailed analyses of a few selected cases.

At the outset it should be noted that much of the difficulty surrounding the interpretation of natural hybrid zones stems from the fact that the "solution" of the problem of each particular interaction really involves the solution of four separate questions:

1. What is the geographical relationship of the ranges of the inter-acting populations?
2. What is the history of the interaction?
3. What will be the probable outcome of the interaction?
4. What is the taxonomic status of the interacting population?

Usually only the first of these questions can be answered objectively. To answer the second question one must be able to distinguish between cases of primary and secondary intergradation, a distinction that is usually impossible to make. The answer to the third question stems from detailed studies of the interacting taxa in nature and the laboratory. Our predictive theory is still so inadequate, however, that even when such information is available, it is only rarely that the third question can be answered. Logically, the answer to the fourth question depends on the answers to the previous ones. Yet, in practice, it has often been the first to be "solved." Many of the older controversies involving particular cases of hybridization arose because undue emphasis was placed on the premature solution of the final question.

A second source of confusion must also be noted; several of the words used to describe hybrid phenomena are actually nonoperational because they are defined in terms of "answers" to the second, third, or fourth questions. I have proposed new criteria for describing cases of natural hybridization that overcome the difficulties inherent in schemes based on the history of the interaction (primary or secondary contact) or the taxonomic status of the interacting forms (Woodruff, 1973a). In this paper I will employ the terms allopatric and parapatric hybridization to imply geographical information only.

Pseudophryne hybrid zones in southeast Australia

Pseudophryne semimarmorata, P. bibroni, and *P. dendyi* are small leptodactylid frogs with generally contiguous distributions in south-east Australia (Figure 9.1). Narrow zones of parapatric hybridization have been found wherever the northern border of *P. semimarmorata* comes into contact with the southern borders of *P. bibroni* and *P. dendyi.* These hybrid zones are particularly interesting because they provided direct evidence for the operation of postmating isolating mechanisms in animals in a natural situation (Woodruff, 1972b). They are also of special interest in that they involve anurans whose breeding calls are indistinguishable and therefore may not serve as premating isolating mechanisms. A highly abridged summary of the following account has appeared elsewhere (Woodruff, 1979).

The *Pseudophryne,* commonly called toadlets, differ from typical anurans in a number of ways. They are terrestrial and cryptozoic as adults. They lack a tympanum, middle ear, and expansive vocal sac, and the breeding call is practically identical in all members of the genus. In the three species under consideration breeding is restricted to the autumn months. Several weeks before mating, males migrate to the breeding area and commence vocalization. In southern Victoria these breeding choruses of 20–50 males form in the last third of March in most years. Males maintain territories and prepare nest sites among the grass roots and leaf litter in situations that will be saturated with water during the winter. Females enter the area and begin mating in the last third of April; they lay 70–90 large eggs in discrete batches and may mate on several occasions over a 4–7-week period. After mating a male usually remains with the eggs and resumes calling, mating again if the opportunity occurs. The embryonic phase of development is protracted, and although hatching

Figure 9.1. Map of southeast Australia and Tasmania showing the general ranges of three species of *Pseudophryne.*

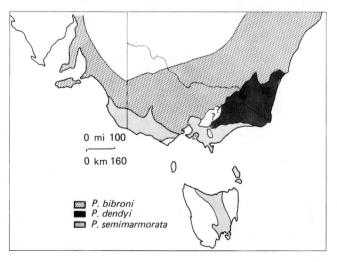

normally occurs after 4–6 weeks, it may be delayed until conditions are favorable for the aquatic larval stage. Metamorphosis occurs in September and October, and toadlets are thought to reach maturity in 2 or 3 years. Adults of these three species have consistently different patterns of coloration. They are, in contrast, very similar in anatomy, karyotype, size, breeding season and site, common male vocalization, pre- and postmating behavior, reproductive rates, mating system, and the pattern and rates of embryonic and larval development. I have supported the above generalizations elsewhere (Woodruff, 1972a, b, 1975c, 1976a, b, 1977, 1978a).

A zone of parapatric hybridization stretches discontinuously over 750 km along the northern border of *P. semimarmorata* (Littlejohn, 1967; Woodruff and Tyler, 1968; Woodruff, 1979). Five diagnostic adult coloration characteristics were used to distinguish the taxa, estimate the hybridity of toadlets of intermediate coloration, and map the hybrid zones.

Figure 9.2. Morphological variation of *Pseudophryne* near Wallan: Hybrid index diagrams illustrate the change from *P. bibroni* (locality A and X) to *P. semimarmorata* (locality G and Z). The hybrid indexes range between 0 (typical *P. semimarmorata*) and 10 (typical *P. bibroni*). The diagrams show all specimens collected at selected localities between 1960 and 1969. The 306-meter and 350-meter (shaded) contour lines are indicated; www is the Wandong-Whittlesea-Woodstock road junction, 3 km east of Wallan.

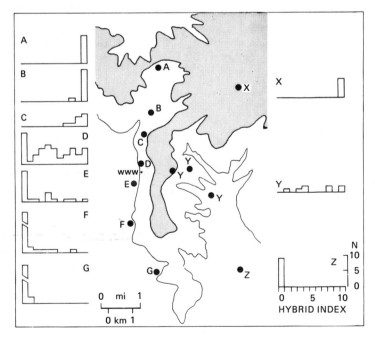

Detailed sampling in seven areas revealed that the interactions between *P. semimarmorata* and *P. dendyi* and between *P. semimarmorata* and *P. bibroni* were very similar in many respects. Morphological variation in two of these areas is summarized in Figures 9.2 and 9.3. The hybrid zone is generally less than 9 km wide, and over 80% of the morphological change occurs in the central 3 km. The diversity of hybrid phenotypes suggests that mating probably occurs at random among toadlets at each breeding site. No evidence for assortative mating based on ecological, ethological, or morphological differences was detected. There are slight asynchronies in the breeding seasons of the interacting taxa, but they do not appear to limit hybridization during the middle 80% of the season. Postmating isolating mechanisms are not sufficient to prevent the occurrence of apparently healthy F_1 and backcross hybrids. In vitro hybridization experiments and studies of field-collected eggs confirmed that whereas hybrids are both viable and fertile, hybrid embryos are less viable than homospecific ones.

Figure 9.3. Distribution and morphological variation of *Pseudophryne* from the hybrid zone between *P. semimarmorata* and *P. dendyi* near Tyers. The map shows the position of collecting sites in relation to roads (heavy lines) and villages. The hybrid index diagrams describe variation at 12 selected localities (A–L): Hybrid indexes range between 0 (typical *P. semimarmorata*) and 10 (typical *P. dendyi*). The fine contour lines are isophenes connecting areas of similar mean hybridity.

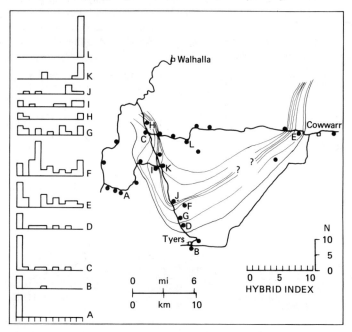

Although natural embryonic mortality rates away from the hybrid zones are typically less than 5% (range 0–10%), many batches of eggs collected from within the hybrid zone suffered levels of mortality two to three times as high (Woodruff, 1972b, 1979). The range of embryonic mortality values seen in field-collected eggs within the hybrid zones indicates that some recombinations may be less viable than others (Figure 9.4).

The histories of these interactions are conjectural. Littlejohn (1967) and Littlejohn and Martin (1974) interpreted them as the outcome of secondary contacts between taxa that diverged from one another in allopatry. Littlejohn suggested that *P. dendyi* was derived from a *P. bibroni*–like ancestor during the last glacial period when it became isolated to the south of the Eastern Highlands. Littlejohn and Martin suggested that *P. semimarmorata* was also derived from the same ancestral stock, but that its differentiation occurred when it became isolated on the continental island of Tasmania during interglacial times. According to these workers, *Pseudophryne* reached Tasmania by way of the land bridge that emerged during the penultimate glaciation and returned to the mainland as *P.*

Figure 9.4. Elevated levels of embryonic mortality were associated with batches of eggs collected in *Pseudophryne* hybrid zones. The *P. semimarmorata–P. bibroni* hybrid zone near Wallan is shown on the left, the *P. semimarmorata–P. dendyi* hybrid zone near Tyers on the right. The upper diagrams show the mean and range of hybrid index scores of samples from localities described in histograms in Figures 9.2 and 9.3. The lower diagrams show embryonic mortality (mean values connected for the Wallan case) in batches of eggs collected at these localities. (After Woodruff, 1979)

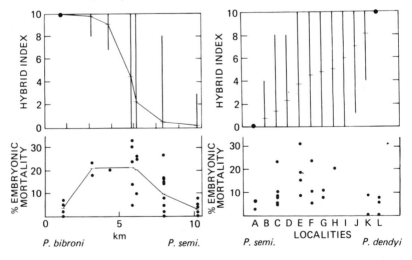

semimarmorata on a second land bridge that emerged during the last glacial phase. According to these scenarios the present hybrid zones could be 10,000–20,000 years old. An alternative scenario involves the hills of the Eastern Highlands as the barrier that facilitated the differentiation of both the southern toadlets. I have argued (Woodruff, 1972b) that present patterns of distribution and morphological variation in *P. semimarmorata* and *P. dendyi* suggest that neither *P. dendyi* nor *P. bibroni* was present in southern Victoria at the various times Tasmania was connected to the mainland. Although the details of the evidence supporting this position no longer seem important, it should be noted that if the highlands rather than Bass Strait served as the barrier facilitating allopatric differentiation of *P. semimarmorata*, then again the present hybrid zone may be at least 10,000 years old.

Two other hypotheses must also be considered. McDonnell et al. (1978) have proposed that the zones (specifically the Wallan contact) may be of more recent origin. Arguing that agricultural activities have increased the area of suitable habitat for *Pseudophryne*, they concluded that such changes would have hastened contact or intensified any existing interactions. It is my subjective impression (based on over 70 field trips) that breeding choruses at undisturbed sites may, in fact, be smaller than those associated with roadside ditches. It seems quite probable that human activities during the last 100 years have had a considerable effect on the local abundance of these animals. That such activities had a marked effect on their overall distribution seems less likely though, for man has tended to modify the habitats originally favored by *Pseudophryne* rather than those where it is not found today even in undisturbed situations. I suspect the hybrid zones may antecede the arrival of Europeans in the region.

A radically different interpretation of these interactions has been offered by Endler (1977), and I have discussed it elsewhere (Woodruff, 1978b, 1979). Although not refuting the allopatric models, Endler has shown how the zones could equally well have evolved by parapatric differentiation. According to this view, they represent complex step clines separating areas characterized by different coadapted modifiers that interact so as to reduce the fitness of the hybrids. The hybrid zone effect is therefore the result of clinal selection patterns between the slightly different habitats of the interacting taxa. The taxa themselves probably differ in relatively few major genes (some of which control coloration); it is the milieu of these genes that is important. In this scenario the present position of the zones may be related to present environmental gradients and partial barriers which affect population density, to the slopes of the selection gradients, and to the history of these variables. Endler's contribution is most important because he has shown how available techniques do not allow us to distinguish between the results of allopatric and parapatric differentiation. Although the various historical hypotheses are all plausible, they are untestable at present and therefore outside of science; the origin of the zones cannot be established yet.

Perhaps we can make some progress by exploring the dynamics of the zones today. If an adequate model can be developed for the existing phenomena, then we may gain new insights into the past and possibly even the future. Three major features of these hybrid zones must be accounted for:

1. There is extensive hybridization between individuals of dissimilar allopatric populations in a restricted narrow zone whose present position is not clearly related to any single environmental factor.
2. Despite the extensive interbreeding and backcrossing, introgression into either homospecific population is apparently very limited.
3. The zone is relatively stable, there being no evidence that fusion is occurring or that premating isolating mechanisms are being reinforced.

One hypothesis to account for these features is suggested by a closer consideration of the intrinsic isolating mechanisms that may operate between these interacting taxa. Let us first review the potential premating isolating mechanisms. No ecological differences were detected that might serve as isolating mechanisms; the interacting taxa appear almost identical in their requirements in the areas where their ranges meet. No morphological differences that might constitute isolating mechanisms were noted; these taxa are very similar in size and shape, and the differences in adult coloration are not visible at night when the toadlets typically mate (Woodruff, 1972b). Behavioral differences were likewise judged to be insignificant. Acoustic signals (key isolating mechanisms in many anurans) are not differentiated. McDonnell et al. (1978) studied breeding calls near the *P. semimarmorata–P. bibroni* contact at Wallan and found overlap in all seven call characters investigated. Pengilley (1971) found the breeding calls of *P. bibroni* and *P. dendyi* to be so similar he suggested the two taxa may be conspecific. Isolating mechanisms based on pheromones remain unknown in anurans (Wilson, 1970). There is a slight asynchrony in the breeding seasons of the interacting taxa, but overlap is significant, and temporal isolation is not apparent during the few weeks when most eggs are laid. The nature of this asynchrony is difficult to quantify (indeed

I overlooked it at the time I was working in the field), but its form is amplified if one compares data on breeding away from the hybrid zones with data from around the contacts. Near sea level more northerly populations of *P. bibroni* and *P. dendyi* breed several weeks later (Fletcher, 1889; Harrison, 1922; Frauca, 1965; Woodruff, 1972b), and more southerly populations of *P. semimarmorata* breed a few weeks earlier (English, 1910; Littlejohn and Martin, 1965, 1974) than they do in the areas of their interaction. Within the hybrid zones the asynchrony is much harder to establish because the composition of the hybrid population at a single site has never been monitored throughout a breeding season. Nevertheless, consistent trends can be seen in the available data, and in view of their possible significance in the *Pseudophryne* case, which is complicated by male territoriality and an unusual mating system, it is worth summarizing some of the evidence here.

Consider the record for the *P. semimarmorata–P. bibroni* contact near Wallan (Figure 9.2). At locality D where 44 adults were collected over six seasons, nine of ten typical *P. semimarmorata* were collected before the end of April and three of four typical *P. bibroni* were found in May. At locality E, 19 adults were collected over two seasons; typical *P. semimarmorata* and hybrids were found in equal numbers during April, whereas the single sample collected in May also contained typical *P. bibroni*. At locality F (46 adults, eight seasons), samples collected in April contained either no hybrids or *P. semimarmorata*–like hybrids (with hybrid indexes, HI equal to or less than three). Two samples collected later in the season both contained hybrids, and they were more *P. bibroni*–like (mean HI = 4.5, range 1–8). Finally, small samples were obtained at about the same level on the north–south transect at the western two localities designated Y. The mean HI ranged from 5.25 on March 16, 1965, to 8.66 on May 2, 1965. Although this evidence for an increase in the frequency of *P. bibroni*–like individuals as the season progresses is somewhat anecdotal, the same pattern was observed in two other areas where the two species hybridize (Woodruff, 1972b). A similar asynchrony has been detected at three transects across the *P. semimarmorata–P. dendyi* hybrid zone (Woodruff, 1972b). Details will be provided elsewhere (Woodruff, in prep.).

At present there is no evidence that the temporal asynchrony in breeding activity is any greater in the hybrid zones than in adjacent allopatry. Although this asynchrony may provide for some assortative mating at the beginning and end of the breeding season, I estimate that it cannot prevent mismating during the height of the season. I conclude that premating isolating mechanisms between these taxa are, at best, highly ineffective, and that mating probably occurs at random among the toadlets present at a site during 80% of the breeding season.

Turning now to the evidence for postmating isolating mechanisms, I could detect no sign that the hybrids suffered sterility or ethological isolation. In vitro hybridization experiments revealed no gametic incompatibility between *P. semimarmorata* and *P. bibroni*. Fertility was 100% in a hybridization experiment between a female *P. semimarmorata* and a male *P. dendyi*, but ova in three reciprocal crosses were infertile. These last results require confirmation and are not supported by field observations, which indicate a high degree of gametic compatibility. In contrast, the evidence for hybrid inviability is good and based on both field observations and laboratory experiments. The major component of this inviability appears to be developmental incompatibility during the embryonic stage. Both natural and artificial hybrids showed higher embryonic mortality than homospecific embryos. I conclude that postmating isolating mechanisms are not sufficiently developed to prevent the occurrence of both F_1 and backcross hybrids. Yet divergence has proceeded to the point where genetic incompatibilities render the hybrids ill-adapted vehicles of gamete wastage.

Although hybrid inviability would seem to provide a basis for selection against hybridization, there is no evidence for the reinforcement of slight differences in premating behavior. [Postmating mechanisms are, of course, beyond the reach of direct selection (Muller, 1942).] All the evidence points to the conclusion that, despite interbreeding for possibly thousands of generations, little progress has been made toward perfecting intrinsic isolating mechanisms. Similarly, there is no evidence that the pairs of interacting populations are fusing, despite demonstrable gene flow in both directions. Three factors (that both parental phenotypes are maintained at the center of the zone, the great variability of the hybrid populations, and the extreme narrowness of the zone) indicate, however, that interpopulation gene flow has been reduced to extremely low levels. The frequency of hybrid phenotypes decreases very rapidly on either side of the zone, and introgression is apparently severely limited. It appears, therefore, that natural selection has been holding its own: The hybrids are ineffective as bridges for gene flow, and the interacting populations are effectively isolated from one another.

Why, then, has selection failed to perfect premating isolating mechanisms to prevent gamete wastage? Is it because of circumstances peculiar to these specific cases? For example, it may be argued that reinforcement of slight differences in breeding season is impossible because the season is strictly circumscribed by unfavorable climatic conditions. Alternatively, perhaps the lack of a middle ear prevents these taxa from utilizing complex acoustic signals that serve as premating isolating mechanisms in many other groups of anurans. Actually, as Bigelow (1965) noted, the evolution of mechanisms that prevent interbreeding is not likely to occur in a narrow hybrid zone at all. The zones of contact involve only a very small fraction of the total range of the interacting populations. The vast majority of each species live outside the zone and have purely homospecific ancestry. Under these conditions, selection will produce mechanisms to inhibit interbreeding between the peripheral populations only incidentally, and by chance. Even if assortative mating genes do appear, and are selected for in the hybrid zone itself, their spread will be severely limited by the slope of the selection gradient, low levels of gene flow into the zone, and the fact that the hybrids suffer developmental problems (Endler, 1977). The last factor tends to keep the hybrid zone narrow as selection operates before migration.

I have concluded that these narrow hybrid zones represent the relatively stable outcome of interactions of unknown origin (Woodruff, 1979). Gamete wastage through hybridization is presumably balanced by gene flow into the zone from adjacent homospecific populations. In the absence of effective premating isolation, homospecific matings in the center of the zone may be too rare to provide a basis for the evolution of reinforcement. The interactions are interpreted, therefore, not as a race between reinforcement and fusion, but rather as a relatively stable balance between the two, in which adequate reinforcement cannot evolve, and where introgression is severely limited.

Empirical evidence to confirm this hypothesis is not yet available. Studies to determine the extent of genetic differentiation between the interacting taxa, the fitness of the various phenotypes across the hybrid zones, the extent of assortative mating, and the direction and magnitude of gene flow have not yet been conducted. In the meantime, considerable support for this interpretation comes from recent advances in population genetics (reviewed by Endler, 1977; Barton and Hewitt, Chapter 7 of this volume). Numerous theoretical treatments point to the conclusion that persistent hybrid zones will form when gene flow balances losses due to hybridization. The width of such stable zones will be related to the dispersal propensities of the interacting taxa and the relative fitness of the hybrids. In theory such zones will migrate across the country if the relative contributions from the homospecific populations are imbalanced, or if the environment changes. These predictions all lead to numerous ideas that might be tested in nature.

McDonnell et al. (1978) have recently reported an independent analysis of the *P. semimarmorata–P. bibroni* interaction near Wallan. In addition to reexamining the specimens collected in the 1960s, they resampled several of the sites in the period 1972–4. Their findings on morphology and embryonic mortality tend to confirm my own. Their discovery that, in this area, *P. semimarmorata* and *P. bibroni* are fixed for alternate alleles of heart lactate dehydrogenase (*Ldh*) adds a new dimension to the story. The distribution of *Ldh* heterozygotes is quite asymmetric about the morphological hybrid zone; whereas the southern limits of the hybrid zone based on morphological, embryonic mortality, and *Ldh* criteria coincide, the northern limits based on each of these criteria differ. This is especially true for *Ldh,* which has a replacement zone about three times as wide as the morphological zone.

McDonnell et al. (1978) cite this asymmetry, coupled with an alleged change in the mean hybridity, to conclude that *P. bibroni* and the zone moved south during the period 1960–74. This point is contrary to my earlier conclusion that the zone was stable; so it deserves comment here. In addition to establishing seasonal within-locality variation, I sought evidence of annual trends in mean hybridity at four localities in this area; the data are summarized in Figure 9.5. McDonnell et al. rescored the hybridity of all the specimens I examined from localities D, F, and G (Figure 9.2) during the period 1960–8 and compared these specimens to those they collected during 1972–4; their data are also summarized in Figure 9.5. [I have extracted these data from McDonnell et al., 1978: Figure 3, and made allowances for errors in that figure, the most significant of which involves locality D (1968), where the cumulative frequency of the animals is shown as 1.5 rather than 1.0.] Also shown in Figure 9.5 are data for 1969 from Woodruff (1972b) that McDonnell et al. ignored. Although there is a good correlation (+0.89) between our estimates for the mean hybridity of the 1960–8 samples, there is some disagreement on the scoring of individual specimens, and some of this could be due to the alteration of color character states following preservation. Nevertheless, as Figure 9.5 suggests, I concluded that the zone was stable during the period 1960–9 and McDonnell et al. reached a different conclusion for the period 1960–74. Unfortunately, the issue cannot be resolved in the absence of information on the collection dates for the 1972–4 samples, the more *P. bibroni*–like nature of which may simply reflect the fact that they were collected later in the breeding season.

The asymmetrical distribution of the genetic change about the morphological hybrid zone is also open to an alternate interpretation; it may be due to differential northward introgression of an *Ldh* gene rather than southward movement of the zone. The imbalance may arise from the fact that *P. semimarmorata* males occupy the breeding sites first and contribute more genes to the interaction. More detailed studies have in fact shown that morphology is a poor indicator of the genodynamics of hybrid zones, and that gene exchange extends into populations well beyond the zones as classically defined (Patton et al., 1979). The well-documented interaction between mice in Denmark involves a morphological zone that has not shifted in 20 years, and the asymmetry of the genetic zone in that case is clearly due to differential introgression (Hunt and Selander, 1973). Until a more critical study is undertaken, the contention that the *Pseudophryne* zone near Wallan is moving south should be regarded as not proven.

I conclude that the *Pseudophryne* hybrid zones of southeast Australia remain problematic. Although their proximity to several universities should guarantee them continued attention, the brief breeding season and

Figure 9.5. Changes in the hybridity of sequential samples from localities D–G in the hybrid zone between *P. semimarmorata* and *P. bibroni* near Wallan (Figure 9.2). Range and mean hybrid index are shown for samples described by Woodruff (1972b) for the period 1960–9 (left) and McDonnell et al. (1978) for the period 1960–74 (right). See text for discussion.

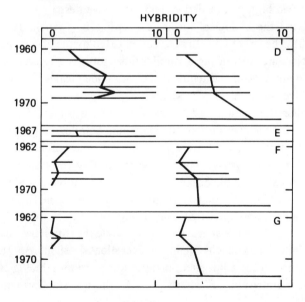

small size of the adult populations will thwart the casual investigator. Nevertheless, the fact that they are now fairly well documented should allow others to make relatively more progress than I was able to in the 1960s; our questions are now more sharply defined, and techniques for the genetic characterization of these zones are now routine. With a thorough genetic analysis of these interactions, more information on the pre- and postmating isolating mechanisms involved, and some measure of the relative densities of toadlets across the zones, we should be in a better position to answer the larger questions posed by these phenomena. The fact that these interactions occur in Bassiana, a natural laboratory of considerable interest to evolutionary biologists (Darlington, 1965; Williams, 1974; Woodruff, 1973b, 1975d), should provide further impetus for the completion of this investigation.

Cerion hybrid zones in the West Indies

Among land snails, the West Indian genus *Cerion* is remarkable for its great display of shell variation. Although intrapopulation variation in shell size, shape, color, and sculpture is not unusual, interpopulation variation is extreme. The dramatic differences often found between adjacent populations have contributed to the recognition of over 600 "species" whose distribution appears to constitute a haphazard crazy-quilt along the coasts of Cuba and the Bahamas. With a very few exceptions the different morphotypes do not occur sympatrically. The group has attained some notoriety because two highly regarded systematists have despaired at the difficulty of applying the biological-species concept to it (Clench, 1957; Mayr, 1963, 1970). The situation is further complicated by the occurrence of numerous hybrid zones separating dissimilar morphotypes. *Cerion* is the classic example, in animals, of the acquisition of morphological differences without reproductive isolation.

Recently, Stephen Jay Gould and I have developed a way of freeing *Cerion* from its taxonomic overburden. By combining detailed mapping of geographic variation in the field with laboratory studies of anatomy, genetics, and shell morphology, we have demonstrated that the systematic problems of these highly variable organisms are not intractable (Gould et al., 1974; Woodruff, 1975a). Gould has sought to characterize the shells of each population morphometrically by taking advantage of two of *Cerion's* special features. First, there is a sharp and recognizable transition between the embryonic shell (protoconch) and the accretionary shell. As the protoconch is retained in the adult shell, its terminus provides a biological criterion for the numbering of whorls and facilitates standardizing measures at various stages of ontogeny. Second, as the shell approaches adult size, the snail changes its direction of coiling, secretes a definitive lip, and ceases growth. This fact permits the sorting of ontogenetic from other variation in the adult shell, something that is impossible in most molluscs. Taking advantage of these attributes, we may characterize any population morphometrically on the basis of the study of covariation among 19 shell variables (Gould et al., 1974). Taking a different approach, I have been able to characterize each population genetically by using standard electrophoretic procedures to survey allozymic variation in foot-muscle extracts (Woodruff, 1975b). I have now examined variation in 20–30 allozymes in about 5000 snails from 250 localities representing nearly 200 "species." *Cerion* populations typically have moderate amounts of genetic variation: mean number of alleles per locus, 1.65–1.70; frequency of polymorphic loci per population, 0.15–0.30; and frequency of heterozygous loci per individual, 0.054–0.128. The frequencies of the different genotypes segregating at polymorphic loci indicate that *Cerion* are outbreeding despite their hermaphroditism.

The coordinated investigation of multivariate morphometrics and biochemical genetics of *Cerion* populations (on the same snails when possible) led to our discovery that the distribution of the various morphotypes was far from haphazard (Gould and Woodruff, 1978; Woodruff, 1978c). In the Bahamas and elsewhere, within situations that previously involved hundreds of "species," we have begun to discern fairly simple biogeographic patterns involving relatively few morphologically variable taxa with allopatric or parapatric ranges. Although our work is far from complete, it appears that these taxa may be characterized on the basis of the different patterns of morphological covariation. The interesting thing is that similar differences are not found in studies of allozyme variation or anatomy. The morphological differences we have studied were acquired without detectable differentiation at a wide range of structural gene loci. Similarly, anatomical studies of various organ systems (including the genitalia) failed to reveal much variation among a range of morphologically distinctive "species" (Chung, 1979). It is becoming clear that the complex differences in adult shell form can be traced to relatively simple differences in developmental rates expressed during ontogeny (Gould, 1977; Galler and Gould, 1979). It is quite possible that all of *Cerion's* diversity may ultimately be under the control of a few regulatory genes. The genetic uniformity was a source of some embarrassment; we had hoped that a molecular technique might succeed in revealing phylogenetic relationships where traditional methods had failed. It was not until we turned our attention to the hybrid zones between the various morphotypes that the approach paid off; the hybrid zones are typically areas of genetic anomaly. The frequency of these zones in Cuba and the Bahamas (where we have found more than 20) suggests that we will not understand *Cerion* until we understand the history and significance of these hybrid zones.

Five hybrid zones will now be described to illustrate the range of phenomena we have encountered. Detailed reports of these and other interactions will appear elsewhere in due course.

> *Hybridization between C. abacoense and C. bendalli near*
> *Rocky Point, Great Abaco, Bahamas*

This interaction involves a ribby morphotype (*C. abacoense*) and a mottled morphotype (*C. bendalli*) and has been described in broad outline by Gould and Woodruff (1978). The original analysis was based on over 400 specimens from 11 localities; our conclusions are supported by the subsequent study of an additional 255 snails. A zone of allopatric hybridization about 500 meters wide separates the parental populations. The snails are continuously and abundantly distributed along the linear transect close to the shore. Morphologically, the geographically intermediate populations display a gradual and continuous transition in mean phenotype. There is no increase in morphometric variability in the hybrid samples. Genetically, the parental species are almost identical (Nei's $I = 0.98$); near Rocky Point 15 of the 20 allozymes studied are monomorphic and fixed for the same allele in each taxon. Variation was detected at five loci: malate dehydrogenase (*Mdh-1*), leucine aminopeptidase (*Lap-1*), glutamic oxalacetic transaminase (*Got-1*), 6-phosphogluconate dehydrogenase (*6-Pgdh-1*), and a nonspecific esterase (*Est-2*). In the case of *Mdh-1*, *Lap-1*, and *Got-1* there was no significant change in allele frequency across the hybrid zone. In addition, there was no significant change in the proportion of heterozygotes of these allozymes in the hybrids. In contrast, both *6-Pgdh-1* and *Est-2* show significant changes in allele frequency associated with the hybrid zone (Figure 9.6). In the case of *6-Pgdh-1*, an unexpected allele, *6-Pgdh-1^b*, was found at high frequency in populations from within the morphological hybrid zone and at decreasing frequency over a distance of 2.5 km to the south. Similarly, in *Est-2* the common allele in both parental species, *Est-2^f*, was replaced by a typically rare allele, *Est-2^d*, over much the same area. In addition, two alleles that were very rare elsewhere, *Est-2^b* and *Est-2^g*, occurred at low but significant frequencies in the genetic hybrid zone.

Hybridization between C. glans and C. gubernatoria on New Providence, Bahamas

A second interaction between ribby (*C. glans*) and mottled (*C. gubernatoria*) morphotypes is being studied on New Providence. Despite the phenotypic and ecological similarities to the previous species pair, the snails on New Providence are anatomically distinct from their counterparts on Abaco (Chung, 1979). The hybrid zone between these morphotypes is also different in several respects from that at Rocky Point. North of the Nassau International Airport we found the hybrids displayed a host of unusual phenotypes and considerable variation in shell size. The hybrids are continuously distributed between the parental types, and the allopatric hybrid zone appears to be less than 500 meters wide. The snails are actually more abundant in the hybrid zone than elsewhere, but this phenomenon may be the result of extensive recent land clearing in the area. Genetically, the hybrids show greater variability than their parental populations, which, as on Abaco, are indistinguishable from one another at the loci surveyed. Morphologically intermediate samples show higher levels of heterozygosity in *Mdh-1* and *Est-2;* in the latter enzyme a fast

Figure 9.6. Allozyme variation associated with the hybrid zone (shaded) between *C. abacoense* (left) and *C. bendalli* (right) near Rocky Point, Abaco. Changes in allele frequency are shown for 6–phosphogluconate dehydrogenase-1^a (upper) and esterase-2^f, -2^b, and -2^g (indicated by a circle) (lower). The morphological hybrid zone is about 500 meters wide.

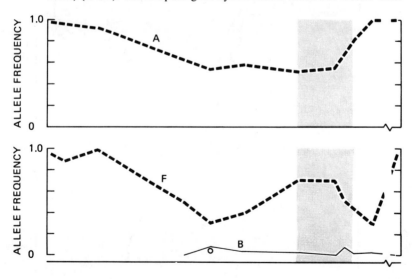

electromorph unique to the hybrid zone rises to a frequency of 0.10. [An earlier statement (Woodruff, 1978c) that this hybrid zone was not an area of genetic anomaly now seems to have been premature.] Whether this unexpected allele is asymmetrically distributed across the hybrid zone and whether the zone of increased genetic variability is wider than the morphological zone are not yet clear.

Figure 9.7. Map of Long Island, Bahamas, showing the position of some of the hybrid zones between dissimilar morphospecies of *Cerion*. Ranges identified are for, from left to right, *C. malonei, C. caerulescens, C. stevensoni,* and *C. fernandina.* Position of the hybrid zone illustrated in Figure 9.8 is indicated by the large arrow.

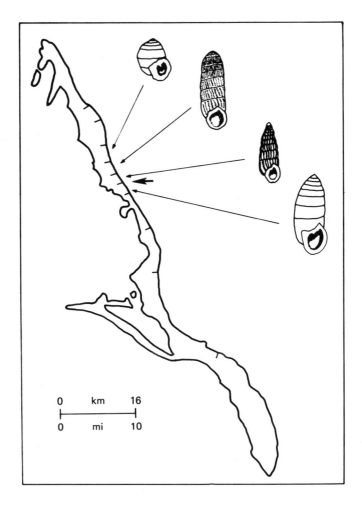

Hybridization between C. stevensoni and C. fernandina on Long Island, Bahamas

On Long Island we have located 12 hybrid zones between the seven distinct morphotypes we recognize (Figure 9.7). Spectacular transitions involve the large white, smooth-shelled *C. fernandina,* the squat, white, smooth-shelled *C. malonei,* and the distinctive member of the subgenus *Umbonis, C. stevensoni.* Defined on the basis of shell morphology, these hybrid zones are typically less than 500 meters wide, and some are only 100 meters wide. Whereas some zones are associated with changes in snail abundance or habitat, others are not. Although the various species involved are very similar to one another genetically, I have again found genetic anomalies associated with all of the hybrid zones examined. The hybrid populations show significantly greater variation in allele frequencies and are characterized by novel genotypes absent in adjacent parental populations. The allopatric hybrid zone between *C. stevensoni* and *C. fernandina* is now reasonably well documented and shows these effects in *Mdh-1, 6-Pgdh-1,* and *Est-2* (Figure 9.8). Furthermore, as at Rocky Point, the area of genetic anomaly is three to four times as wide as the morphological transition and asymmetrically distributed about it; the

Figure 9.8. Allozyme variation associated with the hybrid zone (shaded) between *C. stevensoni* (left) and *C. fernandina* (right) on Long Island. Changes in allele frequency are shown for 6-phosphogluconate dehydrogenase - 1ᵃ, -1ᵇ, and -1ᶜ (upper) and malate dehydrogenase -1ᵇ, -1ᶜ, and -1ᵈ (indicated by a circle) (lower). The morphological hybrid zone is about 300 meters wide.

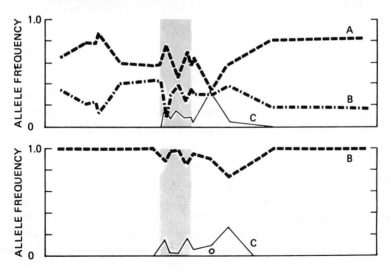

genetic effect was detected abruptly at the *C. stevensoni* end of the morphological hybrid zone but extends about 600 meters into the range of *C. fernandina.*

Hybridization between C. moralesi and C. geophilus on the Banes Peninsula, Cuba

The most widely publicized *Cerion* hybrid zones involve a series discovered by Ernst Mayr in the 1950s, but known only from his map and a single interpretive paragraph (Mayr, 1963:399, 1969, 1970). Recently, Galler and Gould (1979) have published a detailed report on the interaction between *C. moralesi* and *C. geophilus* based on the shells Mayr collected. *C. moralesi* has a smooth, mottled shell; *C. geophilus* has a squat, ribby shell and a spiral sculpture of numerous incised lines that pass over the axial costae. This latter feature is a characteristic of the subgenus *Umbonis* (Clench and Aguayo, 1952). Galler and Gould studied the shell morphology of Mayr's four samples and concluded that the zone of allopatric hybridization was less than 600 meters wide. Mayr's field notes indicated that the snails were abundant and continuously distributed across the hybrid zone. Galler and Gould found no evidence for increased variability in the morphologically and geographically intermediate samples. They demonstrated the existence of several univariate clines and one multivariate cline covering the zone. They concluded that the large morphological differences between the contrasting morphotypes stem from a small alteration in the rate of shell widening during the early postembryonic phase of growth. Nothing is known about the genetics of this interaction.

Hybridization between several morphotypes in the Bimini Islands, Bahamas

Mayr and Rosen (1956) found pronounced but geographically irregular morphological variation among the *Cerion* of six small islands in the Bimini group. They interpreted the pattern to be the result of widespread hybridization between three essentially allopatric and variable morphospecies. Hybrid colonies were shown to be more variable than homospecific ones with respect to sculpture and color, but no such effect was detected for size or shape. Unfortunately, the geographic complexity of this situation precludes a simple discussion of their results along the lines developed in the foregoing accounts. White (1978), for example, has pointed out that one of the highly variable "hybrid" colonies is actually located between two very uniform colonies, which are very similar to one another. A reanalysis of this situation using multivariate and genetic techniques is now in order.

Although the uneven documentation prevents a rigorous comparison of these interactions, they all appear to involve narrow zones of allopatric hybridization (sensu Woodruff, 1973a). They may, or may not, be associated with changes in snail density, or genetic and morphometric variability. In two well-studied cases, the morphological hybrid zone is narrow and lies at one edge of a broader zone of genetic anomaly. Although some of the hybrid zones are associated geographically with environmental changes, others apparently are not. It must be emphasized that no single explanation should be sought to interpret such a range of phenomena; the zones may be of diverse origin and significance.

Mayr concluded that the hybrid zones he studied on Cuba and Bimini were the result of secondary contacts between forms that had differentiated in isolation. The checkerboard distribution pattern of the various morphotypes was presumably the result of infrequent long-distance dispersal by hurricanes. In contrast, our own work suggests that some of the dissimilar populations may have arisen in situ as a result of parapatric differentiation (Woodruff, 1978c). Rather than the checkerboard distribution pattern perceived by earlier workers (results that were partly a consequence of taxonomic malpractice), we find fairly regular patterns of geographic variation over large areas of the Little Bahama Bank (Gould and Woodruff, 1978), the Great Bahama Bank (Woodruff, 1978c), Great Inagua (Gould and Woodruff, in prep.), the Greater Antilles (from Hispaniola to the Virgin Islands; Gould and Paull, 1977), and the Dutch Leeward Islands (Gould, 1969). Wherever we have looked, we have found that forms characterized by the most divergent shell types share a common internal anatomy and set of allozymes. The finding that some major differences in shell morphology arise as a result of relatively simple heterochronous changes in ontogeny also removes one of the objections that earlier workers may have raised to the role of parapatric differentiation in *Cerion*. Perhaps we will find *Cerion* to be capable of very close evolutionary "tracking" of microenvironmental heterogeneity; the analogy with some variable plants, in which highly localized, specially adapted ecotypes replace one another over distances of a few meters, is striking (Bradshaw, 1972). Unfortunately, present techniques do not allow us to discriminate between primary and secondary zones of hybridization. The *Cerion* case is interesting in that it probably contains both types; the challenge is to learn how to distinguish them.

If the diverse origin of these hybrid zones is presently shrouded in mystery, can anything predictive be said of their future? Are they stable or transient phenomena? (And if transient, will the adjacent populations fuse or interact so that some isolating mechanism is reinforced and speciation completed?) *Cerion's* fossil record is of little use beyond establishing that some morphotypes have occupied parts of their present ranges for thousands of years. (We have yet to find a fossil hybrid zone.) Sequential sampling of snails across a hybrid zone is also unlikely to resolve the issues quickly because generation time is relatively long (probably 5 years; Woodruff, 1978c) and sample sites must be located with much greater precision than has been customary in the past. We are also profoundly ignorant of the antihybridization mechanisms that *Cerion* may employ. Statements to the effect that reproductive isolating mechanisms are not easily acquired in this genus are quite plausible but not substantiated. Thus, although our morphological and genetic data indicate that hybrid zones on Abaco and Long Island involve fully compatible morphospecies, the zones are very narrow. There is no evidence for assortative mating, and there is no evidence that fusion is occurring. Although the genodynamics of these interactions remains obscure, it appears that some zones may be relatively stable.

The behavior of the *Cerion* morphospecies would seem to be in clear violation of the biological species concept. Actually, the problem may not be with *Cerion* but with a species concept that overemphasized the importance of reproductive isolation. It is becoming clear that genetic isolation cannot be directly equated with reproductive isolation. Common homeostatic mechanisms and gene flow may protect the integrity of the gene pool even in the absence of reproductive isolation (Bigelow, 1965; Hunt and Selander, 1973; Endler, 1977; Woodruff, 1979). If one allows that semispecies may hybridize without losing their integrity, then these *Cerion* are just extremely variable examples of biological species. The fact that morphologically different populations share a common set of structural genes has no bearing on the speciation question; there is little evidence for the extensive reorganization of gene pools during speciation (Throckmorton, 1977; Nevo and Cleve, 1978). Nevertheless, some genetic differentiation has occurred, as evidenced by the changes in the patterns of covariation among morphometric traits and the genetic anomalies associated with the hybrid zones themselves. The latter phenomena have recently been reported for hybrid zones involving semispecies of *Mus* (Hunt and Selander, 1973) and *Rana* (Sage and Selander, 1979); and Selander and I (in prep.) have discussed the roles of mutation and intragenic recombination in producing these effects.

I do not think that the interpretive difficulties I have encountered with the *Pseudophryne* or *Cerion* hybrid zones are unique. They are merely a reflection of our present scientific inadequacies; specifically, we lack appropriate techniques to test the various alternate hypotheses concerning the origin and future of these phenomena. Progress in this area would appear to require the development of field methods for the characterization of the genodynamics of each situation. If we can monitor the net movement of genes into and across a hybrid zone today, then we should be able to simulate the behavior of that zone at other points in time. I am pursuing this idea with *Cerion,* an organism in which dispersion and dispersal can be estimated both accurately and easily, and I have been conducting pilot studies in populations adjacent to the Rocky Point hybrid zone since 1973. The purpose of these experiments, which involve monitoring the movements of large numbers of snails, is to determine the feasibility of estimating the gene flow parameter. [Gene flow is typically defined as the product of the mean distance traveled in a generation and the square root of probability of leaving a deme or neighborhood (May et al., 1975).] Preliminary results indicate that mean dispersal at a site can be estimated relatively quickly because annual displacements for both juveniles and adults are very similar, and successive annual displacements have not varied appreciably during 5 years; the dispersal estimate for adults for 1 year is nearly the same as the estimate based on a whole generation. Displacements of individual snails at one site are shown for two successive 6-month periods in Figure 9.9. At the two sites monitored it is already apparent that although mean displacement is very small (less than 3 meters), and maximum detected displacement is about 20 meters, gene flow is 50% higher at the northern than at the southern end of the hybrid zone. It is hoped that studies of this type will not only contribute to the explanation of the asymmetrical aspects of some hybrid zones but also lead to a better understanding of their basic genodynamics.

In my opinion one of Michael White's major contributions to evolutionary biology has been the demonstration that the classical sympatric and allopatric models of speciation no longer satisfactorily account for the observed diversity of animals. By very careful attention to cytogenetic details he has helped bring about a change in the way we think about the processes of speciation. With the steady diminution of the alleged role of "cohesive forces" in preventing speciation (Ehrlich and Raven, 1969; Lewontin, 1974; Felsenstein, 1975; Levin, 1979), our focus has shifted to the origin of genetic isolating mechanisms rather than the geographic subdivision of the gene pool as the prime cause of speciation.White's work on the morabine grasshoppers illustrates this well and constitutes one of a small number of examples that set the stage for the ongoing revitalization of the field. I think it is a tribute to the generality of his vision that the processes he elucidated in the morabines are also relevant to the evolution of the *Pseudophryne* and *Cerion,* two groups in which chromosomal rearrangements have not been detected. His continued insistance that speciation represents far more than "a category of biogeographic acci-

Figure 9.9. Dispersal of individual cerion at a study site near Rocky Point, Abaco. The displacement of individually marked snails initially found in a 55 m^2 area is shown for the period November 1974 to May 1975 (upper) and May 1975 to November 1975 (lower). During the former period 227 snails moved, whereas 184 moved during the latter. Snails that did not move during each period (59 during the winter, 32 in the summer) are not shown. Axes show distances in centimeters.

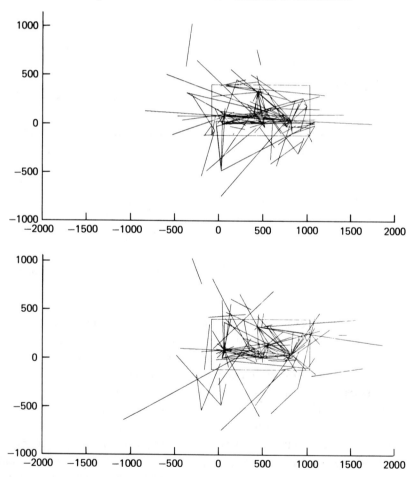

dents" (White, 1959) has helped to bring this crucial evolutionary process back into the mainstream of biological inquiry.

The research was supported by a Commonwealth of Australia Postgraduate Award, a Frank Knox Memorial Fellowship (Harvard University), and grants from the United States National Science Foundation. I am indebted to numerous friends for field and laboratory assistance.

References

Bigelow, R. S. 1965. Hybrid zones and reproductive isolation. *Evolution 19:*449–58.

Bradshaw, A. D. 1972. Some of the evolutionary consequences of being a plant. *Evol. Biol. 5:*25–47.

Chung, D. J. D. 1979. Aspects of the anatomy of *Cerion* (Mollusca: Pulmonata). M.S. thesis, Purdue University, 158 pp.

Clench, W. J. 1957. A catalogue of the Cerionidae (Mollusca–Pulmonata). *Bull. Mus. Comp. Zool. Harv. 116:*121–69.

Clench, W. J. and Aguayo, C. G. 1952. The *scalarinum* species complex (*Umbonis*) in the genus *Cerion*. *Occas. Pap. Mollusks Mus. Comp. Zool. Harv. 1:*413–40.

Darlington, P. J. 1965. *Biogeography of the Southern End of the World.* Cambridge, Massachusetts: Harvard University Press, 236 pp.

Darwin, C. 1859. *On the Origin of Species.* London: John Murray (facsimile of the first edition published by Harvard University Press, Cambridge, Massachusetts, 1964).

Ehrlich, P., and Raven, P. H. 1969. Differentiation of populations. *Science 165:*1228–32.

Endler, J. A. 1977. *Geographic Variation, Speciation, and Clines.* Princeton: Princeton University Press, 246 pp.

English, T. M. 1910. Some notes on Tasmanian frogs. *Proc. Zool. Soc. Lond. 2:*627–34.

Felenstein, J. 1975. The genetic basis of evolutionary change. *Evolution 29:*587–90.

Fletcher, J. J. 1889. Observations on the oviposition and habits of certain Australian batrachians. *Proc. Linn. Soc. N.S.W.* ser. 2, *4:*357–87.

Frauca, H. 1965. Notes on the breeding behavior of a frog (*Pseudophryne*) in central Queensland. *Queensl. Nat. 17:*94–8.

Galler, L. and Gould, S. J. 1979. The morphology of a "hybrid zone" in *Cerion:* variation, clines, and an ontogenetic relationship between two "species" in Cuba. *Evolution 33:*714–27.

Gould, S. J. 1969. Character variation in two land snails from the Dutch Leeward Islands: geography, environment, and evolution. *Syst. Zool. 18:*185–200.

– 1977. *Ontogeny and Phylogeny.* Cambridge, Massachusetts: Belknap Press of Harvard University Press, 506 pp.

Gould, S. J. and Paull, C. 1977. Natural history of *Cerion.* VII. Geographic variation of *Cerion* (Mollusca: Pulmonata) from the eastern end of its range (Hispaniola to the Virgin Islands): coherent patterns and taxonomic simplification. *Breviora 445:*1–24.

Gould, S. J. and Woodruff, D. S. 1978. Natural history of *Cerion.* VIII. Little Bahama Bank – a revision based on genetics, morphometrics and geographic distribution. *Bull. Mus. Comp. Zool. Harv. 148:*371–415.

Gould, S. J., Woodruff, D. S., and Martin, J. P. 1974. Genetics and morphometrics of *Cerion* at Pongo Carpet: a new systematic approach to this enigmatic land snail. *Syst. Zool. 23:*518–35.

Harrison, L. 1922. On the breeding habits of some Australian frogs. *Aust. Zool. 3:*17–34.

Hunt, W. G., and Selander, R. K. 1973. Biochemical genetics of hybridization in European house mice. *Heredity 31:*11–33.

Levin, D. A. 1979. The nature of plant species. *Science 204:*381–4.

Lewontin, R. C. 1974. *The Genetic Basis of Evolutionary Change.* New York: Columbia University Press, 346 pp.

Littlejohn, M. J. 1967. Patterns of zoogeography and speciation in south-eastern Australian Amphibia. In *Australian Inland Waters and Their Fauna: Eleven Studies* (ed. Weatherley, A. H.), pp. 150–74. Canberra: Australian National University.

Littlejohn, M. J., and Martin, A. A. 1965. The vertebrate fauna of the Bass Strait islands: 1. The Amphibia of Flinders and King Islands. *Proc. R. Soc. Vict. 79:*247–56.

– 1974. The Amphibia of Tasmania. In *Biogeography and Ecology in Tasmania* (ed. Williams, W. D.), pp. 251–89. The Hague: W. Junk.

McDonnell, L. J., Gartside, D. F. and Littlejohn, M. J. 1978. Analysis of a narrow hybrid zone between two species of *Pseudophryne* (Anura: Leptodactylidae) in south-eastern Australia. *Evolution 32:*602–12.

May, R. M., Endler, J. A., and McMurtie, R. E. 1975. Gene frequency clines in the presence of selection opposed by gene flow. *Am. Nat. 109:*659–76.

Mayr, E. 1963. *Animal Species and Evolution.* Cambridge, Massachusetts: Belknap Press of Harvard University Press, 797 pp.

– 1969. *Populations, Species and Evolution.* Cambridge, Massachusetts: Belknap Press of Harvard University Press, 453 pp.

– 1970. *Principles of Systematic Zoology.* New York: McGraw-Hill, 428 pp.

Mayr, E., and Rosen, C. B. 1956. Geographic variation and hybridization in populations of Bahama snails (*Cerion*). *Am. Mus. Nat. Hist. Novit. 1806:*1–48.

Muller, H. J. 1942. Isolating mechanisms, evolution and temperature. *Biol. Symp. 6:*71–125.

Nevo, E., and Cleve, H. 1978. Genetic differentiation during speciation. *Nature (Lond.) 275:*125–6.

Patton, J. L., Hafner, J. C., Hafner, M. S., and Smith, M. F. 1979. Hybrid zones in *Thomomys bottae* pocket gophers: genetic, phenetic, and ecologic concordance patterns. *Evolution 33:*860–76.

Pengilley, R. 1971. Calling and associated behavior of some species of *Pseudophryne* (Anura: Leptodactylidae). *J. Zool. (Lond.) 163:*73–92.

Sage, R. D., and Selander, R. K. 1979. Hybridization between species of the *Rana pipiens* complex in central Texas. *Evolution 33:*1069–88.

Throckmorton, L. H. 1977. *Drosophila* systematics and biochemical evolution. *Annu. Rev. Ecol. Syst. 8:*235–54.

White, M. J. D. 1959. Speciation in animals. *Aust. J. Sci. 22:*32–39.

– 1978. *Modes of Speciation.* San Francisco: W. H. Freeman, 455 pp.

Williams, W. D. (ed.). 1974. *Biogeography and Ecology in Tasmania.* The Hague: W. Junk, 498 pp.

Wilson, E. O. 1970. Chemical communication within animal species. In *Chemical Ecology* (ed. Sondheimer, E., and Simeone, J. B.), pp. 133–55. New York: Academic Press.

Woodruff, D. S. 1972a. Australian anuran chromosome numbers. *Herpetol. Rev. 4(6):* 208.

– 1972b. The evolutionary significance of hybrid zones in *Pseudophryne* (Anura: Leptodactylidae). Ph.D. thesis, Zoology Department, University of Melbourne, 373 pp.

– 1973a. Natural hybridization and hybrid zones. *Syst. Zool. 22:*213–18.

– 1973b. Bassiana. *Search – Aust. J. Sci. 4(10):*409.

– 1975a. A new approach to the systematics and ecology of the genus *Cerion. Malacol. Rev. 8(1):*128.

– 1975b. Natural history of *Cerion* V. Allozyme variation and genic heterozygosity in the Bahaman pulmonate *Cerion bendalli. Malacol. Rev. 8(1):*47–55.

– 1975c. Morphological and geographical variation of *Pseudophyryne corroboree* (Anura: Leptodactylidae). *Rec. Aust. Mus. 30:*99–113.

– 1975d. A southern continental island. Review of *Biogeography and Ecology in Tasmania* ed. by W. D. Williams (The Hague: W. Junk). *Science 187:*735–6.

- 1976a. Embryonic mortality in *Pseudophryne* (Anura: Leptodactylidae). *Copeia 1976:*445–9.
- 1976b. North Queensland toadlets of the genus *Pseudophryne. Queensl. Nat. 21:*142–3.
- 1977. Male postmating brooding behavior in three Australian *Pseudophryne* (Anura: Leptodactylidae). *Herpetologica 33:*296–303.
- 1978a. Hybridization between two species of *Pseudophryne* (Anura: Leptodactylidae) in the Sydney Basin, Australia. *Proc. Linn. Soc. N.S.W. 102(3):*131–47.
- 1978b. Mechanisms of speciation. Review of *Geographic Variation, Speciation, and Clines* by J. A. Endler (Monogr. Pop. Biol. 10, Princeton Univ. Press). *Science 199:*1329–30.
- 1978c. Evolution and adaptive radiation of *Cerion:* a remarkably diverse group of West Indian land snails. *Malacologia 17(2):*223–39.
- 1979. Postmating reproductive isolation in *Pseudophryne* and the evolutionary significance of hybrid zones. *Science 203:*561–3.
Woodruff, D. S., and Tyler, M. J. 1968. Additions to the frog fauna of South Australia. *Rec. S. Aust. Mus. 15:*705–9.

PART IV

SPECIATION AND EVOLUTION

10 Stasipatric speciation and rapid evolution in animals

GUY L. BUSH

Species hold a unique and important place in the economy of nature. Each represents a more or less discrete gene pool united into a reproductive community by bonds of common descent, parenthood, and sexual union (Wiley, 1978). Genetic discontinuity between sexually reproducing species is maintained by various kinds of intrinsic barriers to gene exchange usually referred to as "reproductive isolating mechanisms," and the recognition of species on the basis of reproductive isolation is now the widely accepted biological species concept (Mayr, 1963). Species, each of which represents essentially a closed genetic system, are free to evolve and maintain genomes adapted to a specific way of life that makes it possible for them to interact effectively with other species and "track" an ever changing biome. Each species has thus staked out a claim and is well adapted to a portion of the earth's habitable environment. Without such intrinsic reproductive isolating mechanisms, evolution would proceed at a very slow pace with biological diversity arising primarily through anagenic processes in geographically isolated populations.

Reproductive isolation generally arises in two ways. It may result fortuitously as a by-product of adaptation, or it may develop as a direct and integral part of the adaptive process itself (Bush, 1975a). It has long been recognized that when a species is split into two or more geographically isolated populations, reproductive isolation may arise between them simply as an outcome of different stochastic and adaptive genetic changes experienced by each population. If these changes are sufficient to render hybrids between individuals from different populations either inviable or competitively incompetent, then it is generally recognized that the populations represent distinct species. Postmating rather than premating reproductive isolating mechanisms are usually first to develop under these conditions. This scenario is the typical pattern of divergence during *allopatric* (geographic) speciation.

The evidence and reasoning in support of allopatric speciation is so intuitively satisfying and convincing that by the early 1960s evolutionary biologists accepted geographic isolation as a universal prerequisite for speciation in sexually reproducing animals (Mayr, 1963). This broad consensus rested primarily on the belief that a large number of genetic changes were necessary to ensure reproductive isolation between parent and daughter species, and that changes of this magnitude could only be acquired in the complete absence of gene flow.

There are, however, several other ways reproductive isolation can arise between populations in direct partnership with adaptive evolution, which do not involve complete geographic isolation and which may require only minor alterations in the genetic architecture. Recent discoveries in molecular, cellular, population, and theoretical biology have led to a renaissance in our understanding of evolutionary processes, and the universality of geographic speciation has been seriously challenged (for reviews of the subject see Murray, 1972; Bush, 1975a,b; Endler, 1977; White, 1978; Templeton, 1980a,b). Most examples fall under the rubric of *sympatric* speciation (i.e., speciation without complete geographic isolation), although the processes leading to divergence and eventual reproductive isolation are diverse (Maynard Smith, 1966; White, 1968; Bush, 1975a; Endler, 1977; Pimm, 1979; Rosenzweig, 1978; Grant and Grant, 1979). Clearly, a single, all-inclusive model of speciation in animals is oversimplistic. It has therefore become necessary to recast models of animal speciation in more subtle terms of gene flow, selection, stochastic processes, and competition rather than define speciation on the basis of geographic and reproductive isolation alone (Templeton, 1980a).

In this paper I wish to direct attention only to *stasipatric* speciation, a mode of sympatric speciation initiated by chromosome rearrangements which play a primary role in the speciation process because they simultaneously establish a degree of reproductive isolation between parental and daughter species and initiate significant and sometimes novel adaptive change. Appropriately for this volume, this mode of speciation was first proposed by M. J. D. White in 1968 and detailed in his later papers and books. The phenomenon appears to be widespread and may yet prove to be the primary mode of speciation in many groups of animals and plants (Wilson et al., 1975; Levin and Wilson, 1976; Bush et al., 1977; White, 1978). My objective here is to focus on the unique role chromosome rearrangements play in the adaptive process and stasipatric speciation.

Chromosome mutations and novel adaptations

Point and chromosome mutations, which represent the two basic classes of mutations in organisms, differ in many important ways in the amount and complexity of DNA reorganization. Most point mutations, in structural genes, will alter the function or expression of a single protein and its pleiotropic interactions. The effects of a point mutation on a control gene are less well understood, but because it would affect only one control site, its potential for altering phenotypic expression would be limited. Chromosome mutations, on the other hand, rearrange blocks of euchromatic and heterochromatic DNA of varying size and composition, placing them in different linkage arrangements and generating new patterns of regulation and temporal expression (Britten and Davidson, 1971; Davidson and Britten, 1979). Chromosome rearrangements are therefore inherently different in their adaptive and evolutionary roles from point mutations in structural genes which encode for protein.

There is growing evidence that structural genes are not usually directly involved in the speciation process (Avise and Ayala, 1976; Wilson et al., 1977a,b; Nevo, 1979). Novel metabolic changes and major phenotypic evolution depend rather on mutations in control genes (i.e., classical regulatory genes and genes that exert control at various levels). Chromosome rearrangements which alter gene arrangements and associations appear to be more important in this respect than point mutations in structural genes (Britten and Davidson, 1971; Wilson et al., 1977a,b), and recent biochemical studies of chromosome structure and gene regulation support this view. Large-scale duplication and redistribution of genes have occurred between even closely related species (Livak et al., 1978).

From the standpoint of speciation and adaptive shifts, the most profound effects of chromosome rearrangements are at the developmental level. It is now clear they can play an important role in repatterning developmental pathways that lead to striking phenotypic change. Unlike the majority of point mutations which usually express themselves as recessive traits, the changes in the concentration of gene products, cell kinetics, and the timing of their production during development brought about by chromosome rearrangements are frequently expressed even in the heterozygote (Krone and Wolf, 1977). The result may in some cases be a radical phenotypic change, or, in Goldschmidt's much maligned metaphor, "a hopeful monster" (Goldschmidt, 1940). It is the repatterning of the epigenetic architecture that is the key factor in stasipatric speciation.

The effects of chromosome rearrangements on the staging of developmental processes appear to be mediated through changes in the expression of cell surface antigens called *differentiation antigens,* which are involved with cell-to-cell recognition and the induction of differential gene action leading to cellular and tissue diversity as well as cell assembly (Hood et al., 1978). Genes coding for these surface antigens are widely distributed throughout the genome (Weiss and Green, 1967), and are highly conserved during evolution (Hood et al., 1978). The amount and timing of production of antigen corresponds rather closely to gene dosage (Krone and Wolf, 1977), and the position of its gene in the chromosome (Koo et al., 1977) has pronounced effects. Because the course of normal development is directed to a large extent by tissue interactions, a distortion of cell juxtaposition can result in a qualitatively different organ structure, and even a complete restructuring of the local architecture (Wessels, 1977). As stated by Albrech et al. (1979), this phenomenon provides a developmental basis for major adaptational shifts within a few generations, if such heritable mutations do not create insurmountable reproductive barriers or selective disadvantages.

As an example, consider the effects of chromosome rearrangement on the expression of the *H-Y* antigen, a plasma membrane protein associated with the heterogametic sex in all vertebrates. The gene coding for or regulating the expression of this "male-specific" antigen in mammals has been mapped to human satellite III DNA of the Y chromosome (Bostock et al., 1978). In birds and mammals the *H-Y* antigen has been described by Ohno (1979) as a master "regulator" gene that determines primary (gonadal) sex. As such it is at the apex of a regulatory gene series directing the organogensis of sex determination, and affects the expression of many other structural genes of developmental importance.

The *H-Y* antigen when present interacts with the *H-Y* antigen receptor located on the undifferentiated gonad (which is present in both sexes) and initiates its differentiation into the testis. The testis, along with various somatic tissues, then begins producing hormones which induce development of the male duct system and activate the *Tfm* (testicular feminization) locus. Normal functioning of the *Tfm* locus results in the development of secondary male sexual characters. These events are outlined diagrammatically in Figure 10.1.

When the *H-Y* locus is translocated into the X chromosome or one of the autosomes, the expression of the antigen and thus the pattern of sex determination during organogenesis is altered. Individuals bearing such translocations may show a variety of phenotypic effects, such as first and fourth degree hypospadias, intersexuality, Turner's syndrome, and streak gonads in females, depending on where the *H-Y* antigen locus is located in the genome (Koo et al., 1977). Although the *H-Y* antigen structural gene is unchanged in these cases, the timing of its expression during development is altered, resulting in a different pattern of cellular differentiation.

Clearly the expression of genes lower in the sex-determining regulatory hierarchy could be similarly affected with less spectacular phenotypic results if they or their control regions were altered by chromosome rearrangements. Indeed it is reasonable to assume that on rare occasions such mutations may promote a significant shift in an organism's way of life. The striping pattern in zebras, for instance, can be explained by a simple regulatory switch mechanism operating during the third (*Equus burchelli*), fourth (*E. zebra*), and fifth (*E. grevyi*) weeks of development (Bard, 1977).

Because little morphological evolution sometimes accompanies extensive chromosome rearrangements, the suggestion has been made that chromosome rearrangements in evolution should be regarded as neutral changes that simply accompanied speciation (see Ohno, 1974). The most cited examples involves the five species of muntjacs, which range in chromosome number from $2n = 6♀, 7♂$ (XY_1Y_2) (*Muntiacus muntjak*) to $2n = 46$ (*M. reevesi*). Fifteen morphologically distinct subspecies have been recognized in one species alone (*M. muntjak;* Whitehead, 1972). Two of these subspecies have been karyotyped, and they differ in both chromo-

Figure 10.1. Diagrammatic illustration of the influence of the *H-Y* antigen during early embryogenesis on sex determination. See text for details.

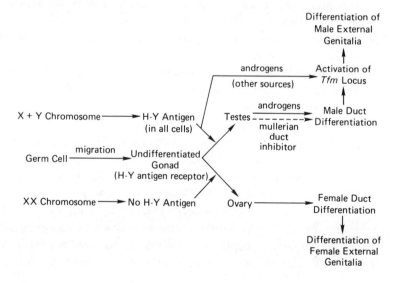

some number and other rearrangements. They probably represent distinct species (Wurster and Atkin, 1972), yet the degree of morphological differentiation from the standpoint of the human observer appears minor.

There is, however, no a priori reason to assume major morphological change is necessary or will always accompany an adaptive stasipatric speciation event involving chromosome rearrangements. Small physiological modifications can have profound adaptive effects. For example, the cyc-1 strain of *Saccharomyces cerevisiae* lacks iso-1-cytochrome c and cannot grow on lactate medium (Sherman and Helms, 1978). However, the CYC7-1 mutation involving a reciprocal translocation between the left arm of chromosome V and the right arm of chromosome XVI causes a 30-fold increase in iso-2-cytochrome c, which normally occurs only at low levels. This overproduction results from the formation of an abnormal regulatory region adjacent to a region that determines the primary structure of iso-2-cytochrome c. As a result of this chromosome rearrangement the CYC7-1 strain can grow on lactate medium. A translocation between chromosome one and four has also been implicated recently in increasing insecticide resistance 100-fold in an aphid, *Myzus persica* (Blackman et al., 1978).

It should be emphasized that chromosome and point mutations are similar in one important respect. Both are usually detrimental and only rarely adaptive in nature, and few produced under laboratory conditions have ever improved fitness. This is a point frequently overlooked by proponents of speciation by more classical modes. They somehow regard point mutations as more adaptively acceptable without much supporting evidence. Chromosome rearrangements, however, may actually provide a broader range of adaptive effects through modification of the regulatory apparatus. Furthermore, they appear to be far more common than previously assumed. In such diverse groups as man and grasshopper, one in 500 individuals karyotyped is heterozygous for a visible chromosome rearrangement (White, 1973).

Rates of chromosome mutation and stasipatric speciation

One feature of chromosome evolution that is also becoming increasingly clear is that chromosome mutation rates vary considerably over time and space within species. Some of the agents producing chromosome mutations in animals have been reviewed by Atkin (1976). These agents include ionizing radiation, chemical compounds, microorganisms (viruses, rickettsias, mycoplasmas, and plasmids), and what has been termed the "chromosome breakage syndrome." To this list can also be added severe protein–calorie malnutrition (Armendares et al., 1971). All of these factors may radically increase the rate of chromosome mutation for various lengths of time.

Two features are shared by these mutator factors. The induced breaks are usually restricted to regions rich in heterochromatin or satellite DNA, and the breaks are not random. Chromosome breaks and rearrangements resulting from ionizing radiation, for example, occur within intermediately repetitive DNA sequences (Lee, 1975). Because the same reiterated sequences are often inserted throughout the genome (Davidson and Britten, 1979; Dover, 1980a,b), breaks in nonhomologous chromosomes that occur within the same repeated sequence can be united and repaired by the same repair enzyme because they share identical DNA sequences. The fact that the breaks do not occur at random and that similar or identical breaks and rearrangements may arise independently and repeatedly suggests that the creation of karyotype phylogenies that rest on the assumed uniqueness of chromosome rearrangements is questionable.

Another outcome of the nonrandom nature of certain types of heterochromatin associated with chromosome mutations is that their pattern of distribution and function will directly influence the type of adaptively acceptable rearrangements within a taxon (Yoon and Richardson, 1978). Thus certain rearrangements that are tolerated in one group may be excluded in another, a feature that White (1973) has called *karyotypic orthoselection*.

Of all the mutator factors inducing breaks, the chromosome breakage syndrome is potentially the most important from an evolutionary standpoint. It may actually be closely linked to and inseparable from the effects produced by certain microorganisms and transposons. The syndrome is expressed following interdeme and interspecific hybridization (Thompson and Woodruff, 1978a,b) and may increase the frequency of chromosome breaks 100-fold. Thus the rate of chromosome mutation may be relatively low in two populations but increase greatly when individuals from the two populations hybridize. The syndrome is also associated with hybrid dysgenesis and sterility, increased mutation in structural genes, and segregation distortion. In eukaryotes the syndrome has been observed in such diverse organisms as maize (McClintok, 1978), man (Atkin, 1976), and *Drosophila* (Thompson and Woodruff, 1978a,b).

The mutator factor resembles in many respects the transposable elements or transposons that cause transposition and inversions in prokaryotes (Starlinger, 1977). Eukaryotic transposons or regulatory elements can not only modify the protein product of the gene but also alter the time during development when the gene product will be made available (Dooner and Nelson, 1977). Furthermore, by means of transposition these elements can regulate a number of gene loci. Most important, they provide a mechanism for rapidly reorganizing the genome, ranging from complex chromosomal translocations to small segments of DNA possibly composed of relatively few base pairs. The ability of these elements to trigger innate systems capable of inducing genome reorganizations may explain the occurrence of "chain" chromosome rearrangement events observed in some groups of animals and plants, resulting in highly divergent karyotypes and rapid speciation (White, 1978).

Hybridization between demes triggering an increase in chromosome mutations accompanied by a period of inbreeding in small demes could thus greatly enhance the rate of chromosome evolution and adaptive speciation. This could occur under either allopatric or sympatric conditions if population structure ensures the maintenance of semiautonomous small demes (as will be discussed).

The role of heterochromatin in adaptive evolution and speciation

Besides the intermediate repetitive sequences of DNA that are the site of chromosome breaks, there are several other classes of heterochromatin or satellite DNA implicated in a diverse array of cellular and genetic activities, ranging from chromosome pairing to gene regulation (Peacock et al., 1978). These various sequences are distributed in a unique pattern even between closely related species and represent libraries or families of related, highly repeated DNA sequences. Evolution of these sequences has occurred by amplification or reduction of a particular family of sequences and by rearrangement throughout the genome (Peacock et al., 1978). Peacock and his group concluded that some or all of the various functions of this type of heterochromatin are attributable to the arrangement of the highly repeated sequences with other DNA sequences that may be interspersed or associated with them, rather than being due to the highly repeated sequences themselves.

Of particular relevance with respect to adaptive evolution is the apparent role these highly reiterated DNA sequences appear to play in gene regulation. Davidson and Britten (1979) have proposed a model of gene regulation based on preliminary evidence in which the formation of repetitive RNA–RNA complexes controls the production of messenger RNA. The sequence organization of these repetitive units in the genome will affect gene expression. Altering the pattern of distribution and the size or diversity of the library by chromosome rearrangement with respect to structural genes could result in a change in the temporal and kinetic aspects of gene control.

Britten and Davidson (1971) have also stressed the evolutionary importance of chromosome rearrangements in dispersion of repetitive sequence families, which appear to arise in a saltatory fashion throughout the genome. A recent example of this phenomenon which involved a multiplication of both structural gene and reiterated DNA, has been observed in mouse melanoma and lymphoblastoid cells (Dolnick et al., 1979; Bostock et al., 1979) selected for resistance to methotrexate. Resistance results in up to a 1000-fold increase in enzyme activity brought about by tandem duplication or other rearrangements of coding and associated repetitive sequences for dihydrofolate reductase enzyme. In one case the new chromosome mutation represents approximately 5% of the total karyotype DNA, only a small part of which must represent unique sequences involved in coding for dihydrofolate reductase itself.

Thus, the repatterning of repetitive DNA sequences involved in the control of gene expression can induce changes and growth in the regulatory system that in turn lead to the origin of evolutionary novelty. This view is consistent with available data on the origin, distribution, and function of repetitive DNA sequences. Although it is clear that other types of gene regulation are possible and highly likely, chromosome rearrangements appear to offer the most reasonable mechanism for providing simultaneously a major adaptive shift through a change in some critical regulatory pathway.

Population size and the fixation of chromosome rearrangements

It has long been known that chromosome rearrangements frequently lower fitness in heterozygotes. This effect may be expressed in reduced fertility or viability and occasionally even phenotypic abnormalities (White, 1973). This lowered reproductive fitness coupled with the extreme rarity of heterozygotes makes it highly unlikely that a new chromosome rearrangement will become fixed as a homozygote in a population unless special circumstances exist to circumvent these problems. It was Sewell Wright (1941) who first pointed out that rearrangements deleterious as heterozygotes could be fixed only in very small populations through inbreeding and drift.

One of the major prerequisites for stasipatric speciation, therefore, is small effective population size (Bush, 1975a; Wilson et al., 1975; Bush et al., 1977). This close relationship between population size and the fixation of chromosome arrangements was illustrated by the computer simulation studies of Bengtsson and Bodmer (1976), who found that even a very small reduction in heterozygote fitness (0.01–0.10) required a relatively small population size ($N < 100$) to ensure fixation. These results were confirmed and treated in greater detail by Lande (1979).

Whether a new chromosome mutation survives and initiates speciation depends on its adaptive role. It is unlikely that a chromosomal mutation expressing negative heterosis would spread throughout the species range or play a significant role in speciation, even if individuals homozygous for the new and old karyotype were of equal fitness. If speciation is indeed primarily an adaptive process, as evidence strongly suggests, then other factors must enter the picture. For instance, if a new rearrangement in homozygous condition improves fitness over the old karyotype, it stands a much greater likelihood of survival once it is fixed in a small deme.

Once fixation has occurred, two possibilities exist. When the new chromosome rearrangement enhances fitness in homozygous condition over the old rearrangement but does not reduce competition with the population bearing the old karyotype, then the latter will eventually become extinct. On the other hand, if the new rearrangement initiates or incorporates genetic changes that lead to a shift in habitat preference and thus reduces competition or competes favorably only in part of the original species range, then the parent and daughter chromosome races under certain circumstances may coexist sympatrically or parapatrically (Bush, 1975a).

Population structure and small population size

If small population size is a prerequisite for stasipatric speciation, how are such populations generated? Small population size has been recognized as a necessary prerequisite for rapid speciation in animals (Wright, 1940; Mayr, 1954, 1963; Carson, 1971), although usually only in the context of geographically isolated peripheral populations established from a small number of founders that undergo a "genetic revolution" due to inbreeding and drift (the founder effect of Mayr, 1954). However, there is a growing body of evidence based on empirical observations and theoretical models that new species can develop rapidly as a result of founder effects in small demes that are *not* geographically isolated and frequently are located near the *center,* not the periphery of the species range (Bush, 1975a,b; Bush et al., 1977; Wilson et al., 1975; White, 1968, 1978; Templeton, 1980a).

This situation arises from the fact that subdivision of a species into many small ephemeral demes throughout its range occurs by various means of population structuring. In vertebrates there is a strong correlation between rates of chromosome evolution and rates of speciation (Figure 10.2). These rates appear to be influenced by the pattern of population structuring rather than generation time or body size (Bush et al., 1977). Thus, in mammals the most rapidly evolving groups often have the longest generation time and largest body size. Certain forms of social structuring, such as the tendency to form harems and clans, for young to settle near parents, strong territoriality, and low vagility, appear to promote the types of population subdivision that cause inbreeding and drift and thus an increase in the rates of chromosome evolution and stasipatric speciation (Bush, 1975a; Wilson et al., 1975; Bush et al., 1977).

For example, in horses of the genus *Equus* (Figure 10.3), which first appeared about 3.5 million years ago, strict and permanent harem formation is the rule in at least four of the seven extant species. The remainder form leks with strong male territoriality. Species have evolved rapidly in

Figure 10.2. Relationship between corrected speciation rate (*S*) and rate of karyotypic change (*r'*) for major groups of vertebrates. Abbreviations: *Ar*, artiodactyls; *Ba*, bats; *Ca*, carnivores, *Fr*, frogs; *Ho*, horses; *In*, insectivores; *La*, lagomorphs; *Li*, lizards; *Ma*, marsupials; *Pr*, primates; *Ro*, rodents; *Sa*, salamanders; *Sn*, snakes; *TC*, turtles and crocodilians; *Wh*, whales (see Bush et al., 1977, for details).

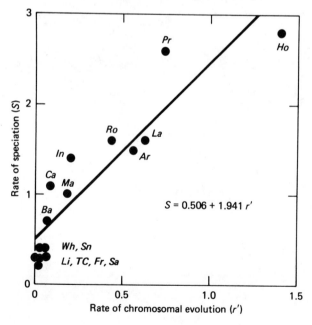

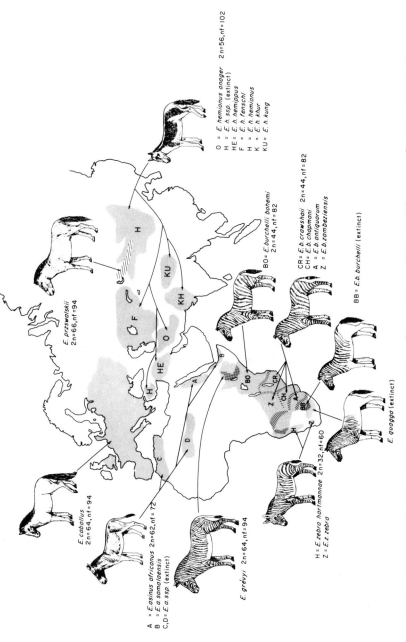

Figure 10.3. Distribution and chromosome numbers of recent horses (*Equus*). The range of each species has been reconstructed from historical records and the distribution of preferred habitats before their destruction by man.

E. przswalskii 2n=66, nf=94

E. caballus 2n=64, nf=94

A = *E. asinus africanus* 2n=62, nf=72
B = *E. a. somalaensis*
C, D = *E. a. ssp.* (extinct)

E. grévyi 2n=64, nf=94

H = *E. zebra hartmannae* 2n=32, nf=60
Z = *E. z. zebra*

E. quagga (extinct)

BO = *E. burchelii bohemi* 2n=44, nf=82

CR = *E. b. crawshaii* 2n=44, nf=82
CH = *E. b. chapmani*
A = *E. b. antiquorum*
Z = *E. b. zambeziensis*

BB = *E. b. burchelii* (extinct)

O = *E. hemionus onager* 2n=56, nf=102
H = *E. h. ssp.* (extinct)
HE = *E. h. hemippus*
F = *E. h. fenschi*
H = *E. h. hemionus*
K = *E. h. khur*
KU = *E. h. kung*

Equus with a karyotypic change and a speciation event occurring about once every 200,000 years (Bush et al., 1977). Diploid chromosome numbers range from 32 to 66 with the fundamental number varying from 60 to 102. A minimum of about 64 major chromosome rearrangements must be postulated to account for the chromosomal diversity in Old World horses. These rates are underestimates because they are based on gross chromosome rearrangements only. When banding data become available, the rates are likely to increase considerably, and speciation rates (which are highly correlated with chromosome evolution rates) may actually approach one every 50,000 years.

That all of these speciation events occurred in small peripherally isolated populations is difficult to reconcile with the recorded high vagility and extensive home ranges, which may cover over 1000 sq mi, of these large animals (Klingel, 1975; Feist and McCullough, 1975). The ability of horses to organize into harems and herds of related individuals provides a great number of small temporarily structured demes throughout the range of a species. From time to time these demes are sufficiently isolated from one another by social structuring to permit the rapid fixation of new chromosome mutations, which in turn may promote genetic isolation and speciation.

Many primate species, including man, were similarly organized into small groups prior to our more recent phase of cultural development. Even today some human groups, such as the Yanomama, sustain relatively high levels of inbreeding as a result of social structuring (Neel, 1978).

Population structuring has also been important in prompting stasipatric speciation of other animal groups. Close inbreeding is widespread in parasitic Hymenoptera (Askew, 1968) and is associated with apparent rapid speciation and chromosome evolution (Goodpasture, 1975). Inbreeding is also characteristic of many flightless insects of low vagility, such as the morabine grasshoppers studied by White (1968) and many insects inhabiting rotting wood (Hamilton, 1978).

Chromosome rearrangements and stasipatric speciation

Stasipatric speciation rests on the assumption that chromosome rearrangements are linked to or promote speciation in some special way. What is the evidence in support of such an assumption?

That chromosome rearrangements usually reduce fitness in the heterozygote has already been discussed. This effect is even apparent in centric fusions which superficially appear to have little deleterious effect (Jacobs et al., 1975). In certain cases they clearly function as the major postmating reproductive isolating mechanism between recently evolved species [e.g., the tobacco mouse, *Mus ponchiavinus,* and the house mouse, *Mus musculus* (Gropp et al., 1972)]. Yet at their inception new chromosome mutations must be at least partially viable as heterozygotes, or they will never pass through the F_1 bottleneck.

From the preceding discussion it is evident that if chromsome rearrangements are directly involved in speciation, they must meet two prerequisites: (1) Chromosome rearrangements must enhance positive assortative mating within chromosome races, and (2) to persist the new homozygote must confer increased fitness in some way over the old karyotype in the parent population. These two criteria are not necessarily independent of one another, as they would at first appear to be.

For instance, any mutation that initiates a shift in habitat utilization could also simultaneously promote assortative mating. A new chromosome rearrangement fixed in a small group of horses might deregulate a digestive enzyme previously present only in low concentrations, thus greatly increasing its level so that the utilization of previously unexploited grasses or other herbs would be possible. The distribution or number of sweat glands might be altered, allowing a shift to a more arid or moist environment; or a change in musculature, bone structure, or striping pattern could provide the innovation that would alter mate choice or permit this group to escape or avoid predators, and to exploit resources and areas previously unavailable to the parent species. The utilization of this new resource would tend to promote assortative mating because individuals with similar habitat needs would congregate.

It is not an unreasonable assumption that a chromosome rearrangement itself through pleiotropic effects might directly result in assortative mating among like karyotypes. Such adaptive rearrangements which simultaneously result in improved fitness and assortative mating have been demonstrated experimentally using sex chromosome translocations in *Drosophila* (Tracey, 1972; Tracey and Espenet, 1976). It can be reasonably assumed that similar mutations occur frequently enough in nature as well, to effect assortative mating and speciation.

Conclusions
Chromosome rearrangements clearly play an important and possibly pivotal role in the evolutionary process of eukaryotic organisms. Evidence is mounting that it is through the reorganization of the regulatory apparatus that major innovative adaptations arise. This reorganization is brought about effectively by a battery of different types of

chromosome rearrangements which may simultaneously initiate reproductive isolation and speciation. Under some conditions this reorganization can occur rapidly by a series of adaptive speciation events that result in a profound change in the way of life.

Speciation and adaptive change mediated by chromosome rearrangements are probably restricted to taxa of a population structure that promotes inbreeding and drift. There is also a clear link between population structure and the rates of molecular evolution in certain kinds of nuclear DNA, particularly those components of the genome involved in structural gene control. Understanding of this interplay between molecular and organismal evolution will undoubtedly unfold in the ensuing years as our knowledge of developmental genetics and gene control progresses. There are exciting discoveries on the horizon that will without question modify our current views of speciation and the evolutionary process.

References

Albrech, P., Gould, S. J., Oster, G. F., and Wake, D. B. 1979. Sex and shape in ontogeny and phylogeny. *Paleobiology 5:*296–317.

Armendares, S., Salamanca, F., and Frenk, S. 1971. Chromosome abnormalities in severe protein calorie malnutrition. *Nature (Lond.) 232:*271–3.

Askew, R. R. 1968. Considerations on speciation in Chalcidoidea (Hymenoptera). *Evolution 22:*642–5.

Atkin, N. B. 1976. Cytogenetic aspects of malignant transformation. In *Experimental Biology and Medicine,* vol. 6. (ed. Wolsky, A.). New York: Kruger, 171 pp.

Avise, J. C., and Ayala, F. J. 1976. Genetic differentiation in speciose versus depauperate phylads: evidence from California minnows. *Evolution 30:*46–58.

Bard, J. B. L. 1977. A unity underlying the different zebra striping patterns. *J. Zool. (Lond.) 183:*527–39.

Bengtsson, B. O., and Bodmer, W. F. 1976. On the increase of chromosome mutations under random mating. *Theor. Pop. Biol. 9:*260–81.

Blackman, R. L., Takada, H., and Kawakami, K. 1978. Chromosomal rearrangement involved in insecticide resistance of *Myzus persicae. Nature (Lond.) 271:*450–2.

Bostock, C. J., Clark, E. M., Harding, N. G. L., Mounts, P. M., Tyler-Smith, C., van Heyningen, V., and Walker, P. M. B., 1979. The development of resistance to methotrexate in a mouse melanoma cell line. I Characterisation of the dihydrofolate reductase and chromosomes in sensitive and resistant cells. *Chromosoma (Berl.) 74:*153–77.

Bostock, C. J., Gosden, J. R., and Mitchell, A. R. 1978. Localization of a male-specific DNA fragment to a sub-region of the human Y chromosome. *Nature (Lond.) 272:*324–8.

Britten, R. J., and Davidson, E. H. 1971. Repetitive and non-repetitive DNA sequences and a speculation on the origins of evolutionary novelty. *Q. Rev. Biol. 46:*111–33.

Bush, G. L. 1975a. Modes of animal speciation. *Annu. Rev. Ecol. Syst. 6:*339–64.

– 1975b. Sympatric speciation in phytophagous parasitic insects. In *Evolutionary Strategies of Parasitic Insects* (ed. Price, P. W.), pp. 187–206. London: Plenum Press.

Bush, G. L., Case, S. M., Wilson, A. C., and Patton, J. L. 1977. Rapid speciation and chromosomal evolution in mammals. *Proc. Natl. Acad. Sci. U.S.A. 74:*3942–6.

Carson, H. L. 1971. Speciation and the founder principle. *Stadler Genet. Symp. 3:*51–70.

– 1975. The genetics of speciation at the diploid level. *Am. Nat. 109:*83–92.

Davidson, E. H., and Britten, R. J. 1979. Regulation of gene expression: possible role of repetitive sequences. *Science 204:*1052–9.

Dolnick, B. J., Berenson, R. J., Bertino, J. R., Kaufman, R. J., Nunberg, J. H., and Schimke, R. T. 1979. Correlation of dihydrofolate reductase elevation with gene amplification in a homogeneously staining chromosomal region in L5178Y cells. *J. Cell Biol. 83:*394–402.

Dooner, H. K., and Nelson, O. E. 1977. Controlling element induced alterations in UDP glucose: flavanoid glucosyltransferase, the enzyme specified by the *bronze* locus in maize. *Proc. Natl. Acad. Sci. U.S.A. 74:*5623–7.

Dover, G. A. 1980a. Problems in the use of DNA for the study of species relationships and the evolutionary significance of genetic differences in chemosystematics: principles and practice. *Syst. Assoc. Ser.* (ed. Binky, F. A., Vaughn, J. G., and Wright, C. A.). New York: Academic Press (in press).

– 1980b. The evolution of "common" sequences in closely-related insect genomes. In *Insect Cytogenetics* (ed. Blackman, R. L., Hewett, G. M., and Ashburner, M.), R. Entomol. Soc. Lond. Symp. 10. Oxford: Blackwell (in press).

Endler, J. A. 1977. *Geographic Variation, Speciation, and Clines,* Monogr. Pop. Biol. 10. Princeton: Princeton University Press.

Feist, J. D., and McCullough, D. R. 1975. Reproduction in feral horses. *J. Reprod. Fert. Suppl. 23:*13–18.

Goldschmidt, R. B. 1940. *The Material Basis of Evolution.* New Haven: Yale University Press.

Goodpasture, C. 1975. Comparative courtship behavior and karyology in *Monodontomerus* (Hymenoptera: Torymidae). *Ann. Entomol. Soc. Am. 68:*391–7.

Grant, B. R., and Grant, P. R. 1979. Darwin's finches: population variation and sympatric speciation. *Proc. Natl. Acad. Sci. U.S.A. 76:*2359–63.

Gropp, A. H., Winking, H., Zech, L., and Muller, H. 1972. Robertsonian chromosomal variation and identification of metacentric chromosomes in feral mice. *Chromosoma (Berl.) 39:*265–88.

Hamilton, W. D. 1978. Evolution and diversity under bark. In *Diversity of Insect Faunas* (ed. Mound, L. A., and Waloff, H.) R. Entomol. Soc. Lond. Symp. 9, pp. 154–75. Oxford: Blackwell.

Hood, L. E., Weissman, I. L., and Wood, W. B. 1978. *Immunology.* Menlo Park, California: Benjamin/Cummings.

Jacobs, P. A., Frackiewicz, A., Law, P., Hilditch, C. J., and Morton, N. E. 1975. The effect of structural aberrations of the chromosomes on reproductive fitness in man. *Clin. Genet. 8:*169–78.

Kedwell, M. G., Kedwell, J. E., and Sved, J. A. 1977. Hybrid dysgenesis in *Drosophila melanogaster:* a syndrome of aberrant traits including mutation, sterility and male recombination. *Genetics 86:*813–33.

Klingel, H. 1975. Social organization and reproduction in equids. *J. Reprod. Fert. Suppl. 23:*7–11.

Koo, G. C., Wachtel, S. S., Krupen-Brown, K., Mittl, L. R., Berg, W. R., Genel, M., Rosenthal, I. M., Borgaonkar, D. S., Miller, D. A., Tantravahi, R., Schreck, R. R., Erlanger, B. F., and Miller, O. J. 1977. Mapping the locus of the H-Y gene on the human Y chromosome. *Science 198:*940–2.

Krone, W., and Wolf, V. 1977. Chromosome variation and gene action. *Hereditas 86:*31–6.

Lande, R. 1979. Effective deme size during long-term evolution estimated from rates of chromosomal rearrangement. *Evolution 33:*234–51.

Lee, C. S. 1975. A possible role of repetitious DNA in recombinatory joining during chromosome rearrangement in *Drosophila melanogaster. Genetics 79:*467–70.

Levin, D. A., and Wilson, A. C. 1976. Rates of evolution in seed plants: net increase in diversity of chromosome numbers and species numbers through time. *Proc. Natl. Acad. Sci. U.S.A. 73:*2086–90.

Livak, K. J., Freund, R., Schweber, E. R., Wensink, P. C. and Meselson, M. 1978. Sequence organization and transcription at two heat shock loci in *Drosophila. Proc. Natl. Acad. Sci. U.S.A. 75:*5613–17.

McClintock, B. 1978. Mechanisms that rapidly reorganize the genome. *Stadler Symp. 10:*25–47.

Maynard Smith, J. 1966. Sympatric speciation. *Am. Nat. 100:*637–50.

Mayr, E. 1954. Change in genetic environment and evolution. In *Evolution as a Process* (ed. Huxley, J., Hardy, A. C., and Ford, E. B.), pp. 157–80. London: Allen and Unwin, 367 pp.

– 1963. *Animal Species and Evolution.* Cambridge, Massachusetts: Harvard University Press.

Murray, J. 1972. *Genetic Diversity and Natural Selection.* New York: Hafner.

Neel, J. V. 1978. The population structure of an amerindian tribe, the Yanomama. *Annu. Rev. Genet. 12:*365–413.

Nevers, P., and Saedler, H. 1977. Transposable genetic elements as agents of gene mutability and chromosomal rearrangements. *Nature (Lond.) 268:*109–15.

Nevo, E. 1979. Adaptive convergence and divergence of subterranean mammals. *Annu. Rev. Ecol. Syst. 10:*269–308.

Nevo, E., and Cleve, H. 1978. Genetic differentiation during speciation. *Nature (Lond.) 275:*125–6.

Ohno, S. 1974. Protochordata, Cyclostomata and Pisces. In *Animal Cytogenetics 4, Chordata 1,* (ed. John, B.). Stuttgart: Borntraeger.

– 1979. *Major Sex-Determining Genes.* New York: Springer-Verlag.

Peacock, W. J., Lohe, A. R., Gerlach, W. L., Dunsmuir, P., Dennis, E. S., and Appels, R. 1978. Fine structure and evolution of DNA in heterochromatin. *Cold Spring Harbor Symp. 42:*1121–35.

Pimm, S. L. 1979. Sympatric speciation: a simulation model. *Biol. J. Linn. Soc. 11:*131–9.

Rosenzweig, M. L. 1978. Competitive speciation. *Biol. J. Linn. Soc. 10:*275–89.

Sherman, F., and Helms, C. 1978. A chromosomal translocation causing overproduction of iso-2-cytochrome C in yeast. *Genetics 88:*689–707.

Starlinger, P. 1977. DNA rearrangements in prokaryotes. *Annu. Rev. Genet. 11:*103–26.

Templeton, A. R. 1980a. The theory of speciation via the founder principle. *Genetics* (in press).

– 1980b. Modes of speciation and inference based on genetic distances. *Evolution* (in press).

Thompson, J. N., Jr. and Woodruff, R. C. 1978a. Chromosome breakage: a possible mechanism for diverse genetic events in outbred populations. *Heredity 40:*153–7.

– 1978b. Mutator genes – pacemakers of evolution. *Nature (Lond.) 274:*317–21.

Tracey, M. L. 1972. Sex chromosome translocations in the evolution of reproductive isolation. *Genetics 72:*317–33.

Tracey, M. L., and Espenet, S. A. 1976. Sex chromosome translocations and speciation. *Nature (Lond.) 263:*321–3.

Wachtel, S. S. 1977. H-Y antigen and the genetics of sex determination. *Science 198:*797–9.

Weiss, M. C., and Green, H. 1967. Human–mouse hybrid cell lines containing partial complements of human chromosomes and functioning human genes. *Proc. Natl. Acad. Sci. U.S.A. 58:*1104–11.

Wessels, N. 1977. *Tissue Interactions in Development.* Menlo Park, California: Benjamin.

White, M. J. D. 1968. Models of speciation. *Science 159:*1065–70.

– 1973. *Animal Cytology and Evolution,* 3rd ed. Cambridge: Cambridge University Press.

– 1978. *Modes of Speciation.* San Francisco: Freeman.

Whitehead, G. K. 1972. *Deer of the World.* London: Constable.

Wiley, E. O. 1978. The evolutionary species concept reconsidered. *Syst. Zool. 27*:17–26.

Wilson, A. C., Sarich, V. M., and Maxson, L. R. 1974. The importance of gene rearrangement in evolution: evidence from studies on rates of chromosomal, protein and anatomical evolution. *Proc. Natl. Acad. Sci. U.S.A. 8:*3028–30.

Wilson, A. C., Bush, G. L., Case, S. M., and King, M.-C. 1975. Social structuring of mammalian populations and rate of chromosomal evolution. *Proc. Natl. Acad. Sci. U.S.A. 72:*5061–5.

Wilson, A. C., Carlson, S. S., and White, T. J. 1977a. Biochemical evolution. *Annu. Rev. Biochem. 46:*573–639.

Wilson, A. C., White, T. J., Carlson, S. S., and Cherry, L. M. 1977b. Molecular evolution and cytogenetic evolution. In *Molecular Human Cytogenetics* (ed. Sparks, R. S., and Comings, D. E.), pp. 375–93. New York: Academic Press.

Wright, S. 1940. Breeding structure of populations in relation to speciation. *Am. Nat. 74:*478–94.

– 1941. On the probability of fixation of reciprocal translocations. *Am. Nat. 75:*513–22.

Wurster, D. H., and Atkin, N. B. 1972. Muntjac chromosomes: a new karyotype for *Muntjacus muntjak. Experientia 28:*972–3.

Yoon, J. S., and Richardson, R. H. 1978. A mechanisms of chromosomal rearrangement: the role of heterochromatin and ectopic joining. *Genetics 88:*305–16.

11 Habitat selection and speciation in *Drosophila*

PETER A.PARSONS

> As applied to the genus *Drosophila,* for example, phenetic classification
> would lump together *D. melanogaster* and *D. simulans,* two quite
> different biological species whose adaptive potential is very different.
>
> White (1977)

The sibling species *D. melanogaster* and *D. simulans* show a remarkable degree of morphological and genetic similarity, and were confused until 1919 when Sturtevant distinguished them. However, the two species are more readily distinguishable at the ecological level (Parsons, 1975a). The term ecological is here defined in the broadest sense to include behavioral factors involved in resources utilized and habitats selected in nature. Such biological divergences in the life histories of the two species must promote isolation between them into discrete habitats (even though they are frequently sympatric in nature as assessed by adult collection data), thus suggesting that microhabitat selection can be invoked as an important isolating mechanism. This discussion commences with a brief account of resource utilization specificity and habitats selected by the Australian *Drosophila* fauna, in order to put the sibling species into biological perspective. From this overview it becomes clear that we are forced to detailed considerations of the ecology, behavior, and genetics of *Drosophila* in natural populations in order to understand the development of heterogeneity among populations, races, and ultimately species. Although the overview leads us to such a conclusion, it is only in "laboratory" species such as *D. melanogaster* that more than rudimentary genetic studies are possible, although speciation studied through cytogenetic processes can be investigated at a much wider level taxonomically within and beyond the genus *Drosophila* (Carson et al., 1970; White, 1978). In the sibling *Drosophila* species under consideration, there are possibilities for genetic analyses that may contribute to an understanding of speciation, but less than common are actual data. Unfortunately, as Bush (1975) comments, "the study of speciation is an *ad hoc* science," the reason being that no one has yet observed the development of a new species from beginning to end in nature. Given this reservation, the comparative approach of detailed studies of life histories in the broadest sense at various taxonomic levels seems to be a necessary approach, especially if habitats preferred in nature are simultaneously considered. As White (in press) recently wrote, "The study of speciation is . . . today a comparative branch of biology."

Although many authors regard animal speciation as primarily the development of reproductive isolating mechanisms, it will be argued here that "an ability to use the resources of the environment in such a way as to be protected against niche competitors" (Mayr, 1977) is an equally if not more important component. To some extent the study of such ecological (and behavioral) characters that may lead to autonomy has been neglected, simply because they are often intrinsically more difficult to study than reproductive isolating mechanisms. *Drosophila* is no exception in this regard.

Resource utilization specificity and habitats selected

Drosophila species depend upon microorganisms bringing about fermentation and decay of plants and fungi (Throckmorton, 1975). Being extremely successful in this role, it is to be expected that the genus is characterized by an incredible diversity of breeding sites (Carson, 1971). To some extent these sites can be assayed by collection methods as shown by a consideration of the Australian endemic and cosmopolitan fauna (Parsons, 1978a; Parsons and Bock, in press). An updated version of the situation is given in Table 11.1. Over half of the known Australian species are from subgenus *Scaptodrosophila,* and are collected by four of the five collection methods listed, whereas subgenus *Hirtodrosophila* species are mainly collected by the fifth collection method, involving fleshy and bracket fungi. In addition, within the subgenus *Scaptodrosophila,* there are taxonomically distinguishable species groups that vary in resources utilized and therefore in habitats chosen. In particular, six of the seven species attracted to fermented fruit baits belong to one species group, *coracina.* Three species groups are identifiable among the swept species: *inornata, barkeri,* and *brunneipennis.* There are greater distributional similarities *within* species groups than *among* species groups. Parsons and Bock (1977) argued that similar distributions imply similar, but obviously not identical, niches (i.e., plant resources) occupied. In other words there is an implied indirect association of morphology with resources utilized. The same argument pertains to subgenus *Sophophora,* all Australian species of which (cosmopolitan and endemic) are attracted by fermented-fruit baits with the exception of three closely related species distinguished by highly unusual sexcombs, which are mainly collected by sweeping.

There is therefore an association between taxonomic and resource utilization divergence at the level of subgenera and species groups. Some of the collection methods are more specific to habitats in the wild than others, but it can hardly be doubted that the various methods yield species with substantially different ecological preferences. In particular, some species

Table 11.1. *Number of species in the Australian endemic and cosmopolitan* Drosophila *fauna classified by resources utilized, as indicated by collection method and subgenus*

	Collection method[a]				
Subgenus	1	2	3	4	5
Endemics					
Scaptodrosophila[b]	7[c]	8		24	2
Hirtodrosophila[b]			10	3	
Sophophora	12[d]			3	
Drosophila	5				
Cosmopolitans					
Sophophora	3				
Drosophila[e], *Dorsilopha*	5				

[a](1) Fermented fruit baiting; (2) rotted commercial mushroom baiting; (3) collected on or near fleshy fungi and/or bracket fungi in the forest; (4) sweeping of foliage, leaf litter, and flowers; (5) demonstrated as utilizing decaying flowers for a larval resource (*D. hibisci* and *D. minimeta*).
[b]Data on resources utilized are unavailable in many rare species in these subgenera.
[c]Includes a minimum of two species (*D. howensis* and *D. specensis*) attracted also to rotted commercial mushroom baits.
[d]Includes *D. dispar,* which is also attracted to rotted commercial mushroom baits, and is obtained by sweeping.
[e]Two introduced cactus-breeding species, *D. buzattii* and *D. aldrichi,* are additionally known in Australia from this subgenus.
Source: Updated from Parsons (1978a).

groups are apparently assignable to habitats as assayed by collection methods.

Species commonly studied in the laboratory are generally attracted to fermented fruits and/or vegetables. Among them, data of Atkinson and Shorrocks (1977) show an association between resource utilization as measured by breeding site items exploited in a market, and taxonomic divergence. For example, a major difference occurs between three subgenus *Sophophora* (*melanogaster, simulans, subobscura*) species, which are fruit feeders, and three belonging to the closely related subgenera *Drosophila* and *Dorsilopha* (*immigrans, hydei, busckii*), which to varying degrees are vegetable feeders in addition.

Of these species most ecological–behavioral data are for the *Sophophora* sibling species *D. melanogaster* and *D. simulans,* and for the subgenus *Drosophila* species *D. immigrans* (Parsons, 1979a). By whatever criterion is selected – larval responses, resources utilized, oviposition, nutritional requirements, temperatures tolerated and preferred, and wasp parasitism – the sibling species are very similar and contrast in total with *D. immigrans* (Table 11.2). The greater diversity of resources utilized by *D. immigrans* has been argued to be consistent with its reasonably frequent occurrence at low population densities in rain forest habitats, which are rarely invaded by other cosmopolitan species in Australia (Bock and Parsons, 1977). In addition, in the Melbourne region (latitude 38°S) *D. immigrans* populations are relatively constant throughout the year; that is, population numbers are virtually independent of macroenvironmental temperature, showing a significant contrast with the two sibling species (McKenzie and Parsons, 1974a). This suggests a greater diversity of resources available for *D. immigrans* in a temperate zone, as compared with the sibling species, whose fruit resources peak in summer and early autumn.

The sibling species *D. melanogaster* and *D. simulans*

Much has been written on the effects of extreme temperatures and desiccation on these species. Although there are known variations among strains *within* the two species for tolerances to these stresses, *D. melanogaster* can normally tolerate a greater range of laboratory temperatures than *D. simulans,* a phenomenon that explains much seasonal and distribution data (McKenzie and Parsons, 1974a). The success of a population obviously depends on its adaptation to climatic conditions. Temperate zone annual cycles provide the greatest range of stresses, but shorter-lived stress periods occur diurnally in addition; in such regions the major stresses are a

Table 11.2. *Comparison of the sibling species* D. melanogaster *and* D. simulans, *with* D. immigrans

	D. melanogaster and *D. simulans*	D. immigrans
Subgenus	Sophophora	Drosophila
Physical environment		
High temperature/desiccation[a]	High resistance, especially *melanogaster*	Lower
Laboratory temperature preferences	High, especially *melanogaster*	Lower
Cold stress[a]	Less resistance than *immigrans,* especially *simulans*	Higher
Ethanol and other potential resources (laboratory)		
Larval responses to ethanol (6%)[a]	High to moderate preference *melanogaster,* minimal response *simulans*	Avoidance
Threshold for ethanol resource utilization	>9% *melanogaster* 3–6% *simulans*	≃ 1.5%
Larval responses to acetic acid, ethyl acetate, lactic acid[a]	High	Moderate
Nutrition		Can develop normally on a low cholesterol diet
Resource utilization (field studies)		
Oviposition sites	Fruit specialists	Fruits and vegetables
Lemons	Adults avoid lemons, and larval survival low, especially *simulans*	Adults select lemons, and larval survival high
Ecological observations		
Occurrence in rain forests	*melanogaster* apparently never, *simulans* rarely	Frequent, rare inhabitant
Wasp parasitism by *Phaenocarpa persimilis* (in sympatric populations in the Melbourne area)	Highly successful	Unsuccessful

[a]Intraspecific geographic differences are known for these items for *D. melanogaster* and *D. simulans* (except for ethanol).
Source: Parsons (1979a), updated and with additional information.

combination of high temperature and desiccation stress, and low temperature. It is normally difficult to breed *D. melanogaster* above about 29°C, and lower for *D. simulans*. But these species must occasionally survive some hours of the day in excess of 30°C; in the field where these and other *Drosophila* species are at high temperatures, they are usually in damp microhabitats (Fellows and Heed, 1972; Cook et al., 1977). That is, there is behavioral selection in the field for damp humid microhabitats. This finding is consistent with Parsons's (1979b) observation that at stressful temperatures in the 30°–34°C range, adults of the sibling species *D. melanogaster* and *D. simulans* readily survive short periods of time at 95% relative humidity and are highly fertile as compared with 0% RH. It shows that provided a humid microhabitat can be found, these species can survive for periods of high temperatures not likely to be exceeded on a diurnal basis in the wild. In addition, the survival and fertility of *D. simulans* are poorer than for *D. melanogaster* – a result consistent with distribution patterns of the two species in the wild. Finally, populations of both species from the temperate climate of Melbourne are more resistant to stresses than those from the subtropical climate of Townsville, north Queensland (latitude 20°S). Such climatic races are related to the direct effects of the greater extremes of temperate compared with subtropical climates on the life histories of the two species (unpub. data).

A second important variable of the physical environment is light intensity. Spieth and Hsu (1950) showed differences in mating success of the species according to lighting conditions; *D. melanogaster* mates freely in both light and darkness, whereas *D. simulans* is inhibited by darkness. In addition, a comparison of the two species dispersing toward a light source and without a light source, showed that *D. simulans* is more dependent upon light than *D. melanogaster* (McDonald and Parsons, 1973). Similarly for phototactic responses along a gradient of light intensities, *D. simulans* shows greater phototaxis than *D. melanogaster* (Parsons, 1975b; Kawanishi and Watanabe, 1978). In both cases therefore, the behavior of *D. melanogaster* is less light-dependent than that of *D. simulans*, arguing that *D. melanogaster* may be regarded as the broader-niched species in this regard. Turning to oviposition, a trait of obvious and direct relevance for fitness, Kawanishi and Watanabe (1978) found that *D. simulans* tends to oviposit in areas of higher light intensity than *D. melanogaster*. In addition, selection of eggs by position in a light-intensity gradient made it possible to segregate a mixed species population into separate species populations because selection for photopositive flies soon eliminated *D. melanogaster,* and selection for photonegative flies soon eliminated *D. simulans*. Therefore it must be expected that these sibling species occupy different niches in nature where varying light intensities occur. Indeed there are Japanese data suggesting that *D. simulans* in some instances is not present inside houses, whereas *D. melanogaster* occurs both inside and outside houses (Watanabe and Kawanishi, 1976).

The third major environmental variable known to differentiate the species at the ecological and behavioral level in nature is ethanol. Beginning with the studies of McKenzie and Parsons (1972), it is a commonplace observation that *D. melanogaster* is more tolerant of ethanol than *D. simulans*, both in the wild and in the laboratory (see also McKenzie and Parsons, 1974b; David et al., 1974; David and Bocquet, 1975). Referring to the exclusion of *D. simulans* from inside buildings (above), this same observation in wineries (McKenzie and Parsons, 1972) can be explained in terms of relative ethanol tolerances.

Because the species are so frequently sympatric, are both attracted to fermented-fruit baits, and utilize fermented and decaying fruits as resources, the implied discrete differences between the two species seem extreme. However, much of the published data concern relatively extreme ethanol concentrations of 6% and up. Indeed it would seem reasonable to suppose that both species could use ethanol at low concentrations, but only *D. melanogaster* could do so at high concentrations. In agreement we have recently shown that in the absence of other food, *D. simulans* longevity increases at 0.5%–3.0% ethanol concentrations (Figure 11.1); this phenomenon is associated with some larval development and also occurs at the most rapid development times (Parsons et al., 1979). In other words the fitness of *D. simulans* is increased by environmental ethanol at low concentrations and is markedly reduced at the high concentrations of 6% and up used in most previously reported studies. The contrast with *D. melanogaster* is that the fitness of this species is increased at all ethanol concentrations up to and including 9%; even so this increase is maximal at the same low concentrations favoring *D. simulans*.

It can be concluded that at low concentrations both species use ethanol, but there is a threshold difference between the two species (Table 11.2; note the even lower threshold for *D. immigrans*, a species apparently utilizing a greater diversity of resources than the sibling species). In the field it would be expected that both species would coexist at low ethanol concentrations but not at high concentrations, as McKenzie and McKechnie (1979) have shown in a study of adults, larvae, and pupae of the two species in a pile of grape residues during different decomposition stages.

On the basis of behavioral and ecological variables of direct significance in determining niches in nature, we therefore arrive at the conclusion that the adaptive potential of these two sibling species is very different. The physical variables of temperature and desiccation and their interrelationships determine the absolute distributions of these and other *Drosophila* species. These variables determine the boundary conditions for resource utilization as directly measured by behavioral variables such as oviposition, sexual behavior, and general activity, which only occur under permissive physical conditions (Parsons, 1978b). Not unexpectedly, McKenzie (1978) has shown that *D. melanogaster* has a somewhat broader range for resource utilization using this type of approach than *D. simulans*. Light intensity is obviously another restricting physical condition, especially for *D. simulans,* although in the field, light intensity variations are often difficult to separate from temperature and humidity variations (Dyson-Hudson, 1956; Carson et al., 1970).

Resource utilization must therefore be studied under permissive physical conditions. Larval responses to possible metabolites are given in Table 11.2, and show that the *melanogaster–simulans* differences are not large except for ethanol. Similarly, with the exception of ethanol, geographical

Figure 11.1. Adult survivorship expressed as LD_{50} (days) calculated by linear interpolation based upon five replicates for *D. melanogaster* and *D. simulans* at each alcohol concentration and temperature. (Modified from Parsons et al., 1979.)

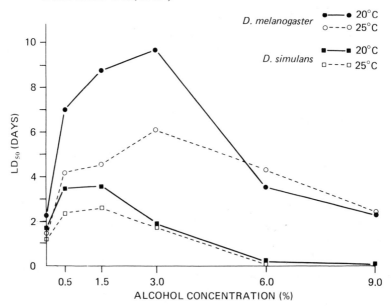

differences in response to the chemicals in Table 11.2 are not large *within* species. In *D. melanogaster* newly hatched larvae in southern Australian (37–38°S) populations have a strong preference for ethanol, whereas the mean preference of a more northern population (19°S) is lower. The lower mean preference of the northern population is due to a high level of variability among isofemale strains whereby some strains are as extreme as those in the south (Parsons, 1977). These results, together with data on adult tolerances to ethanol (David and Bocquet, 1975), show that the differences between the sibling species in relation to ethanol are maximized in temperate compared with more tropical regions.

A general conclusion at this stage, then, is that there is a need for detailed research on these species seasonally, altitudinally, and latitudinally, paying particular attention to microhabitat characteristics. In this way species distributions should be relatable to habitats using multivariate techniques, and the relative importance of the variables so far discussed in the totality of species distributions assessed. As McKenzie and Parsons (1974a) have shown, seasonal species distributions in the Melbourne area are interpretable in terms of temperature, and more important, temperature variations. At the intraspecific level, evidence for climatic races in *D. melanogaster* and *D. simulans* has already been given. In addition Stalker and Carson (1947, 1948, 1949) showed the importance of looking at seasonal, altitudinal, and latitudinal gradients in explaining morphological variation in the widespread North American species *D. robusta,* finding that most variation can be related to mean annual temperature. This is a finding directly suggestive of climatic selection.

The two sibling species are reproductively isolated; the few hybrids that are produced are sterile (Sturtevant, 1919). It is difficult not to conclude that ecological independence preceded reproductive isolation, which evolved as a consequence of differing preferred habitats, especially in sympatric populations (Mayr, 1977). It may well be that selection for utilization of differing light-intensity regimes is an important component of reproductive isolation, given the importance of light-intensity levels in both oviposition site preference and relative mating success. Whether this is true or not, maximum ecological divergence (defined to include associated behavioral traits) between the two species should occur when they are sympatric and less so when they are allopatric – a testable hypothesis. Such character displacement may, as Brown and Wilson (1956) point out, involve morphological, ecological, behavioral, or physiological characters. In the case of sibling species which are morphologically almost identical, being "merely near the invisible end of a broad spectrum of increasingly diminishing morphological differences between species" (Mayr, 1963), differences for morphological characters are known to be slight. The important behaviors are likely to be those associated with habitats selected, rather than those associated with reproductive success, because the level of reproductive isolation between these sibling species is extremely high. The semispecies of *D. paulistorum*, which show more reproductive isolation when sympatric than when allopatric (Ehrman, 1965), would provide a good general test system if ecological–behavioral differences involved in habitats selected can be found differentiating the semispecies.

Intraspecific habitat selection

The cosmopolitan and endemic *Drosophila* fauna of Australia show that habitats selected as assessed by resources utilized become more similar with phenetic convergence (Table 11.1). There is abundant evidence that favorable habitats are selected at the species level; evidence for between–population differences within species is sparse although the latitudinal gradients for larval alcohol preferences are suggestive. But given the stress so far on temperature and humidity, the observations of Waddington et al. (1954), who described differences between mutants of *D. melanogaster* in preferences for environments with different humidities, are important. Where studied, species or geographic strains of *Drosophila* are broadly differentiated in adult preference for various yeasts (Dobzhansky et al., 1956), suggesting that genetic differentiation of olfactory responses has accompanied evolutionary divergence of these species. Geographic larval response differences to ethanol, acetic acid, lactic acid, and ethyl acetate occur in *D. melanogaster* (Table 11.2), and the same occur for adult perception of these odors (Fuyama, 1976). There are few field observations relating to these laboratory-based experiments, although differences in inversion frequencies between *D. melanogaster* breeding on oranges and those breeding on grapefruits have been reported (Stalker, 1976); this is a result suggestive of heterogeneity for microhabitats selected in natural populations, with direct genetic consequences.

There is an ever increasing literature (see, e.g., Vigue and Johnson, 1973; David and Bocquet, 1975; Pipkin et al., 1976) on biogeographic gradients in various *Drosophila* species showing associations of inversions, allozyme frequencies, and fly size with altitude and latitude. As mentioned above, seasonal changes may also occur. In almost all cases temperature is directly or indirectly implicated, although the effects could be through metabolic variations, which themselves are dependent upon temperature. The altitudinal transect of Stalker and Carson (1948) in *D. robusta* covers a short distance geographically, showing a precise correspondence between environment and form. It suggests, in agreement with Ehrlich and Raven (1969), Gould and Johnston (1972), and Endler (1977), that selection is the primary cause of ordering most geographic variation within species so that local differentiation is likely even when the organisms are highly mobile. This possibility can certainly be envisaged for widespread woodland species depending upon natural resources. There is a need for further investigations over short distances (especially across varying altitudes) of cosmopolitan species such as *D. melanogaster,* which can obviously migrate with fruit and vegetable resources. Even so, McKenzie and Parsons (1974b) demonstrated microgeographic differentiation with respect to alcohol in the environment over extremely short distances in the vicinity of a winery.

The findings of Taylor and Powell (1976), who reported on microgeographic differences of chromosomal and enzyme polymorphisms in *D. persimilis,* support the possibility of differentiation of widespread *Drosophila* species into local populations. They found differing gene and inversion frequencies among habitats in a heterogeneous environment made up of various vegetational types associated with differing moisture regimes in a small local study area. Only by invoking habitat selection could they explain their data. In addition, flies were captured from different habitats, marked according to origin, and released from a common site. Flies tended to return to the habitats at which they were originally identified, suggesting the possibility of behavioral differences among genotypes, whereby different genotypes seek out different parts of the environment (Taylor and Powell, 1978). Given the demonstrated importance of temperature and humidity in *Drosophila* population biology, this result is plausible and concurs with the possibility of local genetic differentiation in these mobile insects, especially because on one humid day when conditions were nonstressful, the habitat selection characteristic of the normal warm clear summer days in the study area disappeared.

Results from laboratory studies must therefore be related to natural environments in order to explore the genetics of habitat selection. In nature, habitat selection may be regarded as dependent upon a combination of the physical features of the environment and the nature of resources utilized, provided that the physical features of the environment are permissive. The genetic determination of habitat selection must be extremely complex because the manifestation of the behavior is far removed from its genetic determination. Components such as temperature, light, humidity, and food preferences are analyzable separately in the determination of their genetic basis, but this analysis does not describe the genetics of habitat selection as a "behavioral unit." Such integrated studies have been more fully developed in certain vertebrates such as *Peromyscus* where there is good evidence for overall behavioral phenotypes involved in habitat selection (King, 1967; Ehrman and Parsons, 1976). If habitat selection is common, a consequence is the development of heterogeneity in natural populations, reinforcing the point that the local interbreeding population is in fact the unit of evolution, rather than the species. This will become progressively more testable as we become more familiar with the ecological genetics of *Drosophila*.

A cosmopolitan species such as *D. melanogaster* utilizes a variety of fruit resources. The extent to which populations are differentiated in relation to such resources is not known and is worthy of further investigation. On an annual basis, a variety of resources may be used in urban habitats, but in orchard habitats resource diversity is likely to be lower. A relevant observation in this regard is Manning's (1967) finding that *Drosophila* reared on a medium containing geraniol (a peppermint smell) showed reduced aversion to the odor when adult, which he regards as a form of habituation. It is reasonable that there could be genetic assimilation for such behavior over a number of generations, as suggested by Moray and Connolly (1963). Extrapolating to the wild, if a newly entered habitat is characterized by some sort of resource difference that can be assimilated genetically, then rapid evolutionary change is possible in response to the habitat change. Studies to distinguish learning from habituation are of obvious significance, as are studies on the possibility of learning to utilize different resources. Learning would be important in relation to resource utilization in favoring the development, by genetic assimilation, of local interbreeding populations, specifically adapted to the utilization of local resources. *D. melanogaster* can readily learn a complex discrimination task involving the selective avoidance of an odorant after being shocked in its presence (Quinn et al., 1974). In addition various wild-type strains differ in their abilities to display such learning (Dudai, 1977). These results could be significant in the avoidance of possible noxious resources in nature. If so, then the converse of learning to utilize specific resources is plausible, which could in due course lead to reproductive isolation *following* ecological and behavioral divergence at the habitat level.

Discussion

This discussion is leading to the point that speciation in *Drosophila* could on occasion be a sympatric process, initially involving resource utilization heterogeneity in a population, which under certain circumstances leads to races and possibly even species. Such a scheme has in fact been suggested for two closely related Hawaiian species, *D. mimica* and *D. kambysellisi,* the latter being proposed as a derivative species that may have arisen following the provision of a new rotting plant resource (Richardson, 1974). This explanation is plausible through an altered food preference followed by genetic assimilation. Evidence of such phenomena occurs for those parasitic insects that mate on the host plant or animal; for example, in the true fruit flies (Tephritidae) where species of *Rhagoletis* have evolved over a relatively short time in North America to parasitize domesticated European fruits such as cherries and apples (Bush, 1969, 1975). Bush has stressed the need for an integrated approach involving aspects of ecology, behavior, and genetics for the understanding of speciation in these circumstances. It is argued here that the same applies to *Drosophila.*

Sympatric speciation is almost certainly enhanced if mating occurs assortatively within a niche rather than at random. Indeed, Dethier (1954), when writing on olfactory conditioning, considered that if there is assortative mating among members of a population similarly conditioned, then such behavior could be regarded as an isolating mechanism, which could lead to the development of population heterogeneity at least. Maynard Smith (1966) has shown on theoretical grounds that if there is a positive correlation between mates and niche selection, speciation may ultimately occur by the development of isolation under the control of a polymorphism developed by disruptive selection. [Additional theoretical considerations on the maintenance of polymorphisms in heterogeneous environments appear in Mather (1955), Levene (1953), and Levins (1968).] Thoday and Gibson (1962) have shown that polymorphisms can be developed in this way for sternopleural bristle number in *D. melanogaster,* although as Bush (1975) points out, this is a result in need of generalization. Divergence is enhanced if there is a breeding system incorporating positive assortative mating (Millicent and Thoday, 1960), a point discussed further by Parsons (1962). Later results (Thoday, 1964) showed the development of positive assortative mating in disruptive selection lines, which is reasonable because unselected *Drosophila* populations show positive assortative mating for bristle numbers (Parsons, 1965). Microhabitat selection in the wild would further promote isolation, and is likely once polymorphism is established (Maynard Smith, 1966). Indeed microhabitat heterogeneity implies an array of optima in a population dependent upon interactions between genotypes and environments. Learning associated with habitat selection followed by genetic assimilation would clearly accelerate the development of such population heterogeneity, and in some circumstances could be a major force in the development of species, as was suggested many years ago by Thorpe (1945) in some early explorations of the evolutionary significance of habitat selection. Indeed, he considered that "the positive action of habitat selection works by providing isolation of populations within a species" by a form of "locality imprinting" based upon ecological factors.

Evidence for associations between mating and habitat selection is most readily available for resource-specific species that are "coarse grained" in the sense of Levins and MacArthur (1966). *D. hibisci*, which is restricted to endemic Australian *Hibiscus* plants, has only been found on the petals and corolla tubes of *Hibiscus* species where courtship-type behaviors have been observed (Figure 11.2). Flies spend most of the time in the base of the corolla around the filaments of the anthers; this is a humid microniche in a climate that is otherwise extreme for *Drosophila*. Extreme habitat selection is necessary in this species because the *Hibiscus* flowers themselves remain open for one day only, during which time oviposition occurs because the larvae exploit the decaying flowers as a resource (Cook et al., 1977). Although as yet no definitive laboratory studies have been carried out, host-races within *D. hibisci* could occur, *D. hibisci* having been found on several *Hibiscus* species and subspecies. A second example is the Australian endemic species *D. minimeta*, which has been found to exploit flowers of two introduced species of the family Solanaceae, namely, *Solanum mauritianum* and *S. torvum* in north Queensland (unpubl. observations). The possibility of sympatric races occurs for both of these flower breeding species.

Generally, *Drosophila* species are obtained by one collection method only, implying resources utilization specificity (Table 11.1). However, *D. (Sophophora) dispar* is a widespread endemic species of eastern Australia, which is unique among Australian *Drosophila* in that in many habitats it is collected by the *three* major collection methods of fermented-fruit baiting, mushroom baiting, and sweeping (Bock and Parsons, 1978). Is there any correspondence between the three resources implied by these collection methods and genetic heterogeneity? If so, the heterogeneity so found would assuredly be sympatric. The same question can be asked of two *coracina* group species of subgenus *Scaptodrosophila*, namely, *D. specensis* and *D. howensis*. which come to fermented-fruit and mushroom baits, with some preference for the latter bait (Bock and Parsons, 1978; Parsons and Bock, 1979a).

Two other *coracina* group species, *enigma* and *lativittata*, are mainly

Figure 11.2. *a*. Flower of *Hibiscus heterophyllus* in southern Queensland showing *D. hibisci* on the petals. Note the aggregation of flies in shaded regions. *b*. Adult flies with some courtship-type interactions on the petal of *Hibiscus heterophyllus*.

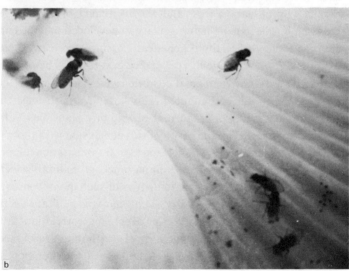

attracted to fermented-fruit baits and are found from southern Queensland to Victoria. These species are reasonably common in urban/orchard areas, especially in southern regions, and have presumably spread subsequent to the introduction of fruit trees after European settlement. These ecologically versatile species, which may be in the process of colonizing new habitats, are therefore of particular evolutionary interest. Presumably colonization of new habitats could involve ecological polymorphisms related to changes in resource utilization with the development of climatic races. To what extent such races are developing in these species is not yet known. However, it can be said that these species are at least as tolerant of temperature/desiccation extremes as *D. melanogaster* and *D. simulans* (Parsons and Stanley, unpub. data), two species with which they are occasionally sympatric in urban/orchard areas; in general, tolerances to such extremes are lower in species restricted to rain forests (Parsons and McDonald, 1978). *D. enigma* and *D. lativittata* therefore have the physical tolerances to environmental extremes necessary for spread, although the possibility of the evolution of resource utilization shifts still remains to be investigated.

The *Drosophila* species discussed here occur in heterogeneous environments that are frequently discontinuous. The environments may be "tracked" by climatic races, and the likelihood of resource-utilization races is high. The genetic heterogeneity so developed may be enhanced by habitat selection, especially if genetically assimilated learned components are present. An array of genotypes can be envisaged as being differentially distributed across environmental conditions, each genotype selecting those habitats to which it is best adapted. This statement is obviously true at the specific level (Table 11.1), and it is now highly plausible at the population/genotype level. The similarity of this conclusion with those of Wright (1969) and others on the importance of fragmentation of populations should be clear. The suggested importance of habitat selection does, however, reduce the need to regard drift as of such major consequence as it is considered by Wright. Formally, such genotypic habitat selection can be regarded as a form of genotype–environment correlation favoring the development of a "coarse grained" array of habitats occupied because clearly an animal selects its environment before its environment selects the animal. Because genotypes may vary in response to a given environment, and conversely a given genotype may have varying responses in different environments, the study of habitat selection requires studies in both "dimensions," perhaps simultaneously, if the experiment itself does not then become excessively large. Possible designs are considered by Parsons (1967).

The formal prerequisites for sympatric speciation appear to exist in the genus *Drosophila* even though the species on which detailed genetic studies have been carried out tend to be "fine-grained" and so do not apparently possess the high level of habitat specificity argued to be most favorable for the development of sympatric speciation (Haldane, 1959; Heed, 1971; Bush, 1969). It must be stressed however, that species collected by identical baiting methods can be defined as occupying different niches at the microhabitat level (Table 11.2). Indeed more studies of the type presented in Table 11.2 will assist in determining how different apparently closely related *Drosophila* species are. It is probably reasonable to suggest that rare species are often extremely habitat-specific in terms of resource utilization, and examples would include many of the numerous *Drosophila* species of the extremely diverse complex mesophyll vine forests of the humid tropics of north Queensland (Bock and Parsons, 1977) and the rain forests of Hawaii (Kaneshiro et al., 1973). A detailed knowledge of speciation in *Drosophila* will necessarily depend upon a knowledge of resource utilization, especially in floristically rich tropical habitats where there may be the greatest chances of discovering species complexes similar to the *D. paulistorum* semispecies complex of Central and South America. Although sophisticated work on reproductive isolation has been done on these semispecies (Ehrman, 1965), a knowledge of resource utilization is equally important in understanding their evolution. Indeed, studies on tropical species where environmental heterogeneity should promote habitat selection should be especially rewarding (Rosenzweig, 1978). Certainly *Drosophila* species diversities increase with decreasing latitude in Australia, becoming apparently greatly magnified in New Guinea (Parsons and Bock, 1979a), and resource utilization diversity increases along the same gradient (Parsons and Bock, 1979b).

White (in press) recently wrote, "We are beginning to understand that to the diversity of population genetics and of chromosomal architecture in the living world, there corresponds a diversity of speciation mechanisms," followed by, "The best strategy would seem to be to concentrate efforts on a limited number of cases so that they can really be investigated in depth." Rather than review chromosomal speciation mechanisms which have been studied in many groups of animals, including the elegant grasshopper studies of White (1973, 1977), this paper attacks the problem by concentrating efforts on the behavioral and ecological biology of a limited number of cases. From a consideration of a diversity of *Drosophila* research, the tentative conclusion is that habitat selection at the intraspecific level is yet another of the interacting components that must be studied in order to understand modes of speciation, especially as habitat selection may be expected to develop in advance of reproductive isolation.

Summary

There is an association between taxonomic and resource-utilization divergence at the level of subgenera and species groups in the Australian endemic and cosmopolitan *Drosophila* fauna.

Although the sibling species *D. melanogaster* and *D. simulans* are morphologically very similar, their adaptive potentials are very different when behavioral and ecological variables of direct significance in determining their habitats in nature are considered. It is suggested that ecological independence preceded reproductive isolation, which evolved as a consequence of differing preferred habitats.

At the intraspecific level, if a newly entered habitat is characterized by a resource utilization shift, then rapid evolutionary change may be possible in response to the habitat change, especially if learning followed by genetic assimilation occurs. Such intraspecific habitat selection is yet another of the interacting components that we must study in order to understand modes of speciation, and probably develops in advance of reproductive isolation. It follows that resource utilization heterogeneity enhanced by habitat selection could under certain circumstances lead to races and even sympatric species.

I thank Michael White for his efforts in convincing me of the importance of chromosomes in evolutionary processes and for sharing with me his recent thoughts on speciation, Hampton Carson for numerous discussions over the years on the evolutionary biology of *Drosophila* as well as introducing me to Hawaiian collecting techniques, and Leanne Weber for helpful discussions on the manuscript. Among others, the writings of and discussions with Guy Bush, Lee Ehrman, William Heed, Ernst Mayr, Dick Richardson, and John Thoday have assisted my thoughts on speciation over the years. Partial financial support for some of the work discussed came from the Australian Research Grants Committee.

References

Atkinson, W., and Shorrocks, B. 1977. Breeding site specificity in the domestic species of *Drosophila*. *Oecologia (Berl.) 29*:223–32.

Bock, I. R., and Parsons, P. A. 1977. Species diversities in *Drosophila* (Diptera): a dependence upon rain forest type of the Queensland (Australian) humid tropics. *J. Biogeog. 4*:203–13.

– 1978. Australian endemic *Drosophila* IV. Queensland rain forest species collected at fruit baits, with descriptions of two species. *Aust. J. Zool. 26*:83–90.

Brown, W. L., and Wilson, E. O. 1956. Character displacement. *Syst. Zool. 5:*49–64.

Bush, G. L. 1969. Sympatric host race formation and speciation in frugivorous flies of the genus *Rhagoletis. Evolution 23:*237–51.

– 1975. Modes of animal speciation. *Annu. Rev. Ecol. Syst. 6:*339–64.

Carson, H. L. 1971. The ecology of *Drosophila* breeding sites. *Harold L. Lyon Arboretum Lecture, No. 2.* Honolulu: University of Hawaii.

Carson, H. L., Hardy, D. E., Spieth, H. T., and Stone, W. S. 1970. The evolutionary biology of the Hawaiian Drosophilidae. In *Essays in Evolution and Genetics in Honor of Theodosius Dobzhansky* (ed. Hecht, M. K., and Steere, W. C.), pp. 437–543. New York: Appleton-Century-Crofts.

Cook, R. M., Parsons, P. A., and Bock, I. R. 1977. Australian endemic *Drosophila* II. A new *Hibiscus*-breeding species with its description. *Aust. J. Zool. 25:*755–63.

David, J., and Bocquet, C. 1975. Similarities and differences in latitudinal adaptation of two *Drosophila* sibling species. *Nature (Lond.) 257:*588–90.

David, J., Fouillet, P., and Arens, M.-F. 1974. Comparison de la sensibilité à l'alcool éthylique des six espèces de *Drosophila* du sous-groupe *melanogaster. Arch. Zool. Exp. Gén. 115:*401–10.

Dethier, V. G. 1954. Evolution of feeding preferences in phytophagous insects. *Evolution 8:*33–54.

Dobzhansky, Th., Cooper, D. M., Phaff, H. J., Knapp, E. P., and Carson, H. L. 1956. Studies of the ecology of *Drosophila* in the Yosemite region of California. IV. Differential attraction of species of *Drosophila* to different species of yeasts. *Ecology 37:*544–50.

Dudai, Y. 1977. Properties of learning and memory in *Drosophila melanogaster. J. Comp. Physiol. 144:*69–89.

Dyson-Hudson, V. R. D. 1956. The daily activity rhythm of *Drosophila subobscura* and *D. obscura. Ecology 37:*562–7.

Ehrlich, P. R., and Raven, P. H. 1969. Differentiation of populations. *Science 165:*1228–32.

Ehrman, L. 1965. Direct observation of sexual isolation between allopatric and between sympatric strains of the different *Drosophila paulistorum* races. *Evolution 19:*459–64.

Ehrman, L., and Parsons, P. A. 1976. *The Genetics of Behavior.* Sunderland, Massachusetts: Sinauer Associates.

Endler, J. A. 1977. *Geographic Variation, Speciation, and Clines.* Princeton: Princeton University Press.

Fellows, D. P., and Heed, W. B. 1972. Factors affecting host plant selection in desert-adapted cactophilic *Drosophila. Ecology 53:*850–8.

Fuyama, Y. 1976. Behavior genetics of olfactory responses in *Drosophila* I. Olfactometry and strain differences in *Drosophila melanogaster. Behav. Genet. 6:*407–20.

Gould, S. J. and Johnston, R. F. 1972. Geographic variation. *Annu. Rev. Ecol. Syst. 3:*457–98.

Haldane, J. B. S. 1959. Natural selection. In *Darwin's Biological Work* (ed. Bell, P. R.), pp.101–49, Cambridge: Cambridge University Press.

Heed, W. B. 1971. Host plant specificity and speciation in Hawaiian *Drosophila. Taxon 20:*115–21.

Kaneshiro, K. Y., Carson, H. L., Clayton, F. E., and Heed, W. B. 1973. Niche separation in a pair of homosequential *Drosophila* species from the island of Hawaii. *Am. Nat. 107:*766–74.

Kawanishi, M., and Watanabe, T. K. 1978. Difference in photo-preferences as a cause of coexistence of *Drosophila simulans* and *D. melanogaster* in nature. *Jpn. J. Genet. 53:*209–14.

King, J. A. 1967. Behavioral modification of the gene pool. In *Behavior – Genetic Analysis* (ed. Hirsch, J.), pp. 22–43. New York: McGraw-Hill.

Levene, H. 1953. Genetic equilibrium when more than one ecological niche is available. *Am. Nat. 87:*331–3.

Levins, R. 1968. *Evolution in Changing Environments: Some Theoretical Explorations.* Princeton: Princeton University Press.

Levins, R., and MacArthur, R. 1966. Maintenance of genetic polymorphism in a hetero-geneous environment: variations on a theme by Howard Levene. *Am. Nat. 100:*585–90.

McDonald, J., and Parsons, P. A. 1973. Dispersal activities of the sibling species *Drosophila melanogaster* and *Drosophila simulans. Behav. Genet. 3:*293–301.

McKenzie, J. A. 1978. The effect of development temperature on population flexibility in *Drosophila melanogaster* and *D. simulans. Aust. J. Zool. 26:*105–12.

McKenzie, J. A., and McKechnie, S. W. 1979. A comparative study of resource utilization in natural populations of *Drosophila melanogaster* and *D. simulans. Oecologia (Berl.) 40:*299–309.

McKenzie, J. A., and Parsons, P. A. 1972. Alcohol tolerance: an ecological parameter in the relative success of *Drosophila melanogaster* and *Drosophila simulans. Oecologia (Berl.) 10:*373–88.

– 1974a. Numerical changes and environmental utilization in natural populations of *Drosophila. Aust. J. Zool. 22:*175–87.

– 1974b. Microdifferentiation in a natural population of *Drosophila melanogaster* to alcohol in the environment. *Genetics 77:*385–94.

Manning, A. 1967. "Pre-imaginal conditioning" in *Drosophila. Nature (Lond.) 216:*338–40.

Mather, K. 1955. Polymorphism as an outcome of disruptive selection. *Evolution 9:*52–61.

Maynard Smith, J. 1966. Sympatric speciation. *Am. Nat. 100:*637–50.

Mayr, E. 1963. *Animal Species and Evolution.* Cambridge, Massachusetts: Harvard University Press.

– 1977. The study of evolution, historically viewed. In *The Changing Scenes in Natural Sciences, 1776–1976,* pp.39–58. Academy of Natural Sciences: Philadelphia, Special Publication 12.

Millicent, E. and Thoday, J. M. 1960. Gene flow and divergence under disruptive selection. *Science 131:*1311–12.

Moray, N., and Connolly, K. 1963. A possible case of genetic assimilation of behaviour. *Nature (Lond.) 199:*358–60.

Parsons, P. A. 1962. The initial increase of a new gene under positive assortative mating. *Heredity 17:*267–76.

– 1965. Assortative mating for a metrical characteristic in *Drosophila. Heredity 20:*161–7.

– 1967. *The Genetic Analysis of Behaviour.* London: Methuen.

– 1975a. The comparative evolutionary biology of the sibling species, *Drosophila melanogaster* and *D. simulans. Q. Rev. Biol. 50:*151–69.

– 1975b. Phototactic responses along a gradient of light intensities for the sibling species *Drosophila melanogaster* and *Drosophila simulans. Behav. Genet. 5:*17–25.

– 1977. Larval reaction to alcohol as an indicator of resource utilization differences between *Drosophila melanogaster* and *D. simulans. Oecologia (Berl.) 30:*141–6.

– 1978a. Habitat selection and evolutionary strategies in *Drosophila:* an invited address. *Behav. Genet. 8:*511–26.

– 1978b. Boundary conditions for *Drosophila* resource utilization in temperate regions, especially at low temperatures. *Am. Nat. 112:*1063–74.

– 1979a. Larval reactions to possible resources in three *Drosophila* species as indicators of ecological divergence. *Aust. J. Zool. 27:*413–19.

– 1979b. Resistance of the sibling species *Drosophila melanogaster* and *Drosophila simulans* to high temperatures in relation to humidity: evolutionary implications. *Evolution 33:*131–6.

Parsons, P. A., and Bock, I. R. 1977. Australian endemic *Drosophila* I. Tasmania and Victoria, including descriptions of two new species. *Aust. J. Zool. 25:*249–68.

– 1979a. Australian endemic *Drosophila* VII. Lord Howe Island, with description of a new species of the *coracina* group. *Aust. J. Zool. 27:*973–80.

– 1979b. Latitudinal species diversities in Australian endemic *Drosophila*. *Am. Nat. 114:*213–20.

– Australian *Drosophila:* diversity, resource utilization and radiations. In *Ecological Biogeography in Australia* (ed. Keast, A.). The Hague: W. Junk (in press).

Parsons, P. A., and McDonald, J. 1978. What distinguishes cosmopolitan and endemic *Drosophila* species? *Experientia 34:*1445–6

Parsons, P. A., Stanley, S. M., and Spence, G. E. 1979. Environmental ethanol at low concentrations: Longevity and development in the sibling species *Drosophila melanogaster* and *D. simulans*. *Aust. J. Zool. 27:*747–54.

Pipkin, S. B., Franklin-Springer, E., Law, S., and Lubega, S. 1976. New studies of the alcohol dehydrogenase cline in *D. melanogaster* from Mexico. *J. Hered. 67:*258–66.

Quinn, W. G., Harris, W. A., and Benzer, S. 1974. Conditioned behavior in *Drosophila melangaster*. *Proc. Natl. Acad. Sci. U.S.A. 71:*708–12.

Richardson, R. H. 1974. Effects of dispersal, habitat selection and competition on a speciation pattern of *Drosophila* endemic to Hawaii. In *Genetic Mechanisms of Speciation in Insects* (ed. White, M. J. D.), pp.140–64. Sydney: Australia and New Zealand Book Co.

Rosenzweig, M. L. 1978. Competitive speciation. *Biol. J. Linn. Soc. Lond. 10:*275–87.

Spieth, H. T., and Hsu, T. C. 1950. The influence of light on the mating behavior of seven species of the *Drosophila melanogaster* species group. *Evolution 4:*316–25.

Stalker, H. D. 1976. Chromosome studies in wild populations of *D. melanogaster*. *Genetics 82:*323–47.

Stalker, H. D., and Carson, H. L. 1947. Morphological variation in natural populations of *Drosophila robusta* Sturtevant. *Evolution 1:*237–48.

– 1948. An altitudinal transect of *Drosophila robusta* Sturtevant. *Evolution 2:*295–305.

– 1949. Seasonal variation in the morphology of *Drosophila robusta* Sturtevant. *Evolution 3:* 330–43.

Sturtevant, A. H. 1919. A new species closely resembling *Drosophila melanogaster*. *Psyche 26:*153–5.

Taylor, C. E., and Powell, J. R. 1976. Microgeographic differentiation of chromosomal and enzyme polymorphism in *Drosophila persimilis*. *Genetics 85:*681–95.

– 1978. Habitat choice in natural populations of *Drosophila*. *Oecologia 37:*69–83.

Thoday, J. M. 1964. Genetics and the integration of reproductive systems. In *Insect Reproduction* (ed. Highnam, K. C.), pp.108–19. Royal Ent. Soc. Symp. 2.

Thoday, J. M., and Gibson, J. B. 1962. Isolation by disruptive selection. *Nature (Lond.) 193:*1164–6.

Thorpe, W. H. 1945. The evolutionary significance of habitat selection. *J. Anim. Ecol. 14:*67–70.

Throckmorton, L. J. 1975. The phylogeny, ecology, and geography of *Drosophila*. In *Handbook of Genetics* vol.3, (ed. King, R. C.) pp.421–69. New York: Plenum Press.

Vigue, C. L., and Johnson, F. M. 1973. Isozyme variability in species of the genus *Drosophila* VI. Frequency–property–environment relationships of allelic dehydrogenases in *D. melanogaster*. *Biochem. Genet. 9:*213–27.

Waddington, C. H., Woolf, B., and Perry, M. M. 1954. Environment selection by *Drosophila* mutants. *Evolution 8:*89–96.

Watanabe, T. K., and Kawanishi, M. 1976. Colonization of *Drosophila simulans* in Japan. *Proc. Jpn. Acad. 52:*191–4.

White, M. J. D. 1973. *Animal Cytology and Evolution.* Cambridge: Cambridge University Press.
– 1978. *Modes of Speciation.* San Francisco: W. H. Freeman.
– Modes of speciation in orthopteroid insects. *Boll. Zool.* (in press).
Wright, S. 1969. *Evolution and the Genetics of Populations,* vol.2, *The Theory of Gene Frequencies.* Chicago: University of Chicago Press.

12 Problems in speciation of *Chironomus oppositus* Walker in southeastern Australia

JON MARTIN AND B.T.O.LEE

In a recent review on animal speciation, Bush (1975) commented upon the significant advances being made as a result of the utilization of independent molecular, organismal, and population studies. However, he pointed out the lack of information on many aspects of the biological properties of most animals and emphasized the need for a multidisciplinary approach to the understanding of the processes involved in speciation. This situation was commented upon also by White (1978), who suggested that a fully documented history of a single case of speciation might include the following data on both forms: (1) a precise map of present distributions; (2) geological and climatological evidence suggesting past distributions; (3) detailed morphological description; (4) detailed information on geographical variation, including multivariate biometrical studies; (5) ecological data on habitats, niches, seasonal cycles, and so on; (6) extensive information on biochemical polymorphisms and the extent of allelic differences between the two forms; (7) detailed descriptions of the karyotypes, based on the most modern techniques (including details of any chromosomal polymorphism in either form, DNA values, information on the types of satellite DNA present, and the amount and distribution of each type of satellite); (8) results of experimental hybridization, and information as to hybridization if any, in nature (this would include a cytological study of meiosis of any hybrids obtained); (9) information regarding any ethological isolating mechanism(s).

Speciation studies that employ a single or few techniques, or those that study a single type of isolating mechanism, particularly ethological ones, generally reveal only one aspect of the speciation process and can give a distorted view. In fact, one can say that these types of studies contribute rather little to the overall concept of speciation theory, and that theories on speciation should only be made in the light of multiple approaches to the collection of data.

The occurrence in *Chironomus oppositus* of at least three and possibly five distinct forms allows the utilization of multiple approaches to data collection within and between the different forms in an attempt to study potential speciation in them.

Chironomus oppositus Walker is one of the most common chironomids of southeastern Australia, particularly during spring and early summer. It has been shown in previous studies to be polymorphic for inversions in all chromosome arms (Martin, 1969; Cragg, 1973). These inversions show seasonal and geographic variation as well as marked nonrandom associations (Martin, 1966, 1967). Some field collections consistently show distributions of inversions markedly different from those found in what is now referred to as form A, to the extent that one can recognize and define two additional forms referred to as form B and form C (Martin et al., 1978). None of the common inversion sequences that are involved in the differentiation of forms B and C is unique to Tasmania, although sequence E_1 is of limited occurrence on mainland Australia (Martin, 1969). The nonrandom association of sequences that occurs in form A in most Tasmanian populations as well as on the mainland is often the reverse of those sequences that occur in populations in which both forms B and C are present, such as those at Bellerive, a major collecting site in Tasmania. Obviously, then, the karyotypic divergence recognized in these populations must be maintained by some restrictions on interbreeding between the groups and also between them and form A, with which they appear to coexist in some localities. Although there is some indication of introgression, no natural F_1 hybrids have been observed, but these hybrids have been produced in the laboratory.

The situation has the potential to allow an investigation of the relative contributions of cytological, biochemical, and morphological characteristics in the speciation process and in sexual differentiation. It also allows an investigation of the relationships of the so far defined forms, and an investigation of their ecological requirements, and allows us to ask the question of whether the forms, particularly B and C, are randomly dispersed within the habitats. We want to attempt to answer the following questions:

1. What is the extent of cytological, biochemical, and biometrical variability in the various populations of *Ch. oppositus?*
2. What is the extent of differentiation between populations, and what level of specific divergence have they reached?
3. Are cytological, biochemical, and biometrical changes equally

involved in such differentiation, or are they perhaps even alternatives?

In our preliminary report on this situation (Martin et al., 1978), we noted that forms B and C appeared only to occur together in the same habitats, sometimes in association with form A. The truth of this assertion, along with further details of the geographic distribution and the frequencies of various inversion sequences throughout Tasmania as well as King and Flinders islands, will be discussed in this review.

The critical point is that these are not clearly allopatric groupings, and so the classical speciation pattern envisaged by Mayr (1970) seems unlikely to be involved. Of the various alternative modes of speciation proposed by Bush (1975), White (1978), or Scudder (1979), this system seems closest to that proposed for host race formation in phytophagous insects (i.e., some form of sympatric speciation; Bush, 1975). However, Bush envisaged this mode as being confined to phytophagous and zoophagous parasites and parasitoids. Therefore the groups for which such a mode of speciation is possible need to be broadened, or else we are dealing with a different and therefore unique mode of speciation. It thus becomes important to discuss the sampling at sympatric localities and to test for the presence of natural F_1 hybrids. This report will also discuss the results of F_1 crosses and backcrosses in the laboratory, hybridization experiments, ecological data, the possible separation of breeding seasons, the presence of swarms, and the ecology of the larvae in an attempt to explain how the differences between the forms are maintained.

Distribution

During earlier studies on *Ch. oppositus* (Martin et al., 1978) it was believed that only form A occurred in mainland Australia, whereas forms B and C were restricted to Tasmania. However, subsequent studies have shown that all three forms occur at Anglesea, southwest of Melbourne (Kuvangkadilok, unpub. data) and that specimens of form B were found at Werribee and the Melbourne Botanic Gardens, all these localities being in Victoria.

Even though extensive collections have now been carried out in Tasmania, the situation there is still not clear. Form A appears to be distributed all across the island although it is rare in some areas such as Bellerive (Tables 12.2 and 12.3). Form B may occur in a band across the north of the island including both King and Flinders islands, reaching the southeast through either the midlands or the east coast. Further sampling of the northern midlands and the northeast is required to clarify the situation.

Form C appears to be restricted to coastal areas, with Queenstown the farthest inland locality at about 35 km from Macquarie Harbour. At least some populations of form C occur on King Island (Table 12.3), but none has yet been identified on Flinders Island.

There is thus no obvious geographic separation of the forms. In fact, two forms can commonly be found together in the same area or even the same pool.

Microdistribution at Bellerive and Queenstown

In order to determine whether the two forms of *Ch. oppositus* are evenly distributed along the Bellerive drain, a 50-meter (approximately) section of it was subdivided into seven regions as shown in Figure 12.1. Samples from Bellerive taken in November 1976 and subsequently, were taken from within these specific regions.

The results for the November 1976 sample clearly indicate that form B is more common in the generally deeper and more polluted left-hand end of the drain (Table 12.1). No pattern is obvious in the next two samples, February 1977 and 1978, but this result is complicated by the low frequency of form B and the lack of water in most areas of the right-hand side of the drain, which forces the larval population into the left-hand side. Further samples during the spring period are required to confirm that the differential distribution of the two forms in the drain is a consistent effect.

There is also evidence from the frequencies of some inversion sequences of further microdistribution within the forms. In form B the sequences of arms C and A show significant heterogeneity of frequencies along the length of the drain. The effect appears to be mostly evident in LHS1 and LHS7, where the frequencies of C_1 and A_2 are higher than in other areas of the drain. Thus although the average frequency of C_1 is 27%, at LHS1 it is

Figure 12.1. Schematic diagram of the drain at Bellerive, Tasmania, to show the relative positions of the sampling sites.

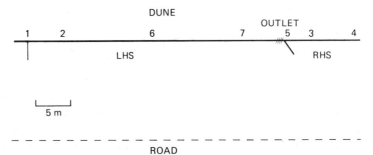

43% and at LHS7, 33%. Similarly the average frequency of A_2 is 58%, but it is 66% at LHS1 and reaches 74% at LHS7. These effects are generally consistent over the three samples analyzed although the higher frequency of C_1 at LHS7 was not evident in the February 1978 collection.

Only arm B shows any heterogeneity in form C. In this case the frequency of B_1 is higher at RHS3 than at other sites along the drain. Because this sequence has been increasing in frequency during the study period (this is discussed under "Changes in inversion frequency"), the frequencies at RHS3 cannot be expressed as compared to an average frequency except in relation to a particular sample (e.g., in November 1976 the average frequency of B_1 was 8%, but at RHS3 it was 19%).

This aspect requires further study to determine what environmental effects these differences reflect.

The Queenstown locality can also be easily subdivided – at present into two main areas, but further subdivisions could be used in the future. This site is at the junction of the Lyell and Zeehan highways. The upper area, near the roads, has been cleared and has a number of shallow rain-filled pools among grass and moss. From this area a marshy drainage channel runs through *Typha* sp. to Pearl creek. Along the edge of this area are a number of holes, which are generally water-filled, some nearly 30 cm deep, and which may be quite shaded from the sun. In February 1977, the samples from these two areas were kept separate and showed quite different frequencies of forms A and C. The upper area was predominantly form A, whereas the more shaded pools were mostly form C (Table 12.3).

Seasonality

Martin et al. (1978) noted that the early samples from Bellerive suggested there may be seasonal differences in the frequency of the two forms. However, with only two samples it was not possible to exclude the

Table 12.1. *Frequency (%) of* Ch. oppositus *form C along the length of the drain at Bellerive, Tasmania (locations in linear order)*

Date	LHS 1	LHS 2	LHS 6	LHS 7	RHS 5	RHS 3	RHS 4
November 1976	17.4	20.2	14.3	37.7	70.5	26.6	58.7
February 1977	96.0		88.9	96.6	—	60.4	—
February 1978	73.0		—	82.6	81.6	—	—

possibility of an actual replacement of one form by another. With additional samples (Table 12.2) it is now clear that there are seasonal differences in the relative abundance of the two forms such that form B is more common in the spring sample, whereas form C is more common in late summer.

In samples subsequent to the original one, some specimens of form A have also been found at Bellerive but at very low frequency (Table 12.2). The numbers found have been too low to indicate any seasonality of occurrence.

At least three other *Chironomus* species also occur at Bellerive, and they must also be considered for possible interactions with, or influence on, the populations of *Ch. oppositus*. *Ch. tepperi* Skuse was recorded only once, from an egg mass collected in November 1976. This is in fact the only record of *Ch. tepperi* in Tasmania. *Ch. alternans* b* also occurs at low frequency and apparently only in February (Table 12.2). The third species is a previously unknown species for which only the larvae are known. The larvae appear to be similar to those of *Ch. oppositus*. The banding sequences of the polytene chromosomes would also indicate a close affinity to that group of species because it appears to have sequences B_2, F_1, D_4, E_1, and A_4. The sequence in arm C is not immediately recognizable, and it is uncertain whether it corresponds to a *Ch. oppositus* sequence. Arm G is quite different from that of *Ch. oppositus* and renders the species immediately recognizable. The species is presently referred to as *Ch. alternans* group species c (*Ch. alternans* c). This species appears to occur mainly in

Table 12.2. *Frequency (%) of species at Bellerive, Tasmania*

| | Ch. oppositus | | | | |
	Form A	Form B	Form C	Ch. alternans b	Ch. alternans c
October 1971	—	71.8	28.2	—	—
February 1975[a]	0.9	11.8	82.7	0.3	4.3
November 1976	0.9	63.4	35.2	—	0.5[b]
February 1977	0.5	15.5	83.4	—	0.5
February 1978	2.4	1.4	78.6	0.7	16.9

[a]Additional data to that in Martin et al. (1978).
[b]Includes a hybrid individual.

* This species will ultimately be described as *Ch. pseudoppositus* (see Martin, 1966).

February samples, and it may be quite common (Table 12.2). Slides made by the Botany Department, University of Tasmania indicate that this species has occurred at Bellerive for at least 15 years.

Samples taken at different seasons have also been analyzed from the Queenstown locality and an area near Smithton. The Smithton samples were taken after a 5-year gap and are not from the same pools, so that any differences may result from causes other than seasonal changes. The major difference is an increase in the frequency of form A at the expense of form C in the February 1977 sample, compared to that taken in October 1971. The frequency of form B was only minimally lower in the February sample (Table 12.3). The Queenstown samples are from the upper levels of this

Table 12.3. *Frequencies (%) of species at various Tasmanian and King Island localities*

	Ch. oppositus			
	Form A	Form B	Form C	*Ch. alternans* b
Three Tree Lagoon, K.I.				
1-iii-77	0.0	100.0	0.0	0.0
Porky Lagoon, K.I.				
2-iii-77	3.9	90.2	5.9	0.0
Sea Elephant R., K.I.				
2-iii-77	0.0	0.0	100.0	0.0
Pearshape Lagoon, K.I.				
3-iii-77	0.0	100.0	0.0	0.0
35 km Smithton, Tas.				
26-x-71	5.8	10.2	84.0	
8-ii-77	33.7	9.6	56.6	0.0
Gordon River area, Tas.				
12-ii-79	100.0 (Lake)	0.0	0.0	0.0
Queenstown, Tas.				
13-xi-76 (upper level)	81.6	13.8	4.6	0.0
7-ii-77 (upper level)	80.1	0.5	18.3	1.0
7-ii-77 (channel pools)	21.1	0.8	78.1	0.0
Lake Crescent, Tas.				
15-x-72	100.0 (Lake)	0.0	0.0	0.0
Campbell Town, Tas.				
1-xi-71	80.4	19.6	0.0	0.0
Lake Dulverton, Tas.				
3-ii-77	0.0	100.0	0.0	0.0
Ida Bay, Tas.				
10-ii-79	77.6	0.0	22.4	0.0

locality taken in November 1976 and February 1977, and clearly show that form B is less frequent in February, whereas form C is more frequent; the frequency of form A is virtually unchanged in the two samples (Table 12.3).

Changes in inversion frequency

Changes in the frequency of inversions can be of two types: changes with time or season at one locality, and geographic variations. Both types are being investigated in the present study, at least for forms B and C.

A relatively simple way of quantifying the amount of frequency variation within forms and at the same time allowing a comparison with the variations between forms, is to calculate Nei's normalized identity measure (I) (Nei, 1972). This measure, called the genetic similarity, has previously been applied to inversion polymorphisms by Carson et al. (1975). In the present instance each chromosome arm, excluding arm G, for which many of the relevant frequencies are missing or inaccurate, was considered as a single unit, so that I is summed over six "loci." Although this is adequate for the present purposes of making inter- and intraform comparisons, it must be realized that this particular approach will tend to maximize the differences because it makes no allowance for those sections of the karyotype not involved in inversions and hence shows no difference between forms or populations.

Form A. As can be seen from the I values (Table 12.4) there is variation in inversion frequencies between the various populations. This variation can involve arms B, D, C, and A. Some of it is due to ecological variants of this form, notably the Lake type, which is completely homozygous for sequences B_1, F_1, D_1, C_2, E_2, A_2, and G_1. This type occurs only in large permanent water bodies. The larvae appear to be somewhat larger than other *Ch. oppositus,* but as yet little is known of this type.

In general, populations of form A in Tasmania have a high frequency of B_1, as in the coastal regions of the mainland (Martin, 1969). However, the population of Queenstown showed 61% B_2, which is an unexpectedly high value and one of a number of unusual features found in this population. In arm D, D_1 is the most common sequence, ranging from 54% to 100%. Sequence D_2 may be quite common, 30%–46%, as at Campbell Town or Smithton, or almost absent as at Queenstown, where D_4 has a frequency of 45%. In arm C, again Queenstown is unusual in having 64% C_2 and 36% C_1, whereas the other form A populations (other than Lake) have higher

Table 12.4. *Genetic similarity (I) for various populations of* Chironomus oppositus

	W'dyte V A	S'fras V A	S'ton T A	Q'town T A	L. Crsct T A (Lake)	C'town T A	3 Tree KI B	S'ton T B	C'town T B	Bell. T B	Sea Ele. KI C	S'ton T C	Q'town T C	Bell. T C
Warrandyte, V A	(0.730)						0.088	0.135	0.638	0.525				
Sassafras, V A	0.853	(0.888)					0.107	0.123	0.449	0.364				
Smithton, T A	0.938	0.878	(0.727)				0.063	0.115	0.602	0.511				
Queenstown, T A	0.844	0.844	0.852	(0.652)			0.268	0.308	0.560	0.503				
L. Crescent, T A	0.884	0.622	0.822	0.790	(1.000)		0.030	0.094	0.607	0.467				
Campbell Tn, T A	0.982	0.770	0.913	0.787	0.911	(0.853)	0.064	0.127	0.666	0.550				
3 Tree Lagoon, KI B							(0.914)				0.067	0.063	0.170	0.194
Smithton, T B							0.974	(0.784)			0.119	0.116	0.190	0.207
Campbell Tn, T B							0.602	0.598	(0.761)		0.518	0.511	0.404	0.344
Bellerive, T B							0.633	0.685	0.971	(0.816)	0.425	0.425	0.357	0.268
Sea Elephant R., KI C	0.817	0.675	0.898	0.853	0.873	0.802					(0.788)			
Smithton, T C	0.800	0.653	0.894	0.833	0.850	0.787					0.998	(0.801)		
Queenstown, T C	0.742	0.685	0.821	0.831	0.653	0.697					0.886	0.895	(0.753)	
Bellerive, T C	0.687	0.639	0.723	0.879	0.717	0.632					0.881	0.872	0.925	(0.810)

The figures in brackets on the diagonal give the average homozygosity of that population. A, B, and C are forms of *Ch. oppositus.* V = Victoria; T = Tasmania; KI = King Island.

frequencies of C_1. On King Island a new sequence, tentatively called C_7, appears common, but its breakpoints and frequencies have not yet been clarified because form A appears to be rare on the island.

Sequence A_5 is generally the most common sequence of arm A with frequencies of 60% and higher, followed by A_2 as the next most common sequence. Sequences A_1, A_3, and A_4 occur at low frequencies in various populations. The only exception to this situation other than the Lake types is the sample from Campbell Town, which is 94% A_2 and the rest A_1.

Form C should be considered next because it is in many ways similar to form A, as indicated by the I values (Table 12.4). Indeed, in many populations where they occur together, it is very difficult to separate individuals as being definitely one form or the other. The most consistent difference between the forms is the occurrence of sex linkage in arm A of form C. In a number of cases some specimens must be sorted into the "most probable" form. At Queenstown, for example, this involved about 12% of the sample.

Some variability occurs in the inversion frequencies of arms B, D, and C, but the greatest variation occurs in arm B. In this arm there is a marked geographic change in frequency of B_1 from about 5% in the south of the island to about 80% in the north or on King Island, as well as an increase in frequency of B_1 with time at Bellerive – from 5% in October 1971 to 13% in February 1978.

Sequence D_1 has similarly tended to increase in frequency at Bellerive, from 40% in October 1971 to 65% in February 1978. At all other localities D_1 is less frequent than D_2 with no obvious geographic pattern.

Sequence C_1 is generally very rare in form C with the exception of the Queenstown population. This, along with the peculiarities of some inversion frequencies of form A at this locality, could indicate some degree of misclassification. However, the frequency differences are too great to be attributed to the relatively small number of doubtful individuals and so seem more likely to reflect either introgression between the forms or even real ecological differences in this environment.

The genetic similarity (I) values of form B (Table 12.4) also indicate considerable differences in inversion frequency between populations. Here the major effects are seen in arms B, F, C, and A, which all show north–south changes in frequency. In arm B the frequency change is the reverse of that seen in form C, with B_1 dropping from 96%–100% in the south to 21% at Smithton and less than 10% on King Island. In arm F there is a dramatic increase in F_2 from only about 5% at Bellerive to 82% at Smithton and 90% on King Island. For arm C there is a replacement of C_1

in the south by C_4 in the north, with little change in frequency of C_3, the most frequent sequence. In arm A the A_1 sequence increases from about 40% in the south to over 90% in the north and on King Island.

At Campbell Town there is an unusually high frequency of D_1 (36%), which is not seen in other populations. At Bellerive the frequency of this sequence shows heterogeneity, with frequencies from 4% to 25%, but no obvious pattern of temporal or seasonal change. The upper value may be due to sampling error, but the variation up to 14%, excluding this sample, is still significant.

One problem that arises with respect to form B is the status of a group of larvae carrying either the standard sequences in all arms or with A_2 instead of A_1. Such larvae have been noted in a number of populations in Tasmania and on the mainland. These larvae could be considered to be form B because they have E_1, and the other sequences are known to be present as polymorphisms in various populations of form B. However, the situation at Queenstown where some typical form B larvae (i.e., homozygous D_4/D_4 and C_3/C_3) were found as well as this group with the standard sequences, but with no D_1/D_4 or C_1/C_3 heterozygotes, suggests that some other explanation may be necessary because it implies that the two groups do not hybridize.

Hybridization

Natural hybrids. It is difficult to determine the extent to which hybridization between the various forms occurs in nature because the results of some hybridization crosses would be difficult to distinguish, even in the F_1. Hybrids would normally be recognized by distinctive combinations of inversion heterozygosity, which occur either in individual larvae in a population sample or among the offspring of a single field mating. The results of field matings may be assessed by collecting inseminated females and assessing the resulting egg masses, or by collecting any egg masses available in the pools under study. The former method is to be preferred because wild-collected egg masses are liable to harbor unrelated larvae, which feed on the gelatinous coating of the mass. Hybrid egg masses should be more easily recognized than an individual hybrid larva, but difficulty may still be encountered because of the high incidence of heterozygosity in many populations, and where form A is involved, the great heterogeneity of sequences present.

The easiest F_1 hybrid to recognize would be that of form B with form C. Such an individual would generally be heterozygous for C_2/C_3, D_1 or D_2/D_4, and E_1/E_2, but other combinations such as D_1/D_1, D_1/D_2, C_1/C_1, or C_1/C_2, which could be scored as normal form A or form C individuals, are possible. However, no individuals or egg masses suggesting F_1 hybrids between form B and form C have been found at Bellerive or any other Tasmanian population.

The occurrence of hybridization involving form A is probably impossible to detect using single individuals and very difficult even with egg masses, because of the variety of sequences that can occur in this form. A hybrid would probably be considered as a normal form A individual.

Although no F_1 hybrids have been recognized, there are some results suggesting that hybridization does occur or has occurred at some time in the past. It has already been noted that the unusual inversion frequencies at Queenstown could be the result of introgression between the forms. At Bellerive there is also some evidence of possible hybridization. In particular the presence of low frequencies of C_2, A_5, and E_2 in form B (Martin et al., 1978) could be the result of introgression from either form A or form C. Similarly, the low frequencies of D_4 in form C could result from introgression following hybridization with form A or form B. It is hoped that electrophoretic studies will help to clarify the extent to which natural hybridization and introgression occur.

There is also the question of whether any hybridization occurs between *Ch. oppositus* and any of the other species of *Chironomus* in Tasmania. Martin (1966) has previously indicated that there is no evidence of hybridization between *Ch. oppositus* and any other species on the mainland. For this reason it is not surprising to find no evidence of hybridization between those species in Tasmania. However, for *Ch. alternans* c, which has not been found on the mainland, a single F_1 hybrid between this species and *Ch. oppositus* was found in the November 1976 sample from Bellerive. The chromosomes prepared from this specimen were poor, but it was possible to determine that the sequences contributed by the *Ch. oppositus* parent were B_2, F_1, D_3, E_2, A_2, and G_1. Arm C was not clear but was probably C_2. From the sequences it would seem that the parent was form A. The actual extent of hybridization and introgression between *Ch. alternans* c and the various forms of *Ch. oppositus* must await a better knowledge of the karyotype of *Ch. alternans* c. One can hypothesize that the limited distribution of *Ch. alternans* c may be due to lack of isolating mechanisms between this species and *Ch. oppositus* A.

Laboratory hybrids. It is possible to breed at least some forms of *Ch. oppositus* in the laboratory using mating cages that are approximately 30-cm cubes, in a constant temperature room at 20°C with a 16-hour light cycle including 1 hour each of artificial dawn and dusk. Form C is the most readily bred, probably because it is smaller (Martin et al., 1978), but even so approximately 20 males are required before matings are achieved (as evidenced by the laying of fertile egg masses). Form A from Tasmania also breeds quite readily in the laboratory, but form B only rarely. Victorian populations are variable, some breeding readily but others breeding rarely or not at all. This means that at least some hybridization crosses are possible in the laboratory. Those that have been attempted are shown in Table 12.5. Crosses involving form C ♂ × form B ♀ or form A (Tas) ♂ × form C ♀ are readily obtained and show no reduction in fertility or egg hatchability. The reciprocal of the latter cross, form C ♂ × form A (Tas) ♀, is classed as uncertain because although no fertile egg masses were obtained, form C self-crosses performed at the same time were also infertile, leading to the loss of the form C stock.

Crosses of form A (Tas) ♂ and form A (Vic) ♀ did not produce any fertile egg masses. However, at that time form A (Vic) self-crosses were also unsuccessful. Later Victorian self-crosses have produced some fertile egg masses. Consequently, further crosses are required to determine whether a difference in laboratory mating behavior exists between the Tasmanian and mainland populations of form A.

Sex determination

Examination of natural populations has only indicated the location of the sex-determining loci in form C, on arm A. All inversions in form

Table 12.5. *Laboratory crosses with* Chironomus oppositus

	Female parent			
	Form A (Vic)	Form A (Tas)	Form B	Form C
Male parent:				
Form A (Vic)	+ + to − [a]	n.t.	n.t.	n.t.
Form A (Tas)	−	+ +	n.t.	+ +
Form B	−	n.t.	+	−
Form C	n.t.	?	+ +	+ +

[a]Symbols: −, no fertile laboratory crosses; ?, results inconclusive; +, few fertile crosses even if >20 males; + +, fertile crosses readily obtained if >20 males; n.t., not tested.

A and form B appear randomly distributed with respect to sex. However, as Rosin and Fischer (1972) have demonstrated, examination of segregation patterns in larvae from single egg masses can sometimes reveal sex linkage in inversions that are randomly distributed in population samples. Attention has therefore been given to analysis of egg mass data showing segregation of inversions. Egg masses of form C consistently show the expected pattern of sex linkage of arm A sequences, with all females homozygous and all males heterozygous.

A number of egg masses of form B also showed segregation of sequences in arm A (A_1 and A_2). None of them, however, gave any clear indication of sex linkage. On the other hand, certain egg masses that were segregating for arm G (G_1 and G_2) had most males heterozygous G_1/G_2 and most females homozygous G_1/G_1 (Table 12.6). Other egg masses segregating for the same sequences showed random distribution between the sexes. This finding, however, is still consistent with the assumption that the sex-determining loci are on arm G in this form. All such egg masses are the result of a $G_1/G_1 \times G_1/G_2$ cross, although we do not know which parent is which. If the female is heterozygous, then G_1 or G_2 chromosomes should be equally distributed to offspring regardless of sex. If, on the other hand, the male parent is the heterozygote, one sequence, the X, will go only to female offspring, whereas the other sequence, the Y, will go only to male offspring except where recombination has occurred between the inversion and the sex-determining locus. In egg masses 9 and 10 (Table 12.6) it would seem that G_1 is on the X chromosome, whereas G_2 is on the Y chromosome, with about 5.5% recombination occurring between the inversion and the sex-determining loci.

Table 12.6. *Segregation of sequences in arm G in egg mass offspring of* Ch. oppositus *form B from Bellerive, Tasmania: parental cross* $G_1/G_1 \times G_1/G_2$

Egg mass no.	Sex	G_1/G_1	G_1/G_2	Total
4	Female	6	5	11
	Male	8	6	14
9	Female	77	1	78
	Male	—	51	51
10	Female	92	7	99
	Male	8	55	63
15	Female	7	6	13
	Male	5	7	12

The egg masses of form A so far examined have shown segregation only for arms B, C, or D, with no apparent sex linkage. It is therefore not possible at this stage to decide where the sex-determining loci of this form may be located.

Electrophoretic data

The biochemical and electrophoretic analyses of various protein and enzyme systems has been very much hindered by the death of our colleague Dr. R. Ananthakrishnan. We are therefore not in a position to present detailed results in this paper. Our results to date indicate definite differences in the frequency of electromorphs for alcohol dehydrogenase and an esterase locus between forms B and C. Study of egg mass data and laboratory hybrids suggests these variations are of genetic origin. It is not yet completely clear whether we are simply dealing with differences in allele frequencies between the forms, or whether there are actual differences in the alleles present. Other enzyme systems, such as isocitrate dehydrogenase, appear to be monomorphic. It remains to be proved just what proportion of systems that can be investigated will show differences between the various forms.

Discussion

The situation with respect to *Ch. oppositus,* particularly in Tasmania, is more complex than originally believed (Martin et al., 1978). Rather than a situation involving three closely related forms that interact in only one or two areas, it now seems that interactions between the forms occur commonly through large sections of Tasmania and the Bass Strait islands, and at least in limited areas of the mainland. The interactions may be between forms A and B, A and C, B and C, or even all three forms. It is also possible that an additional two forms are present at some localities.

The data now available indicate that the distribution of the forms is not allopatric but broadly sympatric. Within this broad sympatry the larvae may live in somewhat different niches, either in different parts of the same pool or in different pools. The adults, however, will emerge to mate, feed, and rest in essentially the same habitat, providing ample opportunity for hybridization unless behavioral or cyclical (seasonal) differences serve to keep their mating swarms separate. Evidence of seasonal differences, at least between forms B and C, has been presented, as has circumstantial evidence of behavioral differences with respect to mating.

At least at Bellerive, it seems that investigation will be required of hybridization between the forms of *Ch. oppositus* and the taxon called here *Ch. alternans* c. Although the adult of this taxon is not yet known, it is thought that it will probably resemble *"Ch. alternans,"* that is, be similar to *Ch. oppositus* (Martin, 1966, 1969) because the larva is morphologically indistinguishable from larvae of *Ch. oppositus*. However, the modified arm G indicates that it is cytologically more differentiated than the forms of *Ch. oppositus*.

The Lake type of *Ch. oppositus* does not seem to be involved in interactions with the other forms, perhaps partly because of the different habitat preference of the larvae, but, presumably, behavioral or seasonal factors must also be involved. The adults could rest and mate in the same area as adults of other forms in cases where shallow pools are adjacent to a lake.

It may be, then, that we are looking at a form of sympatric speciation, similar in some respects to that reported for some Hawaiian *Drosophila* (Richardson, 1974) or for host races of phytophagous and zoophagous parasites and parasitoids (Bush, 1974, 1975). The major questions, as White (1978) has pointed out, are: (1) What is the status of these forms? (2) Was the speciation process initiated under conditions of sympatry as the forms are found at present? We cannot answer either of these questions as yet.

Our studies have, however, indicated that form B is differentiated to a greater degree than forms A and C. This is shown by the genetic similarity values (Table 12.4) where comparisons involving form B give lower values than the comparisons of forms A and C. Form B appears to have distinctive inversion polymorphisms such as F_2 and C_4, and certainly has a different sex-determining mechanism from that of form C. These features would suggest that it has existed for some considerable time. Whether or not it should be considered a distinct species is not certain. Hybrids between form B and form C can be readily obtained in the laboratory and show no indications of infertility or inviability, even though no natural hybrids have been found. The presence of the different sex-determining mechanisms is not a definite indication of distinct species. A reorganization of the sex-determining mechanism is a common feature of speciation in Simuliidae (Rothfels, 1979; White, 1978) and Chironomidae (Martin, unpub., quoted in White, 1978). However, the work of Beermann (1955) and Thompson (1971) with *Chironomus tentans* indicates that different sex-determining systems can exist in the same species, sometimes even in the same population.

The data indicate that forms A and C do not differ greatly except that there is sex linkage of inversion sequences of arm A in form C. As indicated above, this is no indication of specific status. Natural hybrids between forms A and C would be difficult to recognize, whereas laboratory hybrids can be readily obtained – a further problem in defining the status of these forms.

The broad distribution and multiplicity of sequences in form A would suggest that this is the ancestral taxon. However, it is possible that it has not always been in Tasmania. Two lines of evidence suggest this: First, the Tasmanian populations do not contain all the inversion sequences present in mainland populations. This could indicate that only a restricted sample of the mainland polymorphisms was carried across Bass Strait by the founders of the Tasmanian populations, or that new sequences were established on the mainland only after the two areas were separated by the subsidence of Bass Strait. The latter possibility seems unlikely because one of the sequences is A_4, which occurs in the related species, *Ch. australis* and *Ch. tepperi* (Martin, 1971, 1974), and must therefore be of considerable age. It is also in *Ch. alternans* c, which appears to occur only in Tasmania. A further possibility is that some sequences have been unable to establish themselves in Tasmanian ecological conditions. This cannot be ruled out even though A_4 occurs in Tasmania in *Ch. australis* and *Ch. alternans* c, in which it is homozygous. In *Ch. oppositus* the persistence of the A_4 sequence will depend on its adaptive values compared to other arm A sequences.

The second line of evidence comes from the results of the laboratory crosses and the theory of asymmetrical isolation (Kaneshiro, 1976), which states that females of a derived isolated species will be less discriminating than the ancestral females in accepting males. This theory was elaborated for *Drosophila* and has yet to be demonstrated for other groups even though the basis suggested by Kaneshiro would seem to be of general application. It has yet to be shown that Tasmanian form A females will mate with males of other forms, but they do mate with their own males. Females of Victorian form A, however, apparently will not mate with Tasmanian males. This behavior would not seem to be due to reluctance on the part of the Tasmanian form A males because they will mate with females of form C. Therefore it could be suggested that the Victorian form A females are more discriminatory than those from Tasmania and consequently the older form.

If form A only reached Tasmania recently, there is a possibility that form B was already present, which would mean that these two forms had not differentiated under conditions of sympatry. However, form C, being obviously derived from form A, may have arisen only since form A reached Tasmania and may represent a case of sympatric speciation in which the larvae may have slightly different ecological preferences. This is indicated by the microgeographic variation seen at Bellerive and Queenstown. The different microdistributions could be the result of differences in female oviposition preferences, differential survival of the larvae of the various forms in different pools, or a combination of both. The possibility of preferential oviposition can be tested by collecting egg masses from each area. Although the larvae are spatially separated, this does not apply to the adults, for which the possibility of interbreeding exists unless there are seasonal or behavioral barriers.

Although seasonal differences have been found, they are not complete enough to prevent considerable hybridization; therefore, behavioral differences must also be involved. The size differences of the larvae and the difference in laboratory mating behavior suggest that form B may form mating swarms at a different level from at least form C. Other possibilities also exist, namely, the use of different swarm markers or swarming at different times of the day (Downes, 1969; Oliver, 1971). Little attempt has been made to study swarming at Bellerive because of the interferences caused by human settlement, in particular the presence of street lights, which are likely to alter swarming heights and locations.

So far there is no evidence that hybridization is still occurring, except between *Ch. oppositus* and *Ch. alternans* c. There is some evidence of introgression from the occurrence of unusual inversion sequences in some populations such as Bellerive and Queenstown. This aspect requires further study, particularly the investigation of genic rather than just chromosomal introgression.

The interesting situation revealed by our studies of *Ch. oppositus* leaves many questions still to be answered. Future work will aim to clarify the status of the various forms, including the Lake type and the populations with virtually all the standard sequences of the old terminology (Martin, 1969). The extension of electrophoretic data will be a major thrust, but other areas will also be investigated, such as obtaining suitable egg mass material to locate the sex-determining loci of form A, further laboratory hybridization of the forms, biometrical analyses of larvae and adults from various populations and seasons, and the application of multivariate analyses to determine relationships and distances as has been done on Ceratopogonidae by Atchley (1971; Hensleigh and Atchley, 1977).

The presence of parasitic nematodes in high frequency in some populations (e.g., Bellerive) provides a further characteristic on which to evaluate differentiation between the forms. Parasitic mermithids are generally adapted to the life cycle of their hosts (Hominick and Welch, 1971), thus restricting the species they can successfully use as hosts. If it can be shown that the various forms of *Ch. oppositus* are differentially subjected to attack, this will be further evidence of their distinct nature.

More sophisticated techniques of chromosomal analysis may also be applied. Lentzios et al. (in press) have evidence of C-banding differences between form A and form C. However, these results are based on small samples, and larger samples, including form B, from a number of localities are required to establish this point. DNA estimates of the whole complement should also be measured. The production of laboratory hybrids permits a comparison of the amount of DNA in single homologous bands as was done for the sibling species *Chironomus riparius* (= *thummi*) and *Ch. piger* by Keyl (1965).

However, this study shows the importance of recognizing that from the evolutionary point of view this organism has two phases in the life cycle, the aquatic immature stages and the nonaquatic (airborne) adult stage, during which different forms of selection may be operating. Organisms with life history cycles of this type may represent a further category in which sympatric speciation may be occurring.

Summary

Chironomus oppositus populations from Tasmania and southern Victoria have been shown to comprise at least three forms, A, B, and C, which differ in their inversion sequences, sex-determining mechanisms, and size. These forms are broadly sympatric in their distribution but show differences in microdistribution of larvae, seasonality, and mating behavior. The taxonomic status of these forms is not yet certain, nor is it certain that they arose under conditions of sympatry. No definite evidence of recent hybridization between the forms was found, although there is evidence of introgression. An F_1 hybrid was found between *Ch. oppositus* and a related taxon *Ch. alternans* c. In addition to these three well-documented forms there is evidence of at least two more forms.

It is a pleasure to dedicate this paper to Michael White on the occasion of his seventieth. birthday, for it was he who first suggested *Chironomus* to us as a study organism. We also dedicate it to our colleague R. Ananthakrishnan, who died suddenly on November 11, 1978. This work has been supported by grants from the Australian Research Grants Committee D67/16638, D71/17815, and D1-77/15599 as well as an Emergency Grant from the University of Melbourne. We are grateful to Professor W. D. Jackson and Dr. P. A. Tyler, Botany Department, University of Tasmania for telling us of the Bellerive and many other localities, as well as for assistance in many ways. Our thanks also to Chaliow Kuvangkadilok, who made and scored many of the chromosome preparations.

References

Atchley, W. R. 1971. A comparative study of the cause and significance of morphological variation in the adults and pupae of *Culicoides:* a factor analysis and multiple regression study. *Evolution 25:*563–83.

Beermann, W. 1955. Geschlechtsbestimmung und Evolution der genetischen Y-chromosomen bei *Chironomus. Biol. Zentralbl. 74:*525–44.

Bush, G. L. 1974. The mechanism of sympatric host race formation in the true fruit flies (Tephritidae). In *Genetic Mechanisms of Speciation in Insects* (White, M. J. D. ed.), pp 3–23. Sydney: Australia and New Zealand Book Co.

– 1975. Modes of animal speciation. *Annu. Rev. Ecol. System. 6:*339–64.

Carson, H. L., Johnson, W. E., Nair, P. S., and Sene, F. M. 1975. Allozymic and chromosomal similarity in two *Drosophila* species. *Proc. Natl. Acad. Sci. U.S.A. 72:*4521–5.

Cragg, E. 1973. Population dynamics of inversions in *Chironomus oppositus.* Unpublished honours report, University of Melbourne.

Downes, J. A. 1969. The swarming and mating flight of Diptera. *Annu. Rev. Entomol. 14:*271–98.

Hensleigh, D. A., and Atchley, W. R. 1977. Morphometric variability in natural and laboratory populations of *Culicoides variipennis* (Diptera: Ceratopogonidae). *J. Med. Entomol. 14:*379–86.

Hominick, W. M., and Welch, H. E. 1971. Synchronisation of life cycles of three mermithids (Nematoda) with their chironomid (Diptera) hosts and some observations on the pathology of the infections. *Can. J. Zool. 49:*975–82.

Kaneshiro, K. Y. 1976. Ethological isolation and phylogeny on the *planatibia* subgroup of Hawaiian *Drosophila. Evolution 30:*740–5.

Keyl, H.-G. 1965. A demonstrable local and geometric increase in the chromosomal DNA of *Chironomus. Experientia 21:*191.

Lentzios, G., Stocker, A. J., and Martin, J. C-banding and chromosome evolution in some related species of Australian Chironominae. *Genetica* (in press).

Martin, J. 1966. Population genetics of chironomids: cytogenetics and cytotaxonomy in the genus *Chironomus.* Ph.D. thesis, University of Melbourne.

– 1967. Non-random association of unlinked inversions in *Chironomus (Chironomus) oppositus. Can. J. Genet. Cytol. 9:*661.

– 1969. The salivary gland chromosomes of *Chironomus oppositus* Walker (Diptera: Nematocera). *Aust. J. Zool. 17:*473–86.

– 1971. A review of the genus *Chironomus* (Diptera, Chironomidae). IV. The karyosystematics of the *australis* group in Australia. *Chromosoma (Berl.) 35:*418–30.

– 1974. A review of the genus *Chironomus* (Diptera, Chironomidae) IX. The cytology of *Chironomus tepperi* Skus. *Chromosoma (Berl.) 45:*91–8.

Martin, J., Lee, B. T. O., and Connor, E. 1978. Apparent incipient speciation in the midge *Chironomus oppositus* Walker (Diptera: Chironomidae). *Aust. J. Zool. 26:*323–9.

Mayr, E. 1970. *Populations, Species and Evolution.* Cambridge, Massachusetts: Harvard University Press.

Nei, M. 1972. Genetic distance between populations. *Am. Nat. 106:*283–92.

Oliver, D. R. 1971. Life history of the Chironomidae. *Annu. Rev. Entomol. 16:*211–30.

Richardson, R. H. 1974. Effects of dispersal, habitat selection and competition on a speciation pattern of *Drosophila* endemic to Hawaii. In *Genetic Mechanisms of Speciation in Insects* (ed. White, M. J. D.), pp. 140–64. Sydney: Australia and New Zealand Book Co.

Rosin, S., and Fischer, J. 1972. Polymorphismus des realisators für männliches Geschlecht bei *Chironomus. Rev. Suisse Zool. 79 (Suppl.):*119–41.

Rothfels, K. H. 1979. Cytotaxonomy of black flies (Simuliidae). *Annu. Rev. Entomol.* *24:*507–39.

Scudder, G. G. E. 1979. The nature and strategy of species. In: Canada and its insect fauna (ed. Danks, H. V.). *Mem. Entomol. Soc. Can. 108:*533–47.

Thompson, P. E. 1971. Male and female heterogamety in populations of *Chironomus tentans* (Diptera: Chironomidae). *Can. Entomol. 103:*369–72.

White, M. J. D. 1978. *Modes of Speciation.* San Francisco: Freeman.

13 Chromosome change and speciation in lizards

MAX KING

Recent discussions concerning rates of chromosome change in reptiles imply that these are chomosomally conservative organisms. Wilson et al. (1975) suggested that lower vertebrates (fish, amphibians, and reptiles) in general have five times fewer established chromosomal changes in their evolutionary history than have placental mammals. This view was to some extent supported by Morescalchi (1977), who felt that the bimodality present in many reptilian karyotypes was an indication of their conservative evolutionary processes.

In the light of recent chromosomal studies on lizards (suborder Sauria), which are summarized in this chapter, three questions can be posed:

1. Are lizards and other reptiles indeed chromosomally conservative as the earlier data purport to show?
2. What forms of chromosome reorganization can be demonstrated in lizards and other reptiles, and what relationships do these forms have to their modes of speciation?
3. Is the ancestral lizard karyotype based on an all-acrocentric or an all-metacentric format?

Are lizards and other reptiles chromosomally conservative?

The classic view of lizards, and indeed most reptiles, is that they are intrinsically conservative in both their rate and their degree of chromosome change. Although Morescalchi (1977) acknowledged that occasional "unexpected explosions" of chromosomal rearrangements, associated with rapid speciation and population radiation, occurred in some lizards, he argued that "the karyological conservativeness of most sauropsids might demonstrate the intervention of rigid selective factors acting on new, probably unfit, karyotypes which are displaced from the typical bimodal formulae of these vertebrates." There is little doubt that karyotypic orthoselection could, by consistently favoring the same type of rearrangement, produce a particular karyotype that could be interpreted as the end product of such an orthoselected series. Nevertheless, there is clear evidence that karyotypes differing from a standard format are not "unfit" in Morescalchi's sense, for in reality an extraordinary array of different karyotypes is known in three of the major reptilian groups, namely, Crocodilia (Cohen and Gans, 1970), lizards (Gorman, 1973), and snakes (Beçak and Beçak, 1969). It can be generally argued that it is not so much karyotypic orthoselection that provides us with the apparent conservatism found in some reptiles (e.g., the 2n = 36 bimodal karyotype of many snake species), but that many of these reptiles do not have the necessary lifestyle characteristics that facilitate chromosome changes. There is evidence suggesting a relationship between the vagility of a species and its ability to establish karyotypic rearrangements (i.e., high-vagility species are chromosomally more conservative than low-vagility species; see White, 1973). Indeed, there was no need for Morescalchi (1977) to regard chromosomal rearrangements in some lizard taxa as "unexpected explosions," because they occur within species, or groups of species, that have lifestyle characteristics suitable for extensive karyotypic reorganization.

Studies by Wilson et al. (1975) and Bush et al. (1977) have provided estimates of the rate of evolutionary change in the karyotypes of a series of "higher" and "lower" vertebrate genera. These authors argue that placental mammals have a much higher rate of chromosome change than other vertebrates. Wilson et al. (1975) concluded that the difference observed between "higher" and "lower" vertebrates could be related to particular lifestyle characteristics. To use their own words, "reptiles of a given body size have experienced far slower rates of evolutionary change than placentals of comparable body size. The difference lies in the degree of socialization and how the social behaviour affects the potential for inbreeding and the effective population size" (Wilson et al., 1975). That is, placental mammals have socially structured populations, whereas other vertebrates do not, and this is the factor determining the different rates of chromosome change between the groups.

Social structuring of populations in certain mammal species such as horses may have played a significant role in speciation and in establishing numerous chromosome changes. Nevertheless, to make social structuring the criterion that distinguishes the rates of chromosomal change between higher and lower vertebrates is not supported by the existing data. The very high levels of chromosomal reorganization in a number of lizard

Table 13.1. *Intraspecific chromosomal variation in bisexual lizards*

Species	Variation	Reference
Gekkonidae		
Diplodactylus tessellatus	2 races (2n = 38, 28)	King (1975)
Diplodactylus vittatus	5 races (2n = 38, 38, 38, 36, 34)	King (1977a)
Phyllodactylus marmoratus	4 races (2n = 36, 36 ZW, 34, 32)	King and Rofe (1976), King and King (1977)
Gehyra variegata-punctata	5 races (2n = 44, 42, 40, 40, 38), rare mutants	King (1979)
Gehyra australis	4 races (2n = 42, 42, 40, 38), rare mutants	King (unpub.) King (1977b)
Hemidactylus frenatus	2 races (2n = 46, 40)	King (1978)
Xantusidae		
Xantusia vigilis	2 races (2n = 40)	Bezy (1972)
Xantusia henshawi	2 races (2n = 40)	Bezy (1972)
Iguanidae		
Sceloporus clarkii	Locality-specific polymorphisms	Cole (1970)
Sceloporus melanorhinus	Locality-specific polymorphisms	Cole (1970)
Sceloporus olivaceus	Locality-specific polymorphisms	Cole (1970)
Sceloporus spinosus	Locality-specific polymorphisms	Cole (1970)
Sceloporus undulatus	Locality-specific polymorphisms	Cole (1970, 1977)
Sceloporus grammicus	6 races, polymorphisms rare mutants	Hall and Selander (1973)
Anolis monticola	Polymorphism	Webster et al. (1972)
Tropidurus torquatus	Locality-specific polymorphism	Beçak et al. (1972)
Anniellidae		
Anniella pulchra	2 races (2n = 20, 22)	Bezy et al. (1977)
Scincidae		
Scincella laterale	2 races (2n = 30 XY, 30 $X_1 X_2$ Y)	Wright (1973)

species (Table 13.1), all⁵of which have low vagility but lack social structuring of their populations, immediately refutes the argument provided by Wilson and his colleagues which assumes that "vertebrates other than placentals do not organize their populations into small, persistent, social cohesive units." Many lizards in company with some other lower vertebrates species have low vagility, have strong territoriality, and often occur in extensive populations isolated by the "patchiness" of the environment, in which the breeding units are nevertheless remarkably small. Thus, although they lack a precise social structure, they certainly display many of the characters that Bush et al. (1977) readily accept as important in deme formation.

Moreover, although there is no doubt that many amphibian, fish, and reptile species are chromosomally conservative in terms of arm number and chromosome number changes, the presentation of the data by these authors suggests that all lower vertebrates are chromosomally conservative (see Figure 2, Bush et al., 1977). Whereas placental mammals are subdivided into groups in which members have broadly similar lifestyles, such as bats, rodents, carnivores, or whales, the other vertebrates are handled en masse: snakes, lizards, turtles and crocodiles, frogs, salamanders, and teleost fishes. In reality, however, many of these lower vertebrate groupings have animals sharing extraordinary differences in their lifestyle characteristics. It would be just as logical to argue that these organisms too can be subdivided into high- and low-vagility classes to give a more realistic basis for comparison of lifestyle characteristics.

The fact that all reptiles are assumed to be chromosomally conservative is a reflection of the data utilized by Wilson et al. (1975) and Bush et al. (1977), which were in many ways inadequate and consequently biased. Most of their karyotypic information was obtained from a review article by Gorman (1973), which summarized the studies on reptiles available at that time (see p. 5061, Wilson et al., 1975). Unfortunately, many of these data were derived from earlier accounts that failed to make an adequate attempt at population sampling; in many cases a given species was represented only in terms of a single specimen. As a result the data cited by Gorman grossly underrepresent the amount of karyotypic variation within species although they may give an accurate estimate of interspecific variation. More recently, extensive intraspecific chromosomal variation has been recorded in a series of lizard species (see Table 13.1). By contrast, the data used by Wilson, Bush, and their colleagues to estimate chromosomal change in mammals are very extensive, many species were analyzed intraspecifically, and high levels of inter- and intrapopulation variation were detected. Because of this bias the conclusions reached by Wilson et al. (1975) and Bush et al. (1977) concerning lower vertebrates must be regarded as unjustified.

The approach used by Wilson et al. (1975) and Bush et al. (1977) for the handling of their chromosomal data can be criticized on two grounds:

1. They fail to state clearly and precisely what information they are comparing. For example, when they write about correlations between karyotypic evolution and speciation, it is not apparent whether they refer to chromosome changes that determine speciation or those that may accompany it without playing a causal role. Many pericentric inversion differences may play little or no part in producing effective reproductive isolation (contra Bush et al., 1977), yet they are one of the major classes responsible for arm number changes used by these authors.

2. They do not distinguish between chromosome changes that are polymorphic and those that are fixed. Balanced polymorphisms have no role in the speciation process, and they are disproportionately common in some groups such as the rodents (Yonenaga, 1972, in *Akodon;* Patton, 1970, in *Thomomys*). Thus, we have no real idea of the extent to which polymorphisms have contributed to their estimates of karyotypic evolution.

Any results estimating the rate of chromosomal evolution in lizard species that have been grouped together regardless of lifestyle characteristics are also immediately suspect. Families as diverse as the Varanidae, which have both highly vagile and highly mobile species in the sense in which the term is used by White (1978), and the Gekkonidae, in which the majority are sedentary low-vagility forms, are grouped together by Bush, Wilson, and their colleagues. In fact, these two groups of lizards exhibit markedly different types of chromosome change. The varanid species so far analyzed (20 of the 33 extant species) lack any fixed patterns of intraspecific chromosome variation, all having 2n = 40. Even so, in the species that have been karyotyped (King and King, 1975; and unpub. data), extensive differentiation has occurred between species as a result of a series of pericentric inversions. The low-vagility species, such as the fossorial *Anniella* studied by Bezy et al. (1977), on the other hand, have differentiated into a series of chromosome races by a fusion/fission system. Similarly, locality-specific chromosome polymorphisms and rare mutants were found by Cole (1970) in his detailed analysis of the *spinosus* group of *Sceloporus* (Iguanidae). An extensive intraspecific analysis of *Sceloporus grammicus* by Hall and Selander (1973) has shown a remarkable complexity of chromosome races, chromosome polymorphism, and rare mutants. Perhaps the real magnitude of intraspecific chromosome variation possible in lizards can be best seen in the Gekkonidae, a group that has 650 known species, of which only a few have been karyotyped with adequate population sampling. In a study of seven Australian forms my colleagues and I have encountered considerable karyotypic variation (see Table 13.1) in the form of chromosome races, sex chromosome differences, rare mutants, and probable polymorphisms.

Most of the chromosome races are almost surely the product of fusions, and in many cases they appear to be associated with a morphometric differentiation of the animals involved. For instance, specimens from the chromosome races of *Diplodactylus tessellatus, D. vittatus, P. marmoratus, G. variegata,* and *G. punctata* have recognizable size and back-pattern differences, and forms with these differences have definable and extensive geographic distributions. Although the necessary hybrid analyses are not available, there is some basis for arguing that most are biological species, for they are chromosomally uniform throughout their distributions, and in those instances where two distributions appear to make contact, there is no evidence of any introgression of one chromosome race into the other. It is therefore likely that they have a narrow contact zone and are behaving as biological species. If so, the "speciation rate" in these reptiles is remarkably high (contra Wilson et al., 1975; Bush et al., 1977). Additionally, the 163 specimens of *Gehyra variegata-punctata* analyzed could be subdivided into five races, three of which are heterozygous for previously unobserved "new" chromosome changes. Similary four chromosome races have been detected in the 40 specimens of *G. australis* so far analyzed, and six animals were heterozygous for novel rearrangements. These data suggest an extraordinarily high rate of chromosome change in the Gekkonidae. To make it clear that this situation is not peculiar to Australia, it is worth considering the known chromosome number variation in the genus *Hemidactylus* (see King, 1978), which has a circumglobal distribution, and to note also that of three specimens of the Mexican gekko *Phyllodactylus lanei* karyotyped, one was heterozygous for a chromosome fusion (see Figure 13.1).

The extent of chromosomal differentiation in lizards is especially well demonstrated by the variety and nature of sex chromosome systems that have evolved. Whereas male heterogamety is present in many iguanid species and some Scincidae, female heterogamety is present in some *Varanus* and some gekkonids. Many lizards lack any recognizable sex chromosomes. Chromosomal mechanisms establishing these heteromorphisms have involved the gain or loss of chromatin and pericentric

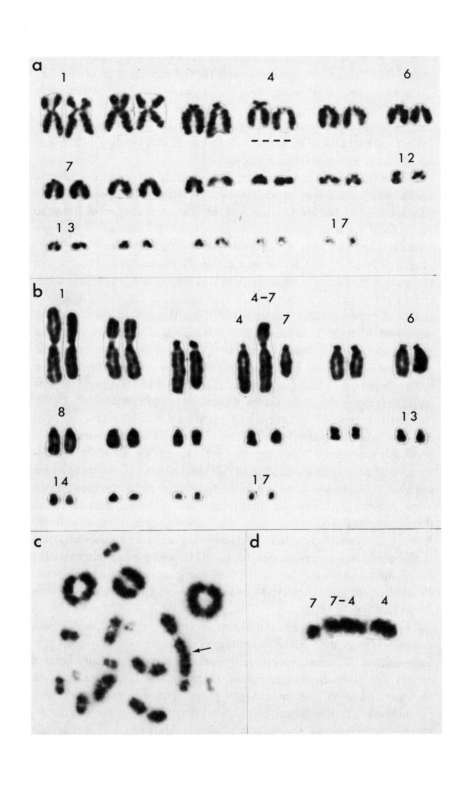

inversions. In some cases sex chromosome heteromorphisms appear to have been established on a number of independent occasions in the one lineage (see King, 1977b, for review).

"All" lizards are thus not chromosomally conservative, though, equally, "all" lizards are not chromosomally "mobile." There are extremes in karyotypic variation between high- and low-vagility reptiles as there are between high- and low-vagility mammals. The important feature is that if a reptile of any description is chromosomally conservative, it is not so because it is a "lower" vertebrate. It is conservative because of the particular lifestyle characteristics possessed by that organism, or because of the karyotypic organization that it possesses (see next section).

Chromosome reorganization and modes of speciation

Whereas the degree of variation can in some cases be related to the animals' lifestyle (high- or low-vagility forms), in other groups of lizards different factors must be involved. Thus a number of lizard species in the Iguanidae and the Australian Agamidae are characterized by pronounced chromosomal stability at least in terms of their external chromosome morphology, yet many of these forms have low vagility and have speciated extensively. Many of the iguanid species have a stable, all-metacentric, 12V (metacentrics) + 24m (microchromosomes) karyotype (see Paul et al., 1976), whereas most of the agamids have an all-metacentric 12V + 20m karyotype (Whitten, 1978; and Figure 13.3b). Many of these species thus appear to have attained a stable karyotypic configuration and, in consequence, may be incapable of further change regardless of their lifestyle characteristics. Such stable configurations may have been derived by a series of fusions from an all-acrocentric ancestral form as a result of karotypic orthoselection (White, 1975). Consequently, once complete metacentricity has been attained, the basis for karyotypic change is automatically terminated. If the organisms involved do not have the capacity for subsequent fission, then the only means of further change available involve inversion, reciprocal translocation, deletion, or addition of chromatin. If the propensity for any of these changes is low, the

Figure 13.1. *a.* The chromosomes of *Phyllodactylus lanei* (♀) from Guerrero, Mexico (2n = 34). Note the heteromorphism in chromosome pair 4. This is a possible ZW sex chromosome system and was present in the only two females examined. *b.* The karyotype of a male *Phyllodactylus lanei* (2n = 33), which is heteromorphic for a fusion of chromosome pairs 4 and 7. *c.* Meiotic metaphase cell of male *Phyllodactylus lanei* heteromorphic for 4–7 fusion. Note fusion trivalent (arrowed). *d.* Enlarged fusion trivalent. Note orientation of elements.

karyotype will be unable to change further, and what may once have been a chromosomally "mobile" group will now appear conservative. Speciation within such a conservative complex can then occur only by changes at the genic level.

The concept of karyotypic orthoselection can be explained in one of two ways. Either there is a differential susceptibility to different types of chromosome change in an organism, or, alternatively, there is differential survival of particular types of chromosomal rearrangements through selection. The former view has been rejected by White (1975). Nevertheless, there is some evidence suggesting that some organisms do indeed show a differential capacity for the induction of certain types of chromosome change. For instance, the frequency of pericentric inversion is far higher in the Orthoptera than in *Drosophila,* where paracentric inversions predominate (see White, 1973). Although it could be argued that this is a product of selection, it is difficult to accept the idea that selection alone has led to a situation in *Varanus* in which only pericentric inversions have survived (King and King, 1975), or where only fusions have survived in the *Mus musculus* complex (see White, 1979) without even some of the other forms of rearrangements having been observed in occasional individuals. Studies on the C-banding characteristics of a series of orthopteran species (King and John, 1979) suggest that certain karyotypes do have a propensity for establishing specific types of chromatin reorganization within most members of their complements, and that this may vary markedly in its expression between species. Analysis at this more refined level of resolution of chromosome change suggests that different karyotypes have a structural predisposition for certain types of reorganization, and not for others.

Table 13.2. *Estimates of chromosome number and arm number changes*

	Wilson et al.	Bush et al.
Arm number changes		
Lizards	1.3[a]	0.031[b]
Snakes	2.1	0.041
Chromosome number changes		
Lizards	1.1	0.027
Snakes	0.5	0.007

[a]Karyotypic changes/lineage/10^8 years.
[b]Karyotypic changes/lineage/million years.
Source: Data from Wilson et al. (1975) and Bush et al. (1977).

One of the more interesting findings made by Wilson et al. (1975), subsequently corroborated by Bush et al. (1977), was that lizards had marginally fewer chromosome arm changes than snakes, but had a slightly higher frequency of chromosome number changes (see Table 13.2). Such a result is not surprising, for many snake species are large carnivores, having high vagility, low population density, and extensive home range when compared to lizards. These lifestyle characteristics would promote higher levels of gene flow and reduce the possibility of small, isolated demes being formed. Consequently, one might expect less emphasis on the differentiation of species or populations into chromosomally monomorphic races. The chromosome differences established between such organisms are often pericentric inversion monomorphisms or polymorphisms, and in many cases appear to have been fixed after allopatric speciation. This is not to imply that all snakes have followed this pattern of evolution, for high levels of chromosomal differentiation, involving both pericentric inversion and fusion/fissions, may well be found in some of the smaller low-vagility forms. However, it is probable that changes such as pericentric inversions play a less significant role as postmating isolating mechanisms than do at least some of the fusion/fission changes. Thus while the latter sometimes reduce the fertility of heterozygotes by nondisjunction at first metaphase (Winking and Gropp, 1976), in those cases where adequate meiotic studies have been made, reverse loops are not usually formed in pericentric inversion heterozygotes (White, 1973:243). Straight pairing with modified chiasma localization prevents any ill effects at meiosis. As previously mentioned, the high-vagility varanid lizards, which have many of the characteristics of the snakes described above, have also established numerous monomorphic inversion differences and at least one known polymorphism (unpub. data). Because these differences are characteristic of species groups it is likely that speciation has occurred allopatrically within these groups. A preponderance of pericentric inversion in snakes would account for the higher incidence of chromosome arm changes than in lizards (Table 13.2). That this is the case is confirmed by the studies of Singh (1972), who found a preponderance of inversion differences between snake species.

The fact that chromosome number changes are more frequent in lizards than in snakes suggests quite a different mode of evolution in these animals (see Table 13.2). Nevertheless, there are good grounds for believing that the frequency of occurrence of changes in chromosome number proposed by Wilson et al. (1975) and in Bush et al. (1977) is a significant underestimate (see preceding section). Many organisms that have low vagility, extensive distribution, isolated populations, and small home ranges have a high incidence of fusion/fission changes established between and within species. Fossorial rodents of the genus *Spalax* (Wahrman et al., 1969) and the *Mus musculus* complex in Italy and Switzerland (see White, 1979 for a review) are two cases in point. In these situations the fusion/fission changes observed are intimately associated with the differentiation of recently evolved forms. In many cases these changes may well be a contributory factor in the fixation of these differences. Lizards that share these lifestyle characteristics, such as some of the gekko species and *Sceloporus* species (see Table 13.1), also share a high incidence of fusion/fission polytypism, whereas many lizard species express equivalent differences between species.

Perhaps the most bizzare claim made by Wilson et al. (1975) was that in vertebrates other than placentals, speciation usually occurs by geographic isolation (allopatric speciation) with very little chromosome involvement – whereas the major mode of speciation in placental mammals has been a stasipatric process with very rapid chromosome change. In those species of lizards in which chromosome races have been detected, these races have defined geographic distributions, are consistent chromosomally throughout this distribution, in some cases have morphological differentiation accompanying this distribution, and in some cases too have narrow, parapatric, contact zones within which there is very little chromosomal introgression (Hall and Selander, 1973). On the basis of the distribution of these chromosome races there is evidence for both stasipatric and allopatric speciation in lizards.

It is important to make the distinction between three basic types of allopatric speciation observed in lizards. These are:

1. *Classical or genic allopatry* (Mayr, 1963), in which geographically isolated populations of a species accumulate a series of genic differences that eventually result in the isolates becoming separate species. This mode of speciation may be seen in the chromosomally conservative Australian Agamidae, in which extensive speciation has occurred without apparent chromosome change in gross terms. This does not mean that populations of a species that become isolated geographically will necessarily become separate species; they could presumably remain unchanged and interbreed freely if ever reunited.

2. *Secondary chromosomal allopatry*. This second form of allopatry occurs in high-vagility species, wherein populations are isolated geographically, speciate genically, and subsequently establish a "new" chromosome rearrangement throughout the new species

distribution. This chromosome change would presumably start as a balanced polymorphism and end up as a monomorphic state. Because it has little effect on fertility the change most likely to occur would be pericentric inversion, and it could explain the abundance of these rearrangements in high-vagility snakes and varanid lizards. These monomorphic differences are only observed between species and are not involved in race formation. In many situations they would give the false impression of a founder effect, with related species having a geographically correlated array of chromosome changes.

3. *Primary chromosomal allopatry.* This quite distinctive and rapid form of chromosomal speciation is based on the revolutionary situation experienced during a colonizing radiation. It occurs within a species or species complex, involves fusion/fission rearrangements, and has, as its principal tenet, the founder principle. That is, a new chromosome form is established in a peripheral deme on the edge of the range of a chromosome race. Specimens with this new rearrangement take part in colonizing radiations into new territory, producing a race with an extensive geographic distribution. Subsequently, new chromosome changes can occur in peripheral populations, which also take part in colonizing radiations, producing a linear array of phylogenetically related chromosome races (A → B → C → D). This mode of differentiation is seen in the Australian gekko *Phyllodactylus marmoratus* (King and King, 1977) where three fusion races are linearly arrayed (2n = 36, 34, 32) (see Figure 13.2a). Similarly, Wahrman et al. (1969) found a series of chromosome races (2n = 60, 58, 54, and 52) in the mole rat *Spalax ehrenbergi* that were distributed both clinically and parapatrically and suggested a similar mode of differentiation. Patton (1969) too, in a chromosomal analysis of the pocket mouse, *Perognathus goldmani,* found a series of chromosome races that he argued had evolved in a series of colonizing radiations.

It is important to make a distinction between this dynamic form of primary chromosomal allopatry and stasipatric speciation. In stasipatric speciation a new chromosome form becomes established within the range of a species, within a small deme, and then displaces the primordial form to produce a new chromosome race. In primary chromosomal allopatry the new chromosome morph is established in a peripheral deme and colonizes new territory in a colonizing radiation. The most significant feature suggesting that this form of allopatry is a dynamic and rapid form of speciation, is that the rate of movement of the new race is directly proportional to the vagility of the species involved, and to the prevailing environmental conditions. This form of evolution may well be more rapid than a stasipatric process because the distribution of the new race relies on population movement, whereas the distribution of the new race in stasipa-

Figure 13.2 *a*. The presumed distributional changes occurring during the evolution of *Phyllodactylus marmoratus* chromosome races. The fusion races 2n = 34 and 32 appear to have evolved during colonizing radiations into Western Australia. They exhibit a linear distribution and are thought to be examples of primary chromosomal allopatry. *b*. The presumed distributional changes occurring during the evolution of the *Diplodactylus vittatus* chromosome races. This is thought to be an example of stasipatric speciation because in eastern and Western Australia independent fusion races 2n = 34 and 36 have displaced the ancestral races, bisecting their present-day distribution.

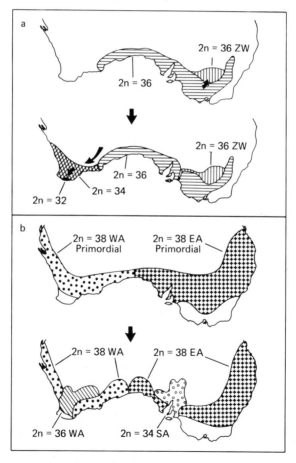

try relies on the rate of movement of a "tension zone" between the ancestral and derived forms.

The major problem that we face in interpreting the mode of speciation in a complex is that we need to determine past events in terms of the present-day distribution of the chromosome races. Such an attempt is obviously fraught with danger, for patterns of distributions may be modified by agricultural and climatic changes. Nevertheless, a number of distributional criteria can be used to determine the likely mode of evolution from the present-day distribution. For example, if the chromosome races are geographically linearly arranged in ancestor–descendant order, there is a reasonable basis for arguing that a series of colonizing radiations based on a dynamic form of primary chromosomal allopatry has occurred. Similarly, White (1979) suggests that in cases of stasipatric speciation "derived taxa occupy central or interior areas and ancestral taxa peripheral or external ones." The distribution of the chromosome races of *Mus musculus* in Italy is a good example because stasipatric distributions of a series of fusion races occur as isolated central populations in a continuous 2n = 40 race of mice. A similar situation is seen in the gekko *Diplodactylus vittatus* (King, 1977a), in which two independently derived fusion races, one in eastern and one in Western Australia, each subdivide the distribution of their ancestral races (see Figure 13.2b).

There are other cases that are not so readily interpreted. Gekkos of the *Gehyra variegata-punctata* complex (King, 1979) are distributed throughout Australia with the presumed ancestral populations in some cases occupying peripheral areas, whereas at least some of the assumed derivative forms occupy central areas, suggesting evolution by a stasipatric sequence. Nevertheless, many of the populations are clearly relictual, occupying rock outcrops isolated by deserts. In fact these distributions could have just as easily been produced by a series of colonizing radiations following extinctions, so that one might then have concluded that differentiation had in the first instance occurred by primary chromosomal allopatry with subsequent changes occurring within the chromosome races by classical genic allopatry.

The distribution of the chromosome races of *Sceloporus grammicus* (Hall and Selander, 1973, Figure 2) is also complex. Here, the ancestral forms are found in three isolated peripheral populations, whereas the chromosomally derived forms occupy more central areas of the distribution. Intuitively, this distribution is suggestive of the stasipatric process. However, because the peripheral populations are found as isolates, it is impossible to say whether this isolation occurred before or after the chromosome race differentiation. If it occurred after differentiation, the distributions would suggest stasipatry, but if it occurred before differentiation, the chromosomally derived races could have evolved in a series of radiations that resulted in the recolonization of territory previously occupied by the locally extinct ancestral forms. Indeed, the linear array of these races may support such a conclusion.

Clearly, the assumptions made by Wilson et al. (1975) on the mode of speciation in lizards and other reptiles are not in accord with interpretations of the observed data. Speciation in lizards is both complex and variable, with, as we have shown, at least four modes possible. Indeed, the type of speciation is determined by the vagility of the species, the nature of the chromosome change, the distribution of populations or races, and the chromosome structure of the organisms involved.

The ancestral karyotype dilemma

One of the great imponderables facing those who study karyotypic relationships in any group of organisms, is the difficulty of determining the direction of chromosome change in a fusion/fission system. Is a single metacentric derived from the fusion of two acrocentrics (or telocentrics), or is it ancestral with subsequent fission producing two uniarmed chromosomes? One way of resolving such a dilemma is to ascertain the phylogenetic position of that organism by other than karyotypic means, and then to use this additional information in deciding which members of a complex can be regarded as ancestral and which are derived. Unfortunately, this approach is rarely used, and we are left with the common means of predicting the ancestral karyotype by its frequency of occurrence in a range of organisms. Although this approach may produce a quite valid result when a great number of samples are taken, an inherent bias exists if the area of sampling is unrepresentative. Such a situation exists in those lizard species that have been analyzed chromosomally and where most studies have centered on North and South American families, with many of the Old World groups remaining untouched.

Despite this obvious overemphasis on the New World species, a number of authors (Gorman, 1973; Williams and Hall in a subsection of Paul et al., 1976) appear determined to make the New World the center for saurian evolution. Both Gorman (1973) and Williams and Hall (1976) propose that the most commonly observed karyotype in reptiles is 12V (metacentrics) + 24m (microchromosomes) and that this therefore represents the "primitive" condition. It is true that species from the families Iguanidae, Agamidae, Chamaeleontidae, Teiidae, Gerrhosauridae, Anguidae, and

Amphisbaenidae do indeed often have this karyotype although others are certainly present. To accomodate these others it was argued that chromosomal evolution has been away from the 12V + 24m format by fissioning and that other karyotypes are thus derived forms. These authors also felt that families such as the Scincidae, Helodermatidae, and Varanidae, which although they retain total metacentricity have structurally very different karyotypes from the 12V + 24m format, could be readily derived from this "ancestral" karyomorph with minor modifications. A number of other families (Lacertidae, Gekkonidae, Pygopodidae, Xantusidae, Anniellidae) were either not considered or else were regarded as "deviant" by Williams and Hall (1976).

Because the vast majority of lizard species that have been karyotyped are New World iguanids (see Gorman, 1973 for review), it is hardly surprising that the most common karyotype in this family was regarded as the ancestral form. It is equally likely that the families Agamidae and Chamaeleontidae might also share this presumed ancestral karyotype, for all are members of the infraorder Iguania. Indeed, a strong case for evolution by fission from an ancestral 12V + 24m karyotype supported by a morphological analysis has been made for members of the Iguanidae by Webster et al. (1972), although an alternative model utilizing fusion was advocated by Cole (1970). However, although it is possible that a 12V + 24m karyotype is indeed ancestral in the infraorder Iguania and some associated families, and that evolution in some cases has proceeded by fission, the evidence is by no means conclusive.

A major weakness of the 12V + 24m ancestral karyotype hypothesis when applied to the Iguanidae is the status of the genus *Polychrus*. Williams and Hall (1976) suggest that animals from this genus form one of the seven lineages directly derived from the iguanine basal stock. In six of the seven lineages, some species at least are seen to have a 12V + 24m complement. However, the chromosomes of *Polychrus* (2n = 20 to 30) have no resemblance to this format and are not readily derived from it. These authors dismiss *Polychrus* as "deviant" without offering any explanation of how such a karyotype could have evolved from an ancestral 12V + 24m format. On the other hand, if the karyotype ancestral to all lizards were acrocentric and evolution proceeded predominantly by fusion, there would be no difficulty in explaining the presence of *Polychrus* or the 12V + 24m karyotypes in basal iguanines. That is, different chromosomes fused to provide the derived karyotypes in each case. The massive reduction in the number of microchromosomes in *Polychrus* could be the result of their loss or their incorporation into larger elements.

a   1   5

                              6   8

9   20

b   1   6

7   16

c   1   6

7   12

13   19

Perhaps the greatest flaw in the ancestral 12V + 24m model with evolution by fission is that it has been forced onto all lizards when in fact it can only apply to a very few if any. Many regard the suborder Sauria as a polyphyletic assemblage, a view to which I subscribe. Indeed, Gans (1978) has recently suggested that the Amphisbaenidae, a family used by Williams and Hall (1976) to support their model, are not lizards and that they should be placed in a separate order. Undoubtedly, the lizards from the infraorders Platynota (Varanidae, Lanthanotidae, Helodermatidae), which are among the most primitive and have many affinities with snakes, and the Gekkota (Gekkonidae, Pygopodidae), which are also among the oldest lizards known, have doubtful affinities with the other infraorders. It is important to realize that a predominant karyotypic form within a lineage or group of lineages is not necessarily the ancestral form; it may itself be the product of karyotypic orthoselection (White, 1973), or else have been secondarily derived from the ancestral basal stock during the initial divergence of a lineage. There is no reason to expect that any or all subsequent karyotypic change will show the same pattern in each lineage.

One of the major problems associated with the assumption of a 12V + 24m ancestral karyotype is that this hypothesis is based on very little evidence. Of the seven families used in support of the argument by Williams and Hall (1976), one (the Amphisbaenidae) is no longer regarded by Gans (1978) as being a lizard. Of the remainder, fusion and fission are equally likely as possible modes of evolution. Thus in the Chamaeleontidae, Teiidae, Anguidae, and Iguanidae many of the species have all-acrocentric karyotypes or intermediate numbers to the 12V + 24m. Incidentally, the Gerrhosauridae can hardly be relied on as evidence because only one species has been karyotyped, and that by Matthey in 1933. Of the 14 agamid species described by Gorman (1973), only three had 12V + 24m, the rest having all-acrocentric or partially modified karyotypes. In Australia, where one of the largest radiations of agamids has occurred, most species have 12V + 20m (Whitten, 1978; Figure 13.3). Sokolovsky (1975) found that many of the Russian species have a 2n = 46 all-telocentric karyotype, and argued that evolution had occurred by

Figure 13.3 *a*. The chromosomes of *Varanus acanthurus brachyurus* (♀) (Varanidae) from Cloncurry, Queensland. This specimen has 2n = 40 (16V + 24m). This specimen is heteromorphic for a ZW pair 9. *b*. The chromosomes of *Amphibolurus caudicinctus* (♂) (Agamidae) from Opalton, Queensland. This species has 2n = 32 (12V + 20m). *c*. The chromosomes of *Bavayia* sp. (♂) from New Caledonia. This species has 2n = 38. Note the enlarged short arms on all chromosomes of this acrocentric karyotype.

fusion. Here, too, the ancestral karyotype is in question. To the author's knowledge there are no published chromosome data on the families Cordylidae, Dibamidae, Feyliniidae, Anelytropsidae, and Xenosauridae; so there is a substantial gap in our knowledge of the entire group.

A model that envisages an acrocentric or telocentric karyotype as the ancestral form with evolution proceeding predominantly by fusion, interspersed with episodes of inversion, addition, or duplication, and in some cases by fission, can produce any of the saurian karyotypes that we see today. Such a model is much more acceptable and definitely more parsimonious than one based on the fissioning of a 12V + 24m karyotype. The great problem associated with making one particular form of metacentric karyotype ancestral is that it is inherently difficult to produce other metacentric karyotypes with differing morphological characteristics from this ancestral karyomorph. Although it is known that a group of species such as the *Chironomus* complex analyzed by Keyl (1962) can undergo numerous whole-arm reciprocal translocations and still retain a superficially uniform metacentric karyotype, this is an exceptional situation. It is in those species that differ markedly in their chromosome morphology but still maintain metacentricity that the problem is most acute. For example, to produce an evenly graded metacentric scincid karyotype decreasing in size from largest to smallest element from a morphologically disjunct 12V + 24m iguanid-like karyotype, a series of complex reciprocal translocations would be required. Reciprocal translocations are simply not common cytogenetic phenomena, and to my knowledge there are very few cases that can be interpreted as reciprocal translocation in lizards.

Williams and Hall (1976) argued that the karyotypes of the Scincidae, Varanidae, Helodermatidae, Xantusidae, Pygopodidae, and Anniellidae were all easily derivable from a 12V + 24m format. This is simply not true. Any of these karyotypes could be produced from an ancestral all-acrocentric form by a series of fusions including different chromosomes, but to produce any of them from the 12V + 24m would require a series of complex translocations and more twists and turns than a python in a revolving door. The derived *Bavayia* species (Gekkonidae), whose karyotype is shown in Figure 13.3c, could only be derived from the 12V + 24m base by complete karyotypic fissioning, loss of microchromosomes, and the addition of chromatin onto the short arms of all these acrocentric chromosomes. In contrast, if they evolved from an acrocentric ancestor, they would only have to add chromatin to the short arms to attain their present-day morphology. Similarly, the varanid karyotype shown (Figure 13.3a) could only be derived from the 12V + 24m if the longer chromosomes first underwent fission and then fused in a different order. To produce this karyotype from an ancestral acrocentric karyotype would only require a series of fusions.

Interestingly enough, many of the groups that Williams and Hall (1976) failed to mention (Lacertidae, Gekkonidae, Pygopodidae, Xantusidae, Anniellidae) or passed off as "deviant" (p. 18) were those that argued most strongly against their case. The karyotypes of these forms either are acrocentric or else require a complex series of chromosome changes to get to the 12V + 24m karyotype. It is noteworthy that the vast majority of the Lacertidae are acrocentric with 2n = 38, as indeed are the morphologically primitive diplodatyline gekkonids. Although I do not suggest a direct relationship between these groups, it is quite possible that both these karyotypes are relics of the ancestral saurian karyotype.

We have some evidence that evolution has occurred by fusion in the Gekkonidae because those chromosome races regarded as being phylogenetically most recent in terms of morphometric characteristics also possess chromosomal fusions (*Diplodactylus vittatus*, King, 1977a; *Phyllodactylus marmoratus*, King and Rofe, 1976; King and King, 1977; *Gehyra variegata-punctata*, King, 1979). In each of these species complexes, different chromosomes have been involved in the fusions concerned. If these changes were in fact fissions, the acrocentric karyotypes we now see in the Gekkonidae would necessarily be a polyphyletic assemblage.

It should be pointed out that the generality of the 12V + 24m ancestral karyotype with the clear implication of evolution by fission has not received general acceptance even in the past. In addition to the gekkonid evidence presented here, Bezy (1972), in a detailed chromosomal and morphometric study on the Xantusidae, argued that fusion had played a predominant evolutionary role. Lowe et al. (1967) also argued for fusion in the *Sceloporus clarkii* complex (Iguanidae) as did Lowe et al. (1970) in the teiid genus *Cnemidophorus*.

Although I advocate chromosome change occurring primarily by fusion with minor episodes of inversion, fission, and addition or duplication of chromatin from an ancestral acrocentric karyotype, it is difficult to establish the diploid number of this ancestral form. Whereas a 2n of 38 is found in the Gekkonidae and Lacertidae, certain gekkonid species also have 2n = 46. This number is also present in the Iguanidae, Agamidae, and Chamaeleontidae. Clearly, the addition or loss of microchromosomes from the postulated ancestral stock could forever conceal the exact number involved.

If an all-acrocentric karyotype is regarded as ancestral, how does one explain the preponderance of 12V + 24m forms in the iguanids and a number of other families? First, the preponderance of species with a 12V + 24m format, or for that matter a 12V + 20m (Agamidae) or 16V + 24m (varanid), may simply be a reflection of karyotypic orthoselection for an all-metacentric karyotype, for either cytomechanical or recombinational reasons. The fact that different chromosome numbers are involved simply reflects the different levels of karyotypic fusion reached, the different elements fused, and the significance of microchromosome gain or loss in a given evolutionary lineage.

A second explanation for the preponderance of 12V + 24m forms in the Iguania could be that this karyomorph was selected for during initial divergence from the ancestral stock (which was presumably all acrocentric), producing the lineage leading to the Iguania (see Figure 13.4). Indeed the same argument can be used to explain the production of the characteristic karyotypes of each of the lineages. It is the subsequent divergence within each of these lineages that then appears to operate independently. Clearly, fusion, addition or duplication of chromatin, and inversion have all occurred within the Gekkota (King and Rofe, 1976; King, 1977b). Inversion is involved in the evolution of the Platynota (King and King, 1975), whereas both fusion and fission are detectable in the Iguania. I have used these major lineages as examples, but clearly a degree of independence in the mode of chromosome change is likely in all of the families.

Figure 13.4. The possible mode of derivation of some of the common karyotypic formats from an ancestral all-acrocentric complement.

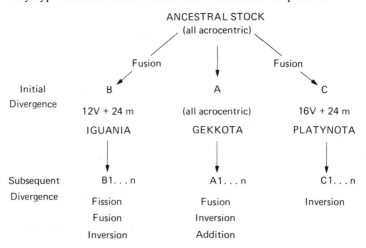

In conclusion, it is clear that there are grave doubts about the acceptability of a 12V + 24m karyotype as the ancestral lizard complement. The difficulty of deriving certain karyotypes from this format, the very biased sampling from which the model was produced, and the need to force chromosome fissioning as the primary evolutionary mechanism all make such a model unacceptable. The most parsimonious model for chromosome evolution in lizards argues for an acrocentric ancestral karyotype with evolution primarily by fusion, but with inversion, addition, and fission playing significant roles. This is clearly the path of least resistance, removing any of the difficulties outlined above.

Concluding remarks

There is little doubt that the lizards are one of the most chromosomally interesting and diverse of the surviving vertebrate suborders. In many respects they pose a series of contradictions:

1. Some groups are chromosomally conservative, whereas others are extraordinarily diverse.
2. Specialized chromosome changes have evolved in some lineages, whereas others have not changed at all, at least in gross terms.
3. Male heterogametic sex chromosomes are present in some families, and female heterogamety is present in other families, whereas many species lack any sexually heteromorphic chromosomes at all.
4. There is evidence for the orthoselection of one karyotypic format in some lineages and markedly different formats in others.
5. Chromosomal fissioning has been the predominant mode of evolution in some lineages, fusion has predominated in others, and inversion has dominated still others.

Most significantly, lizards vary markedly in their lifestyle characteristics and in their modes of speciation. Whereas some species may have differentiated stasipatrically, others appear to have changed by primary or secondary chromosomal allopatry, and yet others by genic allopatry. They tell us quite clearly that "the time seems to have come for us to realize that speciation has followed different patterns in different groups" (White, 1974).

What this essay has attempted to show is that the term group is not applied simply to major classes of phyla as Wilson et al. (1975) would have us believe, but rather to collections of organisms that share common properties at any level.

I would like to thank Professor Bernard John for his critical reading of the manuscript. I am most grateful to Dr. William J. Mautz, who went to much trouble in collecting specimens of

Phyllodactylus lanei from Mexico for me. I am also indebted to the late Peter Rankin, who with Ross Sadlier collected the New Caledonian *Bavayia*, and to John Wombey for collecting *Amphibolurus caudicinctus*.

References

Beçak, W., and Beçak, M. L. 1969. Cytotaxonomy and chromosomal evolution in serpents. *Cytogenetics 8:*247–62.

Beçak, M. L., Beçak, W., and Denaro, L. 1972. Chromosome polymorphism, geographical variation and karyotypes in Sauria. *Caryologia 25:*313–26.

Bezy, R. L. 1972. Karyotypic variation and evolution of the lizards in the family Xantusiidae. *Contrib. Sci. 227:*1–29.

Bezy, R. L., Gorman, G. C., Kim, Y. J., and Wright, J. W. 1977. Chromosomal and genetic divergence in the fossorial lizards of the family Anniellidae. *Syst. Zool. 26:*57–71.

Bush, G. L., Case, S. M., Wilson, A. C., and Patton, J. L. 1977. Rapid speciation and chromosomal evolution in mammals. *Proc. Natl. Acad. Sci. U.S.A. 74:*3942–6.

Cohen, M. M., and Gans, C. 1970. The chromosomes of the order Crocodilia. *Cytogenetics 9:*81–105.

Cole, C. J. 1970. Karyotypes and evolution of the *spinosus* group of lizards in the genus *Sceloporus. Am. Mus. Novit. 2431:*1–47.

– 1977. Chromosomal aberration and chromatid exchange in the North American fence lizard, *Sceloporus undulatus* (Reptilia: Iguanidae). *Copeia 1:*53–9.

Gans, G. 1978. The characteristics and affinities of the Amphisbaenia. *Trans. Zool. Soc. Lond. 34:*347–416.

Gorman, G. C. 1973. The chromosomes of the Reptilia, a cytotaxonomic interpretation. In *Cytotaxonomy and Vertebrate Evolution* (ed. Chiarelli, A. B., and Capanna, E.), pp. 349–424. New York: Academic Press.

Hall, W. P., and Selander, R. K. 1973. Hybridization of karyotypically differentiated populations in the *Sceloporus grammicus* complex (Iguanidae). *Evolution 27:*226–42.

Keyl, H. G. 1962. Chromosomenevolution bei *Chironomus*. II. Chromosomenumbaten und phylogenetische Beziehungen der Arten. *Chromosoma (Berl.) 13:*496–514.

King, M. 1975. Chromosomal studies on Australian lizards. Ph.D. thesis, University of Adelaide.

– 1977a. Chromosomal and morphometric variation in the gekko *Diplodactylus vittatus* (Gray). *Aust. J. Zool. 25:*43–57.

– 1977b. The evolution of sex chromosomes in lizards. In: Evolution and reproduction, *Proc. 4th Int. Conf. Reprod. Evol.*, pp. 55–60. Aust. Acad. Sci.

– 1978. A new chromosome form of *Hemidactylus frenatus* (Duméril and Bibron). *Herpetologica 34:*216–18.

– 1979. Karyotypic evolution in *Gehyra* (Gekkonidae: Reptilia). 1. The *Gehyra variegata–punctata* complex. *Aust. J. Zool. 27:*373–93.

King, M., and John B. 1980. Regularities and restrictions governing C-banding variation in acridoid grasshoppers. *Chromosoma (Berl.) 76:*123–50.

King, M., and King, D. 1975. Chromosomal evolution in the lizard genus *Varanus* (Reptilia). *Aust. J. Biol. Sci. 28:*89–108.

– 1977. An additional chromosome race of *Phyllodactylus marmoratus* (Gray) (Reptilia: Gekkonidae) and its phylogenetic implications. *Aust. J. Zool. 25:*667–72.

King, M., and Rofe, R. 1976. Karyotypic variation in the Australian gekko *Phyllodactylus marmoratus* (Gray) (Gekkonidae: Reptilia). *Chromosoma (Berl.) 54:*75–87.

Lowe, C. H., Cole, C. J., and Patton, J. L. 1967. Karyotype evolution and speciation in lizards (genus *Sceloporus*) during evolution of the North American desert. *Syst. Zool. 16:*296–300.

Lowe, C. H., Wright, J. W., Cole, C. J., and Bezy, R. L. 1970. Chromosomes and evolution of the species groups of *Cnemidophorus* (Reptilia: Teiidae). *Syst. Zool. 19:*128–41.

Matthey, R. R. 1933. Nouvelle contribution a l'etude des chromosomes chez les Sauriens. *Rev. Suisse Zool. 40:*281–316.

Mayr, E. 1963. *Animal Species and Evolution.* Cambridge, Massachusetts: Belknap Press of Harvard University Press.

Morescalchi, A. 1977. Phylogenetic aspects of karyological evidence. In *Major Patterns in Vertebrate Evolution* (ed. Hecht, M. K., Goody, P. C., and Hecht, B. M.) pp. 149–67. New York: Plenum.

Patton, J. L. 1969. Chromosome evolution in the pocket mouse *Perognathus goldmani* Osgood. *Evolution 23:*645–62.

– 1970. Karyotypic variation following an elevation gradient in the pocket gopher, *Thomomys bottae grahamensis* Goldman. *Chromosoma (Berl.) 31:*41–50.

Paul, D., Williams, E. E., and Hall, W. P. (1976). Lizard karyotypes from the Galapagos Islands: chromosomes in phylogeny and evolution. *Brevoria 441:*1–31.

Singh, L. 1972. Evolution of karyotypes in snakes. *Chromosoma (Berl.) 38:*185–236.

Sokolovsky, V. 1975. Comparative and karyological study of lizards in the family Agamidae. II. Karyotypes of five species of the genus *Agama. Cytologia (U.S S.R.) 17:*91–3.

Wahrman, J., Goitein, R., and Nevo, E. 1969. Geographic variations of chromosome forms in *Spalax,* a subterranean mammal of restricted mobility. In *Comparative Mammalian Cytogenetics* (ed. Benirschke, K.). New York: Springer-Verlag.

Webster, T. P., Hall, W. P., Williams, E. E. 1972. Fission in the evolution of a lizard karyotype. *Science 177:*611–13.

White, M. J. D. 1973. *Animal Cytology and Evolution,* 3rd ed. Cambridge: Cambridge University Press.

– 1974. Speciation in the Australian morabine grasshoppers: the cytogenetic evidence. In *Mechanisms of Speciation in Insects* (ed. White, M. J. D.), pp. 57–68. Sydney: Australia and New Zealand Book Co.

– 1975. Chromosomal repatterning – regularities and restrictions. *Genetics 79:*63–72.

– 1978. *Modes of Speciation.* San Francisco: Freeman.

– 1979. Chain processes in chromosomal speciation. *Syst. Zool. 27:*285–98.

Whitten, G. J. 1978. A triploid male individual *Amphibolurus nobbi nobbi* (Lacertilia: Agamidae). *Aust. Zool. 13:*305–8.

Williams, E. E., and Hall, W. P. 1976. Primitive karyotypes. In: Lizard karyotypes from the Galapagos Islands: chromosomes in phylogeny and evolution" (Paul, D., Williams, E. E., and Hall, W. P.) *Brevoria 441:*1–31.

Wilson, A. C., Bush, G. L., Case, S. M., and King, M. C. 1975. Social structuring of mammalian populations and rate of chromosomal evolution. *Proc. Natl. Acad. Sci. U.S.A. 72:*5061–5.

Winking, H., and Gropp, A. 1976. Meiotic nondisjunction of metacentric heterozygotes in oocytes versus spermatocytes. *Serono Symp. 1975.* New York: Academic Press.

Wright, J. W. 1973. Evolution of the X_1X_2Y sex chromosome mechanism in the scincid lizard *Scincella laterale* (Say). *Chromosoma (Berl.) 43:*101–8.

Yonenaga, Y. 1972. Chromosomal polymorphism in the rodent *Akodon arviculoides* ssp. (2n = 14) resulting from two pericentric inversions. *Cytogenetics 11:*488–99.

14   Chromosomal tracing of evolution in a phylad of species related to *Drosophila hawaiiensis*

HAMPTON L.CARSON

A chromosomal mutation may serve as a precise and indelible marker that can reflect details about the evolutionary history of a species. The exciting field of chromosomal evolution was virtually founded by M. J. D. White in pioneer research carried out more than 40 years ago. His pivotal paper (1940) dealt with translocations and thus carried the study of chromosomal evolution for the first time beyond the inferences from polyploidy that were fashionable at the time. White's work was immediately extended in a landmark series, "The evolution of the sex chromosomes."

Although extensive use was first made of inversion polymorphisms in construction of phylogenies (e.g., Dobzhansky and Sturtevant, 1938), exploitation of the polytene chromosome technique soon led in a different direction, namely, to intraspecific studies of the genetics of natural populations of a number of species. Realization that the gene orders of full biological species could yield useful inferences about genetic relationships between species stems largely from work initiated by Patterson et al. (1940) on the *virilis* group of *Drosophila*. The technique was later refined by Wasserman (1954), who extended polytene analyses to species that cannot readily be hybridized.

Analyses based on gene sequence, however, have an intrinsic drawback. Thus, the direction of the evolutionary series cannot be inferred from the chromosomal data alone. In most cases, indirect outside phylogenetic information must be brought in. Indeed, when two species differ by a fixed inverted section, in most cases no chromosomal clue exists as to which of the two sequences is the ancestral condition. The result is that the phylogenetic schemes generated, in the absence of the necessary outside information, must be considered only as relationship diagrams based on a sensitive means of determining genetic similarity. Direction of evolution, however, remains indeterminate.

Beginning with the work of Carson et al. (1967), the polytene and metaphase chromosome relationships of a large group of drosophilids endemic to Hawaii have been worked out. Currently, data are available on 103 species of a closely related but archipelago-wide group, informally denoted the "picture-winged" flies (Carson and Yoon, 1980).

Outside information on the evolution of this group of flies in Hawaii is extremely powerful, especially with regard to the age of the islands on which the species are found. Most of the species are endemic to single islands, and it now appears conclusive that the islands have been thrust up successively at perforations of the Pacific tectonic plate. Thus, ages of the current high islands range from 5.6 million years (Kauai) to less than 700,000 years (Hawaii). Accordingly, the species and chromosome arrangements endemic to the newer islands may be safely designated as newly evolved in situ, permitting evolutionary direction to be established (Carson and Kaneshiro, 1976). Complementary to the above information is the discovery of Stalker (1972) that *D. primaeva,* a species endemic to Kauai, serves as a chromosomal link between the continental *robusta* group and the Hawaiian flies endemic to the newer islands.

The present chapter provides detailed phylogenetic information on a series of 14 species that may be designated the *hawaiiensis* phylad, included within the larger picture-winged group. Generally, they have been assigned to subgroup I, the *grimshawi* subgroup (Carson and Yoon, 1980).

Phylogenetic use has been made of a hitherto unexploited permanent chromosomal character. Thus, five members of this phylad manifest a striking constricted area at a particular point in polytene chromosome 2. Such areas are well known in polytene chromosomes of a number of species, and sometimes have served as specific and useful cytological markers. They appear to be areas where DNA replication is locally interrupted (Lefevre, 1976). The permanence of the constriction in question (constriction A, chromosome 2), together with information on inversions, heterochromatin differences, and geographical relationships, has permitted the construction of a rather detailed hypothetical scheme for the evolution of these species within the archipelago.

Material and methods

Specimens of the 14 species concerned have been captured, usually at bait streaks of yeasted banana, from many high-altitude locations throughout the islands. Isolated females are permitted to oviposit in the laboratory, and chromosomal studies are carried out on F_1 larvae. Analysis of aceto-orcein smears of the salivary glands of at least seven larvae is done, if possible, from each wild female. The precise localities of collection, the number of chromosomes sampled, and the intraspecific polymorphisms observed have already been reported in scattered earlier publications. The breakpoints of all inversions listed in this paper have been previously mapped and published (references in Carson and Kaneshiro, 1976).

Following a convention established at the beginning of the mapping work, chromosomal sequences on each of the five large polytene chromosome arms were determined relative to the picture-winged standard sequences, as found in the homokaryotypic stock G1 of *Drosophila grimshawi;* the standard sequences are designated simply as X 2 3 4 5 (Carson et al., 1970). Recording of alternate sequences without use of hybridization techniques was facilitated by a projection method involving the use of a compound microscope drawing-tube. This enables the observer to view the unknown sequence under the microscope at table level simultaneously with a photograph of the standard map (see Carson, 1970).

Each inversion relative to the standard is denoted by a lower-case letter following the designation of the chromosome; for example, 2b or Xa^2. The superscript on the latter merely means that the inversions are so numerous (211 in 103 species) that for certain variable chromosomes like the X, the alphabet has been used more than twice (see Carson and Yoon, 1980).

Observations

All the data on which the inferences that follow are based are given in Table 14.1. The metaphase and polytene chromosome information has been compiled from the literature (references in Carson and Kaneshiro, 1976). Annotations to Table 14.1 record the more striking karyotypic variations that have been observed in the metaphases of these species. The X and Y chromosomes are frequently similar in length in ordinary squashes, and in a number of species the sex chromosomes have not been precisely identified. Clayton (1976) and Baimai (1977) interpret this variation from species to species as being due to added heterochromatin. In one case (*D. recticilia*) these authors differ in the way they have interpreted the position of the added heterochromatin relative to the centromere (see Clayton, 1977).

Five of these species show a strong constriction at a precise point in chromosome 2 (constriction 2A). This contrasts strongly with the condition found in the other nine species at this same point on the polytene map. In the latter species, the banding at the relevant spot in the chromosome is uninterrupted and is similar to the rest of the picture-winged *Drosophila*. The 2A constriction is thus lacking in 98 species, including both the standard *D. grimshawi* and the most primitive picture-winged species, *D. primaeva* of Kauai (Carson and Stalker, 1968, 1969). That it is a condition uniquely evolved in an ancestor of these five species appears to be firmly established.

Figure 14.1*A* illustrates the homozygous presence of the standard $(++)$ condition as exemplified by *D. silvarentis* (Q10Q8, Papa, South Kona

Table 14.1. *Chromosomal characteristics of 14 species of the* Drosophila hawaiiensis *phylad*

| Island and species | Metaphase | Constriction 2A | Inversions[a] | | |
			Shared with one or more species		Not shared
Kauai					
D. musaphilia	5R,1D[b]	AA	Xa² 2b 3g 4u		
Oahu					
D. flexipes	5R,1D	AA	2b		41³/+
D. gradata	5R,1D	++	2b 3g		
D. turbata	5Rᶜ,1D[d]	++	Xa² 2b 3g 4u 5g		
Molokai					
D. villitibia	5R,1D	++	2b		
Maui					
D. gymnobasis	5Rᶜ,1D[d]	AA	Xa² 2b 3g 4u		
D. hirtipalpus	5R,1D[d]	++	2b	Xp² 2j	Xq²
D. lasiopoda	6R	++	2b		
D. recticilia	5Rᶜ,1D[d]	++	Xa² 2b 3g 4u 5g/+		3sv/+
Hawaii					
D. formella	5R,1D[d]	AA	2b 4[g]		4j³/4k³ᵍ
D. hawaiiensis	5R,1D	AA	Xa² 2b 3g 4u		3d²/+; 4t²/+
D. heedi	6Rᶠ	++	Xa² 2b 3g 4u		
D. psilotarsalis	5R,1D	++	2b	Xp² 2j	
D. silvarentis	5R,1D[d]	++	Xa² 2b 3g 4u		5k/+

[a]Inverted sequences are entered only if they differ from the standard of *D. grimshawi*.
[b]5R or 6R, five or six rods; 1D, one dot.
[c]One pair of rods more than double length.
[d]Large dots.
[e]Two pairs of rods double length.
[f]One pair of rods double length.
[g]Lacks standard 4.

Figure 14.1. Salivary gland chromosome map positions of constriction
2A. *A*. Central portion of chromosome 2 (gene arrangement 2b) of
D. silvarentis. The 2A constriction is not present at arrow (homozygous
+ + condition). *B*. The same region of chromosome 2 of *D. hawaiiensis*.
Constriction 2A is indicated at arrow (homozygous AA condition). *C*.
The same region of chromosome 2 in a hybrid between *D. gymnobasis*
and *D. silvarentis*. Heterozygosity for the constriction is shown at arrow
(A/ +). The *gymnobasis* chromosome is the one closest to the
arrowhead. Distal ends are to the left. For map reference, the breakpoints of inversion 2a are added to each figure (cf. Carson and Stalker,
1968, Figure 1).

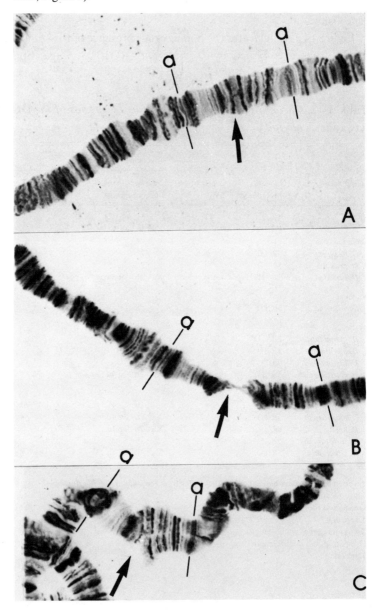

District, Hawaii). The arrow designates the relevant band which is altered in the 2A constriction. Figure 14.1*B* illustrates the homozygous presence of the 2A constriction (AA) in *D. hawaiiensis* (Q57M5, Laupahoehoe, North Hilo District, Hawaii). Figure 14.1*C* depicts the heterozygous state (A/+) in a laboratory F₁ hybrid of *gymnobasis* (AA) ♀ × *silvarentis* (++) ♂ (M64G2 *gymnobasis*, Auwahi, Hana District, Maui; K18A1 *silvarentis*, Humuula, North Hilo District, Hawaii). The constricted homologue (deriving from the *gymnobasis* mother) is on the side toward the arrow in Figure 14.1*C*. In all three photographs, the positions of the breakpoints of inversion 2a are entered. This will facilitate location of the 2A constriction region on the standard chromosome map of *D. grimshawi* (see Figure 1 in Carson and Stalker, 1968).

The presence or absence of the 2A constriction may be easily determined on each salivary gland chromosome preparation. Indeed, it is often possible to observe the condition definitively in a single cell. On the other hand, when the constriction is present, it is ordinarily not possible observationally to confirm that both homologues are indeed constricted. Comparable constrictions, nevertheless, have been studied in species hybrids and in hemi- and heterozygous X chromosomes in *D. setosimentum* (Carson and Johnson, 1975). Additionally, *D. affinidisjuncta* carries two constrictions in the standard homologue of chromosome 4. Both are absent from the 4v inversion, which segregates in the same populations as the standard homologue in natural populations of this species from West Maui (Carson and Sato, 1970; Baimai and Ahearn, 1978). Constrictions are absent in 4v4v homozygotes but are present in 4/4v heterozygotes and 44 homozygotes.

The above cases of constrictions in *D. setosimentum* and *D. affinidisjuncta* are cited here as evidence that constrictions can indeed exist in polymorphic states within a single species. No such polymorphism, however, has been detected for the 2A constriction within any of the species of the *hawaiiensis* phylad. The presence of constriction 2A has been diagnosed in every individual larva examined from wild-caught females of the five species where it is present. The numbers of larvae examined are as follows: *musaphila*, 16; *flexipes*, 60; *gymnobasis*,. 20; *formella*, 70; *hawaiiensis*, 264. Clearly, constriction 2A is a unique and permanent cytological character in the polytene chromosomes of these five species. One may go a step further and claim that these regions are basically chromosome mutations serving as useful visible regional markers.

Table 14.1 also lists the various inversions that are shared by two or more species (central columns) and those that are found within a single species only (extreme right-hand column). All inversions that are shared are fixed in the species concerned with the exception of 5g, which is fixed in *turbata* but has remained polymorphic in *recticilia*. Significantly, all 14 species are fixed for inversion 2b. This inversion is not found in any other of the 89 picture-winged species studied. Accordingly, 2b serves as a definitive cytological marker for the *hawaiiensis* phylad. In one species (*D. villitibia* of Molokai) this is the only chromosomal character separating that species from the standard, *D. grimshawi*.

More than half of the more than 100 picture-winged species are involved in homosequential groupings; that is, there is no detectable fixed polytene sequential difference between them. It should be recalled that the position of every polytene band is known and mapped in these species. Homosequential species are also frequent within the *hawaiiensis* phylad. Thus, half of the species have the same fixed inversion formula (Xa^2 2b 3g 4u). Three more show only inversion 2b relative to the standard. Furthermore, one pair of species (*musaphilia* and *hawaiiensis*) have identical metaphase karyotypes, constrictions, and fixed inversions. Nevertheless, these species manifest strong morphological and behavioral differences and are found at opposite ends of the archipelago. Identity certainly cannot extend to the genic and molecular level, and it may be noted that *hawaiiensis* shows a number of polymorphic inversions not found in *musaphilia*.

Interpretation and discussion

Figure 14.2 diagrams a hypothetical scheme for the allopatric evolution of each of the 14 species of the *hawaiiensis* phylad. This proposal is based wholly on chromosomal inversions and constriction 2A. Each existing species is represented by a solid circle, and certain hypothetical ancestral populations are represented by open circles. What follows is essentially a proposed solution to a jigsaw puzzle in space and time.

Starting in the upper left-hand corner, a curving line represents the arrival of the standard sequences (X 2 3 4 5) on Oahu from Kauai via an interisland founder. In previous papers (Carson et al., 1970; Carson and Yoon, 1980) documentation of the presence of the standard sequences on both islands has been presented. This has been interpreted as requiring an event involving the movement of an interisland founder individual, possibly only a single fertilized female (see Carson, 1971). In past publications each founder event involving the four major island areas (Kauai, Oahu, Maui complex, and Hawaii) has been given an identifying number. This system is continued in this paper. The one just referred to is No. 5 (see

Figure 16, Carson et al., 1970); others may be found on Figure 14.2 and in the figures in Carson and Yoon (1980).

The key inversion for the *hawaiiensis* phylad is thought to have arisen on Oahu after the establishment of the standard founder. This is not difficult to visualize because in addition to *D. grimshawi* itself (formula X 2 3 4 5) there are two other species homokaryotypic for the standard on Oahu (Carson and Yoon, 1980). Indeed, these three species are chromosomally both monomorphic and homosequential. This ancestral 2b population, probably at a very early time after fixation of 2b, acquired constriction 2A in the polymorphic state, a condition that apparently persisted in ancient populations on both Oahu and Maui. From this 2b 2A/+ ancestor, four lines of descent were derived (Figure 14.2). Two of these involve interisland founders No. 34 and No. 35. The former gave rise to a lineage of four species via *villitibia* of Molokai. These species retain the ancestral nonconstricted (+ +) chromosome 2b in homozygous state. A second line of descent leads directly to *formella* of Hawaii, which is homozygous 2A. The third derivative gave rise to the modern *flexipes* on Oahu, a species

Figure 14.2. Evolution of 14 species of the *hawaiiensis* phylad of picture-winged Hawaiian *Drosophila* based on polytene chromosome characters. Solid circles: existing species; open circles: hypothetical populations. Encircled numbers on arrows represent hypothetical founder events. Stepwise fixations of inversions relative to the standard (X 2 3 4 5, founder No. 5) are indicated (e.g., 2b, 3g, etc.): + +, homozygous for the standard condition at polytene chromosome constriction A of chromosome 2; A/+, heterozygous, and AA, homozygous for the constriction.

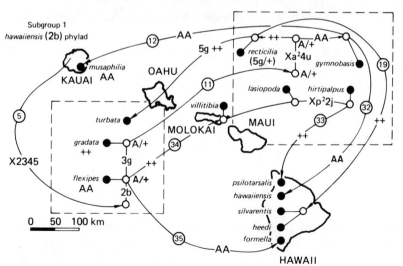

homosequential with *formella* of Hawaii and having the same constriction condition.

The fourth derivative from the 2b 2A/+ ancestor has acquired the important marker inversion 3g, which is fixed in all the other eight species of the phylad. During this period of evolutionary time, 2A appears to have remained in the polymorphic state. That 3g arose on Oahu is suggested by the fact that the 2b 3g formula is found today only on that island, embodied in the modern species *gradata*. It is further inferred that a derivative of an ancestral 2b 3g 2A/+ population reached Maui (founder No. 11, Figure 14.2). Subsequently, two new fixed inversions (Xa^2 and 4u) were added, giving rise to a population differing from standard by four inversions (Xa^2 2b 3g 4u). Seven species have this formula, and it appears to have spread widely in the islands.

It is noteworthy that in some years of collecting no such species has been found on Molokai. Completion of the present-day formulae as observed may have been achieved by the final separation of two lines of descent on Maui. These are the *recticilia–turbata–silvarentis–heedi* group, which lack the 2A constriction, and *gymnobasis–musaphilia–hawaiiensis,* which bear it in fixed condition.

This evolutionary scheme thus makes the assumption that the major sequential karyotype (Xa^2 2b 3g 4u) evolved on Maui. This necessitates interpreting both *turbata* (Oahu) and *musaphilia* (Kauai) as "back migrants," the founders for which moved in the opposite direction from the usual southeastern mode of colonization. A key fact suggesting this is that *recticilia* of Maui carries the standard ancestral chromosome 5 in the polymorphic state. It would therefore be a more ancient species than *turbata,* which carries the newer arrangement 5g in the fixed condition.

D. musaphilia of Kauai is thought to have been founded from Maui directly, as it has the same polytene formula as *gymnobasis* and, indeed, as *hawaiiensis* of Hawaii. It seems unlikely that it could be derived from *turbata* of Oahu, because of the difference at constriction 2A.

Perusal of the data on metaphase karyotypes given in Table 14.1 reveals that they are not useful in constructing phylogenetic schemes. Metaphase variations are often striking. The six-rod (6R) condition represents a case in which the microchromosome (the "dot") has acquired a long heterochromatic segment. The two species showing this condition are far apart according to inversions. Similarly, the rest of the metaphase variation cannot be used phylogenetically, and one is left with the conclusion that heterochromatic additions to the karyotype are unpredictable and sporadic in occurrence. This has been widely observed in the chromosomal data on the picture-winged species, and this observation is not confined to the *hawaiiensis* phylad.

Emphasis should be placed on the fact that the scenario developed in this paper is based wholly on chromosomal characters. Detailed morphological, hybridization, and electrophoretic information has not been systematically collected or published in a form that would add much to the current interpretations. Most particularly, electrophoretic data would be a useful adjunct to the chromosome data. Genetic similarity based on such biochemical characters appears to erode at a steady rate following the cladistic event that occurs as a founder establishes a new population (Carson, 1976). Such data would be very important in evaluating the chromosomally based hypotheses and in dating these past events.

External morphology is not especially reliable for phylogenetic inferences in Hawaiian *Drosophila*. Much of the extraordinary morphological diversity found appears to relate to the complex sexual behavior of these flies and is the product of sexual selection. Abrupt and convergent changes frequently accompany speciation.

This particular phylad includes some species that breed in the slime fluxes of certain endemic forest trees of Hawaii, including *Sapindus, Myoporum, Osmanthus,* and especially *Acacia koa* (Montgomery, 1975). The last-named tree, which has a wide geographical and altitudinal range in the islands, serves as a host for *musaphilia, gradata, turbata, recticilia,* and *hawaiiensis*. Some of the species (e.g., *hawaiiensis*) can complete development on the slime fluxes of several species of host tree. Clearly, the species are quite similar ecologically. Nevertheless, a striking case of niche separation is known between *D. silvarentis* and *D. heedi* (Kaneshiro et al., 1973). The former oviposits on trunk fluxes of *Myoporum sandwicense,* whereas the sympatric *D. heedi* utilizes areas of soil saturated by flux drip. Accordingly, eggs and larvae of the two species are spacially separated, utilizing two quite different variations of the flux niche. Although the metaphases of these two species are different, and *heedi* manifests an extraordinarily large puff in chromosome X, the two could have diverged recently on the island of Hawaii, as indicated in Figure 14.2.

In closing, mention might be made of the opportunity for unusually accurate dating of the evolutionary events recounted here. Precise dating of the lava flows gives evidence that the islands have been formed successively from northwest (Kauai, 5.6 million years) to southeast (Hawaii, less than 700,000 years). The scheme proposed in this paper begins on Oahu, an island that shows no lava flows older than 3.3 million years (Macdonald and Abbott, 1970). Accordingly, the 14 relevant species must have evolved in somewhat less than that time.

The chromosomal data indicate that after an early episode on Oahu, major evolutionary events occurred on Maui, an island having no lava flows older than 1.3 million years. The *villitibia* subphylad colonization of Molokai may have taken place somewhat earlier (1.8 million years). On the other hand, the colonization of the island of Hawaii (five species allopatric to the rest) seems likely to have been quite recent, as no lava flows from that island are known that are older than 700,000 years.

Summary

The *hawaiiensis* phylad consists of 14 large picture-winged species of *Drosophila* distributed over five of the six major islands of the Hawaiian archipelago. Each species is confined to a single island; Hawaii island has the largest number, five. The species are morphologically quite distinct from one another but share to varying degrees a number of fixed inversions and a specific polytene constricted region. Based on these data, a detailed proposal is made concerning the evolutionary pattern of these species. After an early episode on Oahu, perhaps three million years ago, major evolutionary events apparently occurred on Maui about one million years ago. The five species apparently reached the newest island (Hawaii) less than 700,000 years ago. The pattern appears to be largely allopatric, and the founder effect is evoked to explain the formation of most of the species. Although the patterns of metaphase karyotypes and ecological substrates used are complex, they appear to be less useful for phylogenetic interpretation than the polytene chromosome characters.

I acknowledge the great skill of Linden Teramoto and Joyce Kurihara, who prepared most of the salivary gland smears used in this study. The work was supported by Grant BMS 74-22532 from the National Science Foundation.

References

Baimai, V. 1977. Chromosomal polymorphisms of constitutive heterochromatin and inversions in *Drosophila*. *Genetics 85:*85–93.

Baimai, V., and Ahearn, J. N. 1978. Cytogenetic relationships of *Drosophila affinidisjuncta* Hardy. *Am. Midl. Nat. 99:*352–60.

Carson, H. L. 1970. Chromosome tracers of the origin of species. *Science 168:*1414–18.

– 1971. Speciation and the founder principle. University of Missouri, *Stadler Symp. 3:*51–70.

– 1976. Inference of the time of origin of some *Drosophila* species. *Nature (Lond.) 259:*395–6.

Carson, H. L., and Johnson, W. E. 1975. Genetic variation in Hawaiian *Drosophila*. I.

Chromosome and allozyme polymorphism in *D. setosimentum* and *D. ochrobasis* from the island of Hawaii. *Evolution 29:*11–23.

Carson, H. L., and Kaneshiro, K. Y. 1976. *Drosophila* of Hawaii: systematics and ecological genetics. *Annu. Rev. Ecol. Syst. 7:*311–46.

Carson, H. L., and Sato, J. E. 1970. Microevolution within three species.of Hawaiian *Drosophila. Evolution 23:*493–501.

Carson, H. L., and Stalker, H. D. 1968. Polytene chromosome relationships in Hawaiian species of *Drosophila*. I. The *D. grimshawi* subgroup. *Univ. Tex. Publ. 6618:*335–54.

– 1969. Polytene chromosome relationships in Hawaiian species of *Drosophila*. IV. The *D. primaeva* subgroup. *Univ. Tex. Publ. 6918:*85–94.

Carson, H. L., and Yoon, J. S. 1980. Genetics and evolution of Hawaiian *Drosophila*. In: *The Genetics and Biology of Drosophila,* vol. 3b. (ed. Ashburner, M., Carson, H. L., and Thompson, J. N.). New York: Academic Press.

Carson, H. L., Clayton, F. E., and Stalker, H. D. 1967. Karyotypic stability and speciation in Hawaiian *Drosophila. Proc. Natl. Acad. Sci. U.S.A. 57:*1280–5.

Carson, H. L., Hardy, D. E., Spieth, H. T., and Stone, W. S. 1970. The evolutionary biology of the Hawaiian Drosophilidae. In *Essays in Evolution and Genetics in Honor of Theodosius Dobzhansky* (ed. Hecht, M. K., and Steere, W. C.), pp. 437–543. New York: Appleton-Century-Crofts.

Clayton, F. E. 1976. Metaphase configurations in *Drosophila:* a comparison of endemic Hawaiian species and non-endemic species. *Proc. Arkansas Acad. Sci. 30:*32–5.

– 1977. Chromosomes of *Drosophila recticilia. Genetics 86:*499–500.

Dobzhansky, Th., and Sturtevant, A. H. 1938. Inversions in the chromosomes of *Drosophila pseudoobscura. Genetics 23:*28–64.

Kaneshiro, K. Y., Carson, H. L., Clayton, F. E., and Heed, W. B. 1973. Niche separation in a pair of homosequential *Drosophila* species from the island of Hawaii. *Am. Nat. 107:*766–74.

Lefevre, G. 1976. A photographic representation and interpretation of the polytene chromosomes of *Drosophila melanogaster* salivary glands. In: *The Genetics and Biology of Drosophila,* vol. 1a (ed. Ashburner, M., and Novitski, E.), pp. 32–66. New York: Academic Press.

Macdonald, G. A., and Abbott, A. G. 1970. *Volcanoes in the Sea.* Honolulu: University of Hawaii Press, 441 pp.

Montgomery, S. L. 1975. Comparative breeding site ecology and the adaptive radiation of picture-winged *Drosophila. Proc. Haw. Entomol. Soc. 22:*65–102.

Patterson, J. T., Stone, W., and Griffen, A. B. 1940. Evolution of the *virilis* group in *Drosophila. Univ. Tex. Publ. 4032:*218–51.

Stalker, H. D. 1972. Intergroup phylogenies in *Drosophila* as determined by comparisons of salivary banding patterns. *Genetics 70:*457–74.

Wasserman, M. 1954. Cytological studies of the *repleta* group. *Univ. Tex. Publ. 5422:*130–52.

White, M. J. D. 1940. The origin and evolution of multiple sex chromosome mechanisms. *J. Genet. 40:*303–36.

15　Reproductive isolation: a critical review

MURRAY J.LITTLEJOHN

> The grossest blunder in sexual preference, which we can conceive of an animal making, would be to mate with a species different from its own and with which the hybrids are either infertile or, through the mixture of instincts and other attributes appropriate to different courses of life, at so serious a disadvantage as to leave no descendants.
>
> R. A. Fisher (1930:130)

For some 40 years the notion of reproductive isolation has been presented as the keystone of the biological species concept (Dobzhansky, 1937, 1975; Mayr, 1942, 1970; Bush, 1975; Dobzhansky et al., 1977; White, 1978); and it is usually couched in terms of more-inclusive units of biological organization, such as demes, populations, and gene pools, rather than individuals. Thus, Mayr (1970:13) defined a biological species as: "a protected gene pool. It is a Mendelian population that has its own devices (called isolating mechanisms) to protect it from harmful gene flow from other gene pools." Likewise, Dobzhansky (1975:3638) stated that: "Species in sexually reproducing and outbreeding organisms are arrays of Mendelian populations reproductively isolated from other population arrays. Reproductive isolation is thus an earmark of species distinction."

The term "isolating mechanisms" was introduced by Dobzhansky (1937:405–6) as "a convenient general name for all the mechanisms hindering or preventing the interbreeding of racial complexes or species." It is usual to separate extrinsic isolation arising from the imposition of a barrier of impassable habitat (geographic isolation) from other forms of isolation which derive from the genetic, ecological, structural, and behavioral properties of sexually reproducing organisms, which are then termed reproductive isolating mechanisms (Riley, 1952). The concept of reproductive isolation is thus restricted to these latter mechanisms, and is of relevance only in biparental systems.

The idea of reproductive isolation can be traced to Darwin (1859), who, although he considered infertility and hybrid sterility in some detail, gave other aspects only scant treatment. The first detailed accounts of reproductive isolation, although described with a different terminology, were published in the early post-Darwinian period by Romanes (1886, 1897), Wallace (1889), and Gulick (1890, 1891), all of whom saw isolation and segregation as central issues in the process of divergent evolution. Gulick (1890) developed a comprehensive classification of the means of segregation (i.e., reproductive isolation), including all the main categories recognized in the modern schemes that stem from that presented by Dobzhansky (1937). Romanes (1886) was mainly concerned with physiological segregation, which may now be equated with genetic incompatibility, and he later provided a detailed and critical review of the literature on the subjects of isolation and segregation (Romanes, 1897). Wallace (1889) also considered the topic of isolation in some depth, and emphasized the role of natural selection in the origin of reproductive isolation. It was then commonly believed that separation (or geographic isolation), by preventing intercrossing, was important in allowing the differentiation of varieties and species. That infertility and hybrid sterility could not arise through the direct action of natural selection, but only as incidental effects or by-products of other adaptive processes, was realized by Darwin (1859:245); but Romanes (1886, 1897) and Wallace (1889) presented arguments attempting to explain the direct role of natural selection in the origin of infertility and hybrid sterility. Gulick (1890:235–6) clearly outlined the process whereby sexual preference and sexual discrimination (leading to homogamy) could be favored by natural selection, whereas Wallace (1889:174–80) was vague on this point, and seems to have confused the origin of homogamy with that of infertility and hybrid sterility. The evolution of reproductive isolating mechanisms will be explored in more detail in a later section.

As noted by Riley (1952), there was a "dark age" for studies of reproductive isolation during the first quarter of the twentieth century, perhaps because of the rediscovery of Mendelism and the great development of genetics that occurred during this period. Robson (1928) described the forms of isolation operating between species, but did not consider their origins. Fisher (1930) made a significant advance in the provision of a theoretical basis for divergent evolution and homogamy, viewing them in terms of individual fitness, and from a genetic point of

view. Robson and Richards (1936) provided a more advanced description
of the forms of isolation than that presented by Robson (1928). The
subject was further developed by several authors in *The New Systematics,*
edited by Huxley (1940); and H. J. Muller (1940, 1942), in particular,
emphasized the genetic aspects of isolation. It seems, however, that the
concept of reproductive isolation did not attract significant and wide
interest until the formulation of the biological species concept by Dobzhansky (1937, 1940) and Mayr (1942).

Most recent classifications of reproductive isolating mechanisms have
emphasized a dichotomy between two levels of action: premating and
postmating, this division being based on the conservation of gametes, and
on the assumption that only those mechanisms operating at the premating
level are amenable to the direct action of natural selection for their
isolating effect per se (Littlejohn, 1969; Levin, 1978). But this dichotomy
may be arbitrary. Hagen (1967) found it difficult to separate the lack of
hybrid fitness due to ecological incompatibility (a postmating mechanism),
which was a consequence of differential parental adaptation, from the
ecological isolation of the parents (a premating mechanism). V. Grant
(1966, 1971) and Levin (1971, 1978), when considering reproductive
isolation in higher plants, saw the rejection of foreign pollen grains by the
stigma as amenable to natural selection – an interaction that would be
considered at the postmating level (i.e., after the release of gametes) in
animal systems. Coyne (1974) developed a hypothetical case of a viviparous form in which natural selection would favor the earlier recognition
and abortion of a hybrid embryo of lowered fitness, an event clearly
occurring at the postmating level.

If one rejects the concept of biotic adaptation (sensu Williams, 1966)
(i.e., an adaptation that is directly selected for benefits acting only at a
more inclusive level than that of the individual), then all evolutionary
phenomena and processes should first be considered in terms of advantages
only to individuals, and as measured by their consequent increase in
reproductive success and fitness relative to alternative states. Furthermore,
errors of interpretation, such as the confusion of cause and effect, and of
incidental by-product and evolved function, can easily arise in thinking
developed at the more inclusive or group level (Williams, 1966).

It is clear from the statements cited earlier that the concepts of the
biological species and reproductive isolation were developed largely in
terms of supraindividual units of biological organization. It now seems
appropriate to reconsider those devices and mechanisms previously
subsumed under reproductive isolation in terms of their relevance to the

fitness of sexually reproducing individuals. Only then may one usefully appraise the processes leading to the development and maintenance of the distinct biparental systems, generally referred to as biological species. Unavoidably, such an individual-oriented approach thus must have a direct bearing on the biological species concept, for it represents another attempt to reintroduce nominalism into this area of evolutionary biology, a philosophy that was strongly rejected by Mayr on several occasions (e.g., Mayr, 1969). Such a nominalistic approach is the only one that can be applied to uniparental (asexual) systems, in which each "species" is seen as a cluster or cloud of individuals representing an adaptive node or adaptive peak (Dobzhansky, 1940; Hutchinson, 1968).

An immediate difficulty that arises in this nominalistic approach to the species problem is seen in sexual reproduction, which involves the cooperation of two individuals, and generally the reconstruction of their previously successful genotypes and resultant phenotypes (as evidenced by their survival and successful reproduction) through recombination. Hence, the uniqueness of the individual and its associated genotype is lost, and the high level of fitness of each parent is deliberately broken down when its haploid genome is combined with that of another individual. But if the individual is seen as the principal unit of selection in contemporaneous biological systems, then by considering only the effective life span of an individual – between the formation of a zygote and the end of its reproductive life – these conceptual difficulties may be avoided. The evolutionary significance of sex has been discussed by Williams (1975) and Maynard Smith (1978).

Mate selection and sexual preference

A convenient starting point for reexamining the concept of reproductive isolation in terms of individuals rather than groups is to develop a generalized account of mate selection based on the approach of Jones (1928), Burma (1949), and Straw (1955). Of all those individuals detected or encountered by a subject during its period of reproductive activity, some will represent potential mates (i.e., will be of opposite sex, of appropriate size, and also in breeding condition), and thus constitute the critical array from which a "choice" must be made. The range of potential mates may be further reduced by considering only those individuals that produce gametes with which the subject's haploid genome, represented by its gametes, is genetically compatible. That is, a mate must be selected from individuals with sufficiently similar genotypes that the shared and combined genetic information will result in a viable zygote (or zygotes) that will develop into a functional, fertile individual. Furthermore, the close match of adaptation of the progeny to existing or anticipated environments, their ability to efficiently utilize or compete for limiting resources, and their success in avoiding predation are critical if fitness is to be maintained. Hence, optimal mates must exhibit phenotypically evident signs regarding their suitability as reproductive partners. Because of the almost universal occurrence of outbreeding and associated genetic recombination, it must be assumed that it is advantageous to produce progeny that differ to some extent from their parents (Williams, 1975). Thus, there should be some optimal amount of genetic difference between potential parents that will yield a maximum level of reproductive success. If the genotypes of the parents are too similar, then fitness may be lowered because of a lack of sufficient variation among progeny, where such variation may be important for survival of at least some offspring in unpredictable future selective regimes (Williams, 1975). This condition need not apply, however, where the advantage of sexual reproduction rests solely with the level of reproductive success that results from having two parents with a significant genetic interest in the progeny, and thus sharing parental care, as in the alternation of incubation and feeding in brooding birds. If the parental genomes are too dissimilar, then genetic incompatibility could arise, or there may be a loss of adaptation due to the intermediacy of the progeny, both aspects leading to a lowering of reproductive success.

Hence, a breeding individual should select as a mate another individual with some optimal level of genetic difference, and because this is a mutual interaction, the reciprocal condition would apply to the other potential mate. Because anisogamy (gamete dimorphism and the male–female phenomenon) is usual in sexually reproducing forms, one mate, generally the female, may have a greater investment than the other in the reproductive products, in the nurturing and development of the zygote, or in parental care. In promiscuously breeding systems, males may have a much smaller investment in each fertilizing event, and, in contrast to females, the opportunity or potential to experience multiple matings during a breeding season. With anisogamy and associated sexual dimorphism, potential mates of one sex may present themselves through display to a wide range of individuals of the other sex. In contrast, individuals of the opposite sex may exercise discrimination or "choice," such that the outcome is one of maximizing reproductive success, and hence fitness, for both interactants.

For breeding systems in which the two parents are essential for the survival of the progeny, the level of genetic difference between them may be assigned to a secondary role. In such balanced and monogamous systems with sustained pair bonds and parental care, the mutual assessment of potential mates would be more important, and requires that individuals of both sexes have detectable indicators of fitness incorporated into their phenotypes. Such fitness-indicating attributes could be further developed and exaggerated through sexual selection, as outlined by Fisher (1930); but it seems unlikely that a "runaway" process (positive feedback) could proceed very far and independently of other forms of natural selection, if viewed in terms of the elemental and universal currency of energy.

One point that emerges from this consideration of mate selection is its positive nature (i.e., the active selection of an appropriate mate) and its extension to the notions of sexual selection. Thus the assortative mating that results is positive in two senses: statistically, and through active selection. Paterson (1978) stressed this positive nature of mate selection, and described such processes and devices as specific mate recognition systems (SMRS). The correlation between positive assortative mating, which results from the active processes performed by breeding individuals, and sexual or ethological isolation is obvious.

This argument may be extended to flowering plants that utilize animals as pollinators, and to the development of floral structure and associated homogamy. If other constraints are placed on the system, however, such as a limited number of kinds of pollinator, or pollination by passive dispersal by wind, then a temporal displacement of flowering seasons could be the result of natural selection favoring efficient homogamy (McNeilly and Antonovics, 1968; Crosby, 1970). Most other aspects of reproductive biology and anatomy that facilitate homogamy (e.g., habitat separation and morphological differences) may have ambiguous explanations.

Ecological aspects of biotic discontinuity

It is a commonplace observation that a discontinuous array of biological diversity exists, particularly when viewed in a spatially restricted situation. Furthermore, it is clear that the understanding of the mechanisms and processes accounting for the origin and maintenance of this diversity represents a most fundamental area of evolutionary biology and ecology. Of particular relevance to the present essay is the consideration of the role of ecological factors in the processes of divergent evolution in biparental systems.

The significant initial steps in the formalization of a combined ecological and genetic analysis of biological diversity were taken by Fisher (1930) and Wright (1932). Fisher (1930:125–31) advanced the idea of environmental heterogeneity and the associated development of adaptive complexes, and saw the possibility of unstable, though inhabitable states, and the application of a divergent evolutionary stress upon the "cohesive power of the species." Most of his subsequent discussion was concerned with a spatial model of divergent evolution in a continuously distributed array of interbreeding individuals spread along an ecological gradient. The basis for divergence under these conditions was seen as the maintenance of a selective gradient and the reduction of rate of exchange of genetic information through the linking of "heritable variations which will influence diffusion" (Fisher, 1930:127–8), namely, those concerned with dispersal and habitat selection. Wright (1932) advanced the notion of adaptive peaks, wherein biotic responses to a range of environments and associated selective regimes were likened to a topographic landscape on which relative fitness was directly equated with altitude. Thus the elevated areas (peaks) represented successful adaptive complexes, the intermediate altitudes (foothills and plains) lower levels of fitness, and the deepest parts (valleys) unoccupiable conditions in which fitness was negative.

Dobzhansky (1940:316) aptly described the existing state of biological diversity when he noted that: "each species is . . . an adaptive complex which fits into an ecological niche somewhat distinct from those occupied by other species," and: "The adaptive value of such a complex . . . is a property of the genotype as a whole." Simpson (1951, 1961) further developed the adaptive explanations of diversity when advancing the evolutionary species concept in which the separate evolutionary role and tendencies were emphasized. The idea of an adaptive grid on which biotic discontinuity was associated with essentially unstable ecological zones was advanced by Simpson (1953:199–206) and further developed by Van Valen (1976).

Hutchinson (1957) introduced a measure of precision to the ecological niche in the idealized concept of the multidimensional state or hypervolume in which each environmental variable is given a value along a separate coordinate. He subsequently equated the niche hypervolume to the total selective regime acting on a biological system, and saw the evolutionary response (through natural selection) to such a set of factors as a cluster, or cloud of phenotypes. Thus, discrete groups of individuals could exist if integrated responses or phenotypic transforms, and the associated selective regimes, were discontinuous (Hutchinson, 1968). Hutchinson (1968) considered that these clouds of points (sets of individuals) could be equated to the adaptive peaks of Wright (1932), in which case, fitness might be related to the density of points (phenotypes) in a cloud.

In spite of his insistence on the primary role of reproductive isolation in explaining the maintenance of diversity in coexistent biparental systems, Mayr (1957) nevertheless clearly appreciated the importance of ecological factors for uniparental systems. Thus, he commented that:

> Curiously enough there seem to be a number of discontinuities which make taxonomic subdivision possible. The most reasonable explanation of this phenomenon is that the existing types are the survivors among a great number of produced forms, that the surviving types are clustered around a limited number of adaptive peaks, and that ecological factors have given the former continuum a taxonomic structure. [Mayr, 1957:382]

It is curious that he did not then extend this significant conclusion to biparental systems. Later, Mayr (1963:546-7) saw ecological compatibility as an important property of biparental species in addition to reproductive isolation, although still giving it a secondary role, presumably in the context of sympatry. He also saw that the move to a new ecological niche can lead to speciation because of the shift in adaptation and associated genetic reconstruction of a population (Mayr, 1963:575).

Van Valen (1976:233), when introducing an ecological species concept, slightly modified the definition of the evolutionary species advanced by Simpson (1961) to read: "A species is a lineage (or a closely related set of lineages) which occupies an adaptive zone minimally different from that of any other lineage in its range and which evolves separately from all lineages outside its range." Levin (1979), in a review of the nature of plant species, also discussed the ecological approach to species as adaptive and evolutionary units. He concluded that the most operational and utilitarian concept for plants is one of "a mental abstraction which orders clusters of diversity in multidimensional character space" (Levin, 1979:381), a conclusion that is similar to that of Hutchinson (1968), and with nominalistic tendencies.

A basic assumption in this approach, which emphasizes the ecological control of diversity, is that the discontinuities between adaptive sets cannot be occupied by any of the available biotic systems (or their evolved derivatives) because the gaps represent unstable zones (sensu Simpson, 1953) or nonexistent states, or are already occupied by the members of other genetic systems. Thus, biological organization, with its distinctive breaks in ranges of variability, may be explained primarily in terms of the operation of discrete selective regimes (equivalent to the ecological niche, in part), and the evolutionary responses reflected in the associated phenotypic transforms (sensu Hutchinson, 1968). Hence the origin and maintenance of the distinctness of at least some, and perhaps most, clusters of sexually reproducing individuals may be explained without any recourse to reproductive isolation. Ehrlich and Raven (1969) have attributed the phenetic constancy within spatially separated subunits of an adaptive cluster largely to the similarity of selective regimes and the sustained operation of both stabilizing and disruptive selection, rather than high levels of migration and of interbreeding between members of different subunits.

But such a primarily ecological explanation can only account for the basic number of independent and discrete adaptive systems that could exist in the biosphere. This number could be increased by several processes, or perhaps reduced by others. Ecologically similar and stable adaptive peaks may be scattered through space and time. These peaks could be colonized by independent genetic systems which have then achieved an equivalent level of adaptation through convergent evolution, or with an equally effective but independent complex of characters based on the potential and constraints of the phenotypes of the colonizing individuals. Alternatively, the colonists may be derived from the same ancestral stock, and could then achieve an equivalent level of adaptation with eventual phenotypic similarity (parallel evolution), but through a distinctly different sequence of underlying genetic changes.

Polymorphism, wherein one interbreeding system includes individuals of alternative adaptive characteristics, is an evolutionary strategy that allows exploitation of more than one adaptive peak, as might a very general level of adaptation in a simplified biological environment. Two independent but sympatric adaptive systems may share the resources of one adaptive peak under a regime of frequency-dependent selection (Mather, 1955; Ayala and Campbell, 1974). Rosenzweig (1978) also has discussed the situations in which disruptive gaps are not necessary or sufficient for explaining the origin of distinctive genetic adaptive systems.

Acceptance of the plausibility of a primary role for extrinsic factors in the establishment and maintenance of a discontinuous range of biological diversity leads directly to a general ecological species concept in which there is no requirement for reproductive isolation, and which is thus applicable to both biparental and uniparental systems. As Levin (1979:383) has pointed out, if the origin of species is defined by the appearance of reproductive isolating mechanisms, then the process is divorced from divergent evolution because divergence is not dependent on reproductive isolation. An ecological species concept is not a universal one, and does not, of course, alleviate the taxonomic problems associated with the other species concepts. All degrees of gradation between adaptive peaks are to be expected, so that inhabitable saddles may support intermediate phenotypes, thus preventing the nonarbitrary assignment of individuals to one adaptive set or another.

The process of differential ecological adaptation in closely related genetic systems (sets of interbreeding individuals) derived from one common ancestral system also will involve the directly selected evolution of attributes that may have incidental effects on the process of mate selection. A shift in habitat or host that is maintained during the reproductive phase will incidentally segregate the members of the derived systems, thus greatly reducing or eliminating the chance of selecting a member of the alternative adaptive system as a mate (Bush, 1975). Likewise, the complete displacement of breeding seasons of two coexistent sets of individuals due to competitive interactions (Rosenzweig, 1978), or to the same selective factor acting on different sensitive phases of complex life cycles (Alexander, 1968), could result in the elimination of contacts between reproductive individuals of two systems.

The reduction in predation due to more effective background matching, and the consequent restriction in habitat range, may also indirectly lead to positive assortative mating. Tauber and Tauber (1977a,b) developed a model to explain the involvement of habitat shift, displacement of breeding seasons, and predation in the sympatric origin and subsequent maintenance of homogamy in the two interfertile neuropteran systems, *Chrysopa carnea* and *C. downesi,* which occur on evergreen conifers and deciduous broad-leaved trees, respectively.

Fisher (1930) considered the process of differential adaptation that took place along a smooth ecological gradient, that was initiated during a phase of continuous geographic distribution such that there was the potential for free interbreeding between adjacent individuals and for migration from one region to the other. He postulated that, after a time, genetic differences would tend to accumulate in those individuals inhabiting the extreme regions, so that there was a progressive increase in the level of adaptation to those conditions. He saw that two categories of genes could be involved in this process of adaptation – those concerned directly with the adaptation to the particular conditions, and those that influenced the diffusion of genetic information through the system. Thus individuals that tended to move away from the region to which they were maximally adapted would be of lower fitness than those that remained in these conditions. As a consequence:

> The constant elimination in each extreme region of the genes
> which diffuse to it from the other, must involve incidentally the

elimination of those types of individuals which are most apt so to diffuse. . . . The effect of such a progressive diminution in the tendency to diffusion will be progressively to steepen the gradient of gene frequency at the places where it is highest, until a line of distinction is produced, across which there is a relatively sharp contrast in the genetic composition of the species. Diffusion across this line is now more than ever disadvantageous, and its progressive diminution, while leaving possibly for long a zone of individuals of intermediate type, will allow the two main bodies of the species to evolve almost in complete independence. [Fisher, 1930:128]

The reasons for the development of a definite line under these conditions are not clear. If, however, this type of reasoning were applied to a sharp ecological transition, rather than a smooth gradient, then it is possible that a "zone of avoidance" could develop, with a corresponding decrease in density of the two well-adapted forms near the interface. As an incidental effect of this trend, which might be termed the "ecotonal avoidance effect," there would be a progressive reduction in the probability of encounters between reproductive individuals of each adaptive set, so incidentally resulting in a high degree of homogamy.

Categories for distinctive sets of successfully reproducing individuals

Turesson (1922a,b) introduced the terms coenospecies, ecospecies, and ecotype during his pioneering studies of the genecology (i.e., the experimental-genetic approach to ecology) of higher plants, in which he stressed the importance of discrete selective regimes in explaining the distinct phenotypic gaps between Linnean species. Although his definition of the ecotype is clearly stated as: "an ecological unit to cover the product arising as a result of the genotypical response of an ecospecies to a particular habitat" (Turesson, 1922a:112), those of the coenospecies and ecospecies are not so clear. The original ecospecies concept was seen as "an understanding of the Linnean species from an ecological point of view" (Turesson, 1922a:102). The term coenospecies was used "to denote the total sum of possible combinations in a genotype compound" (Turesson, 1922b:345).

The development of the ecospecies concept can be seen in the following quotation:

It is evident that we do not find realized in nature the whole possible range of combinations within such an ecospecies because of the control of living and non-living factors of the outer world. If the ecospecies be subjected to artificial crossing or withdrawn from the close control of some of these controlling factors, as is already accomplished to a certain extent when the species is brought under culture by man, the great number of possible combinations within the ecospecies might be brought to light. Such an extension of the limits ordinarily set by nature might also, as is well known, be attained when different species become crossed. [Turesson, 1922b:344–5]

Thus the potential range of phenotypic variation is greater than that realized in ecospecies because of the constraining influence of discrete selective regimes. The limits to such genetic recombination and variability then are set at the boundary of the coenospecies. Ecotypes would be characterized by intergradations. Turesson held to his initial concept in this statement:

From the point of view of genecology the Linnean species represent a genetically complex community, the distribution and the composition of which is largely determined by the ecological factors and the genotypical constitution of the individuals comprising the species-community. The Linnean species represents as such a much [sic] important ecological unit, to which the name *ecospecies* has been given by the present writer. [Turesson, 1923:172]

Later, Turesson (1929:333) enlarged his definition of an ecospecies to become: "An amphimict population the constituents of which in nature ' produce vital and fertile descendents with each other giving rise to less vital or more or less sterile descendants in nature, however, when crossed with constituents of any other population."

Gregor et al. (1936) interpreted Turesson's classificatory units as follows:

Coenospecies. A group distinguished by morphological, physiological or cytological characters, or a combination of these; separated from all other plants by sterility or by failure of hybrids to produce viable seed.

Ecospecies. A group also distinguished by morphological, physiological or cytological characters, or a combination thereof; separated from other parts of its coenospecies by restricted interfertility or by failure of hybrids to establish themselves in Nature.

Ecotype. A population distinguished by morphological and physiological characters, most frequently of a quantitative na-

ture; interfertile with other ecotypes of the ecospecies, but prevented from freely exchanging genes by ecological barriers.
[Gregor et al., 1936:325]

Gregor (1939:313–14) later defined the coenospecies as: "a group of sexually reproducing plants separated from other groups by sterility or by the failure of the hybrids to produce viable seeds," and saw ecospecies as: "being separated from [one] another by restricted interfertility and by failure of hybrids to establish themselves in nature." Ecotypes were seen as: "populations which occupy particular ranges [of variation] on an ecocline [or ecological gradient]" (Gregor 1939:319).

Clausen et al. (1939) also adopted the three genecological categories of Turesson, and advanced criteria similar to those of Gregor et al. (1936), namely, an absolute genetic barrier for the separation of coenospecies, and free interbreeding between ecotypes. Thus: "If there is no genetic but only geographic or ecologic isolation, the units are considered ecotypes of one ecospecies. . . . The only difference between the two units [ecotypes of one ecospecies] is the lack of a genetic barrier between ecotypes" (Clausen et al., 1939:105). They treated the ecospecies as a unit with partial genetic incompatibility. Thus they made the following comment on the ecospecies: "If the genetic isolation is only partial, we usually find that there is an additional geographic or ecologic isolation" (Clausen et al., 1939:104).

Gregor (1944), in a comprehensive review of the three genecological units, traced their development as evolutionary concepts, and discussed the one most significant addition to the original definitions, namely, that of partial genetic isolation to the ecospecies. Turrill (1946) also reviewed the history of the three units.

Valentine (1948, 1949) separated the ecospecies into two categories: abrupt or *a-ecospecies,* and gradual or *g-ecospecies.* The former category included polyploids and other major structural cytological changes, and hence a high level of genetic incompatibility would be present. Accordingly, at least the polyploid a-ecospecies should be placed in a category of coenospecies. He saw difficulties in distinguishing between ecotypes and some forms of ecospecies because some ecotypes represent early stages in the formation of g-ecospecies. He commented that it is the actual amount of gene exchange between ecospecies under natural conditions, rather than the potential (as measured experimentally), that is important, and that: "the classification should surely take account as well of things as they are" (Valentine, 1948:128). He then advanced a new definition of the ecospecies: "species are ecospecies of a coenospecies when exchange of genes can take place between them, but when the exchange under either natural or artificial conditions is limited in some way" (Valentine, 1948:128).

Dobzhansky (1951) briefly considered the ecospecies and coenospecies, and the implication that the former is an earlier stage of evolutionary divergence. He commented that:

> There is, however, no basis for the assumption that the diverging populations become isolated first ecologically, seasonally, or mechanically, and then hybrid inviability and sterility are added to make the reproductive isolation complete. On the contrary, it is arguable that at least some hybrid inviability or hybrid breakdown appears first, and serves as a stimulus for the development of other forms of reproductive isolation. [Dobzhansky, 1951:273]

Baker (1952) also reviewed the ecospecies concept and the problem of the amount of gene exchange. After discussing the g-ecospecies concept of Valentine (1948, 1949), he noted that: "forms which can produce fully fertile and strong hybrids after artificial crossing are allowed to qualify as separate ecospecies if they do not exchange genes freely in nature" (Baker, 1952:62). He also saw difficulties in making a nonarbitrary separation between ecotypes and ecospecies, and concluded with the following statement: "it is proposed that the dividing line between ecotype and ecospecies should be drawn at the level where the first bar to crossing is imposed which is not purely concerned with habitat-preference" (Baker, 1952:67). Thus he is in accord with the later definition of Turesson.

Cain (1953) compared the genecological terminology of the plant systematists with the biological species concept, which was largely a product of zoologists. He summarized the reviews of Valentine (1949) and Baker (1952) as follows:

> (1) If two forms are distinct in nature, and cannot be made to produce fertile hybrids artifically, then they belong to different coenospecies.
>
> (2) If two closely related, but ecologically distinct forms do produce fertile hybrids occasionally when they meet in nature but both there and under artificial conditions the hybrids show a reduced viability, then these forms are separate ecospecies within the same coenospecies.
>
> (3) If two closely related but ecologically distinct forms either totally intergrade in nature where they meet, or appear completely interfertile under artificial conditions, they are ecotypes within the same ecospecies. [Cain, 1953:79–80]

He also noted that there is a continuum between the three categories. Cain (1953) compared the botanical and zoological approaches to these categories of divergence, and advanced a modified version of the biological species definition which considered only the actual or potential genetic connectedness.

The concepts of ecospecies and coenospecies were not seriously considered in later reviews, where emphasis was placed on the biological species, following the general acceptance of the biological species concept so strongly promoted by Mayr (e.g., Bush, 1975); the emerging evolutionary species concept (e.g., Levin, 1979); or both (e.g., V. Grant, 1971).

In summary, an ideal coenospecies includes the a-ecospecies of Valentine (1948, 1949), and represents a set of sexually reproducing and genetically compatible individuals that are incapable of combining haploid genomes with individuals of other coenospecies, so that fertile offspring cannot be produced under any circumstances. Thus, coenospecies are closed genetic systems. There is a sufficient level of intrinsic genetic incompatibility, which is reflected in infertility, hybrid inviability, or complete hybrid sterility. Nonarbitrary decisions are possible on the assignment of boundaries, based on experimental analysis, or observations of the consequences of crosses occurring under natural conditions. Coenospecies could occupy the same or different adaptive peaks.

The ideal ecospecies might be seen as a set of individuals occupying the same adaptive peak, with the potential to freely combine genetic information with members of other adaptive peaks. That is, no intrinsic genetic incompatibility is present. Thus, the potential for successful combination of genetic information is not realized in sympatry or parapatry because the action of extrinsic selective factors on the recombination products (hybrids) is such that they do not survive or reproduce successfully. Mechanisms of homogamy may or may not be present. The outcome in this ideal case then is to effectively maintain the genetic separateness of the two sets of differently adapted individuals in spite of the intrinsic compatibility of their genomes.

Ecotypes would then be expressions of less distinctive extrinsic selective regimes, occupying saddles or ridges between adaptive peaks, so allowing recombination products with intermediate adaptive characteristics to survive and to reproduce successfully in the appropriate environment of matching intermediacy. In this case, the transfer of genetic information between sets occupying different adaptive peaks is possible. Because the carriers of such nonadaptive information would be of lower fitness than optimally adapted phenotypes, the input would be continuously eliminated by natural selection.

All degrees of intermediacy between ecotypes, ecospecies, and coenospecies are to be expected in the real world, so that arbitrary decisions of classification are unavoidable. Even so, each category has considerable utility and heuristic value in evolutionary ecology, and should be considered independently of the pragmatic demands of taxonomy. For when coupled with the evolutionary and ecological species concepts, they allow a logical approach to, and clearer understanding of, the bases to discontinuity in biparental systems. Thus the gaps between biological species may be due to intrinsic genetic incompatibility, sustained disruptive selection between ecologically discrete states, or a combination of these processes. Hence, the oversimplified and combined approach advanced in the conventional view of reproductive isolation is confusing and does not relate to the underlying basic mechanisms.

Intrinsic genetic incompatibility

The divergence of genetic systems (i.e., sets of interbreeding individuals) may be considered independently of any differential ecological adaptation and its consequent effects on the genotypes of the component individuals. Three possible processes may be envisaged that could develop when the exchange of genetic information between members of two subsets of one previously interbreeding system is prevented by either the imposition of an extrinsic barrier, or the evolution of a system of social organization leading to the formation of discrete inbreeding groups (Wilson et al., 1975). The selective regimes acting on the now spatially separated sets of individuals are postulated to remain virtually identical during the period of isolation, and a secular environmental change is then imposed.

First, the resulting adaptive trends in tracking the change could follow parallel courses of evolution, so that equivalent levels of fitness would be maintained, but through the incorporation of different changes in the underlying genetic information (H. J. Muller, 1940). Thus the genetic systems would progressively diverge because of the improbability of identical genetic changes being incorporated into the genotypes of each disjunct set of interbreeding individuals. Second, the incorporated complexes of adaptive characteristics of each set might follow diverging trends of independent responses, even though subjected to identical selective regimes, because of the availability of different "suitable" mutations and the sequence of their appearance. Third, if a spatially separated set consisted of only a small number of individuals, then chance effects, particularly changes in chromosome structure, could also be important in promoting the divergence of genetic systems (White, 1978).

Should sufficient genetic differentiation have occurred, then the haploid genomes of the members of such pairs of sets, if brought together in experimental matings or artificial fertilizations, would be incompatible. Thus the origin of infertility, or of hybrid inviability and sterility, could be explained either as a purely incidental by-product of the maintenance of the same type of ecological adaptation, or through chance effects.

If the extrinsic barrier were no longer effective, and reproductive individuals of the two previously separated sets came into contact so that interset matings could occur, then no genetically viable progeny would result from these unions. Hence the two systems would maintain their genetic distinctness in spite of any interbreeding. Because the two genetic systems are, a priori, ecologically equivalent, competitive interactions could develop, or their effective mean reproductive rates might differ (de Bach, 1966). Under either of these conditions, there are three possible outcomes: (1) the extinction of the members of one set; (2) the establishment of a parapatric contact at the line of intersection, or at an equilibrium of fitness if slight environmental differences were present; or (3) the evolution of ecological divergence through ecological character displacement (P. R. Grant, 1972), which would shift the systems out of the limited area of consideration in the present section.

A similar array of interactions would be expected with autopolyploidy or major structural changes in karyotypes, which give rise to instantaneous genetic incompatibility but presumably retain the ecological equivalence.

Such modifications of genotypes, leading to complete genetic incompatibility, would be virtually irreversible; and thus, highly stable, contemporaneous, and ecologically equivalent systems could exist.

Extrinsic ecological incompatibility

Differential ecological adaptation need not automatically carry with it the by-product of intrinsic genetic incompatibility. Crosses between members of phenotypically and ecologically divergent systems frequently yield viable, healthy progeny under experimental conditions (e.g., Blair, 1964). If the phenotypic differences have a complex and polygenic basis, as is usually the case (Mayr, 1963; Dobzhansky, 1970), then the recombination products probably will be optimally adapted to an intermediate selective regime, which is limited to an ecotone (C. H. Muller, 1952; W. S.

Moore, 1977) or may not yet be realized. The recombination products also may be of lower competitive ability in the range of either parental taxon (Hagen, 1967), but may be superior to both in an "open" or novel situation (i.e., one that has not yet been fully exploited because of the recency of its formation).

In the absence of such open or intermediate environments, hybrid genotypes would thus be of lower fitness than either parental type within its normal environment. When interacting with individuals of either parental system, the hybrids would not be expected to survive to reproduce successfully. Hence the low level of fitness of hybrids is determined by extrinsic factors, and so the distinctness of the parental sets is maintained regardless of the outcomes of any cross-mating. The situation, however, is an unstable one, because if an intermediate or suitable open habitat were created, then the fitness of the recombination products could increase and exceed those of the parental forms in these habitats.

In the absence of suitable habitats favoring recombination products, interset matings would again result in a serious loss of fitness for the participating individuals. Accordingly, ecotonal avoidance, or increased specificity of mate recognition and attraction systems, should be favored by natural selection. Conversely, where hybrids were fitter than parental individuals, such mechanisms should break down in an ecotonal situation, for there would not be any loss of fitness in crossbreeding in an ecotonal region. This aspect of homogamy will be considered further in a later section.

If intrinsic genetic incompatibility arises as a consequence of the divergence of the genotypes of the members of each system under consideration, then the process of differential adaptation, having a polygenic basis, could lead to genetic incompatibility as a consequence of the divergence, although by an indirect path. Clearly a spectrum of degrees and kinds of genetic divergence is to be expected, and some ecologically distinctive forms also may have intrinsic genetic incompatibility in addition to the extrinsic ecological incompatibility. But where the distinctness of adaptive sets is maintained solely by the continuous operation of disruptive selective regimes, the situation is unstable and will last only as long as the distinctness of selective regimes is maintained (H. J. Muller, 1942:83).

Mate-selecting mechanisms as communication systems

The production and transmission of a characteristic mate-attracting signal by a breeding individual of one sex, and the reception of that signal together with the appropriate response by a breeding individual of the opposite sex of the same adaptive set, constitute the primary stage in the process of mate selection. Competition among the signalers, and discrimination and preference by the receivers, leads to the secondary within-set aspects of mate selection. This interaction between signaler and responder, in which both individuals benefit through the enhancement of reproductive success and fitness, constitutes a biocommunication system (Littlejohn, 1977). The evolution of such systems will be markedly influenced by other coexistent and contemporaneous users of the same communication channels, both within sets (through sexual selection) and between sets (as interference or noise). These problems of communication at the between-set level could operate where adaptive sets were ecologically disjunct, except during overlapping breeding seasons when a common resource was required for reproductive purposes.

The communication system may be exploited by illegitimate signalers and receivers, who then are the only ones to benefit from the interaction (Otte, 1974). Displaying individuals may simultaneously be exposed to a greater risk of predation or reduction in reproductive success through drawing attention to themselves. If such an influence were present in one area and not in another, or if different forms exploited a communication system in different areas of distribution, then natural selection might favor shifts in signal structure and communicative behavior associated with reproduction.

Such an interaction was postulated by McPhail (1969) for western Canadian populations of the threespine stickleback, *Gasterosteus aculeatus,* sympatric with the western mudminnow, *Novumbra hubbsi,* a presumed predator on young sticklebacks concentrated near the breeding male parent at the nest site. Males of most populations of sticklebacks sympatric with the mudminnow have a deep black nuptial color, rather than the normal bright red that is characteristic of allopatric populations, the color having an epigamic function. Furthermore, a male showing an intense black color can fade to a drab mottled color in less than 1 minute, whereas one showing red color may only vary the intensity but not lose the color completely. McPhail (1969) suggested that the black color and its lability evolved in response to predation on the young sticklebacks by the mudminnow. Mate preference tests showed that females from red populations rarely accept a black male in a choice situation. Females from allopatric black populations also prefer red males, but females from sympatric black populations appear to mate randomly. Thus there is evidence for the development of some positive assortative mating. Hence the postulated response to predation may incidentally have influenced the process of mate selection. There is, however, some genetic incompatibility between the two color forms which could also provide a selective component. Hybrid fitness is reduced to about 50% in backcrosses; and where the parental male is an F_1 hybrid, most of the young die because of his defective egg fanning behavior (McPhail, 1969). '

Littlejohn (1959, 1969) advanced the idea of the reproductive environment and its influence on the divergence of homogamic mechanisms in spatially separated subsets of one adaptive system. Consider a situation in which, in different areas, different assemblages of other adaptive sets simultaneously utilize resources of reproductive communication common to those of the set under consideration. Then divergent evolutionary trends, resulting from selection for increased efficiency of communication by reduction of interference and noise, could lead to spatial differentiation in the original adaptive set of the homogamic mechanisms involving signal production and reception. If disjunct allopatric populations are considered to be undergoing divergence according to the classical model of geographic speciation (Mayr, 1963), then regionally distinct communication systems could evolve as a consequence of interactions with these other genetic systems. In an ensuing contact between the subsets, there need be no communicative interference between breeding individuals of each subset (i.e., daughter populations), as a result of the incidental divergence. If coupled with adequate levels of ecological distinctness or genetic incompatibility, then the two divergent sets could assume sympatric or parapatric distributions without any significant interaction because each was "preadapted" to coexist with the other.

Let us consider another situation in which there are two systems, either spatially coexistent (sympatric) or contiguously distributed (parapatric), whose distinctness is maintained by the continuous action of discrete extrinsic selective regimes, so that each set occupies a distinct adaptive peak but there is no intrinsic genetic incompatibility. Furthermore, the reproductive efficiency of the component individuals is maintained by effective homogamic mechanisms based on set-specific communication. If intermediate (ecotonal) or unexploited (open) environments, accessible to members of each adaptive set, were created, such that hybrids were of equivalent or greater fitness than either parental type in these habitats, then the distinctive communication system might no longer constitute noise. If any errors of mate selection occurred in these environments, then the previously set-specific communication systems would rapidly break down as the "mistakes" no longer resulted in reduced reproductive success, but, rather, increased fitness for the individuals concerned (Littlejohn,

1977). Littlejohn and Watson (1973, 1976) have explained the establishment and maintenance of a narrow zone between the leptodactylid frogs, *Geocrinia laevis* and *G. victoriana,* by such a process.

A biparental species concept in terms of individuals

We may commence with the observation that phenotypically distinctive sets of contemporaneous, sexually reproducing organisms exist; and that where such sets are coexistent (sympatric), reproductive pairs are formed between members of the same phenotypic set; that is, mating is nonrandom. Such set-selective, positive assortative mating presumably leads to greater reproductive success for the individuals concerned because it results in the most efficient production of the potentially fittest progeny.

The distinctness of the sets of individuals implies that there are gaps in the potential continuum of phenotypic variability. The presence of such gaps may be explained by either, or a combination, of the following two processes or properties: (1) the sustained action of ecological disruptive selection on the maintenance of adaptive peaks in "discrete niches" or selective regimes; and (2) the presence of genetic incompatibility between the haploid genomes of individuals from different sets, such that a functional, fertile hybrid cannot be produced under existing conditions and circumstances. The reciprocally selective mating by males and females of the same set implies that where individuals are coexistent and in breeding condition at the same time (synchronic and syntopic), such mate-selecting mechanisms will be set-specific and so may categorize all the included breeding individuals. These devices or processes (homogamic mechanisms) are components of what is generally recognized as premating reproductive isolation; however, they must operate only to maximize the fitness of individuals, and any other consequences are purely incidental effects. Such sets of individuals, characterized by either similarity of ecological adaptation (i.e., ecospecies, in part), or level of genetic compatibility (i.e., coenospecies, in part), or both, may be equated to the biparental or biological species as promoted by Mayr (1970, and earlier).

There is also an interaction between ecological divergence and genetic incompatibility. If subjected to different selective regimes, then the recombination products of matings between individuals belonging to different adaptive sets will most probably be adapted to an intermediate, unrealized environment, and hence will be of very low fitness, even if the genomes of the parents are of sufficient genetic compatibility to have the potential to produce a functional hybrid (in terms of ontogenetic development). Such a hybrid might, however, have a very high degree of relative fitness if placed in an intermediate or unexploited (open) environment.

In sum, the observed distinctness of sets of contemporaneous sexually reproducing individuals may reflect adaptation to distinctive and discontinuous selective regimes, the expression of genetic incompatibility in crosses between members of different sets, or both. Where syntopic (sensu Rivas, 1964), individuals of each set will display set-specific mate-selecting (or homogamic) mechanisms. Such sets of individuals may be equated with biparental or biological species.

It now seems appropriate to advance the following statement as a provisional, albeit lengthy, definition of biparental species, framed strictly in terms of individuals.

A biparental species consists of a set of similarly adapted individuals, the members of which can combine genetic information through sexual reproduction. Their phenetic and genetic distinctness is maintained as long as matings between similarly adapted and genetically compatible individuals result in greater reproductive success (fitter progeny) than do those between differently adapted or genetically incompatible individuals (i.e., belonging to different species). The lowered reproductive success of interspecific matings may be due to genetic incompatibility (grading from parental infertility to hybrid sterility), to the ecological incompatibility of hybrids which results from their intermediate adaptive characteristics and the lack of compromising environments, or to a combination of these factors. If the opportunity for interspecific mating occurs, then those factors or processes that directly increase an individual's probability of mating with a member of the same adaptive or genetically compatible set (i.e., homogamic mechanisms) will be favored by natural selection. Accordingly, such set-selective homogamic mechanisms should be characteristic of established, syntopic, synchronously breeding individuals belonging to the same species.

Reproductive character displacement

By building on the comments and criticisms of P. R. Grant (1972), character displacement, or divergence, may be redefined as a process whereby attributes of individuals of one biparental species are changed through natural selection resulting from an interaction with individuals of one or more other biparental species that have similar or overlapping requirements of ecology, reproduction, or both. A distinction usually is made between ecological character displacement and reproductive character displacement (Brown and Wilson, 1956); and it is the latter process that will be emphasized in the present review. The term "reinforcement," introduced by Blair (1955), is a useful synonym for reproductive character displacement.

As discussed earlier, interspecific infertility and hybrid sterility cannot be directly enhanced by natural selection; hence, reproductive character displacement can only apply to processes associated with the increase in efficiency of optimal mate selection (i.e., those factors previously included in the category of premating reproductive isolating mechanisms, particularly ethological or sexual isolation).

The origin of the idea of reproductive character displacement has been attributed to Wallace (1889) by V. Grant (1966) and Murray (1972), and termed the "Wallace Effect." It seems, however, that Wallace (1889:174–5) was attempting to explain the origin of interspecific infertility and hybrid sterility through the direct action of natural selection, rather than positive assortative mating (or homogamy). He only once makes a direct comment on mate selection, namely: "We may fairly suppose, also, that as soon as any sterility appears some disinclination to *cross unions* will appear, and this will further tend to the diminution of the production of hybrids" (Wallace, 1889:179).

The credit for the originality of the idea probably should go to Gulick (1890), who, in a comprehensive essay read to the Linnean Society of London in 1887, made several clear statements concerning the inception of selective mating. Thus, he wrote: "whenever there arises a variety that can maintain itself by crosses within the same variety, any variation of instinct that tends to segregation [i.e., homogamy] will be preserved by the segregation" (Gulick, 1890:235); and: "As long as the groups are held apart by divergent sexual instincts, it is evident that divergent forms of Sexual Selection are almost sure to arise, leading to a further accumulation of the divergence initiated by the previous causes" (Gulick, 1890:236). In this regard it seems that "divergent sexual selection" might be an appropriate term for this form of reproductive character displacement.

Fisher (1930:129–31) considered the development of sexual preference in some detail, and presented his arguments in terms of individual fitness, as is indicated in the quotation at the beginning of this essay. The following extract from his treatment of sexual preference indicates his understanding of the process.

> A typical situation in which such discrimination will possess a definite advantage to members of both sexes must arise whenever a species occupying a continuous range is in the process of fission into two daughter species, differentially adapted to different parts of that range; for in either of the extreme parts certain relatively disadvantageous characters will constantly appear in a certain fixed proportion of the individuals in each generation, by

reason of the diffusion of the genes responsible for them from other parts of the range. Individuals so characterized will be definitely less well adapted to the situation in which they find themselves than their competitors; and in so far as they are recognizably so, owing, for example, to differences in tint, their presence will give rise to a selective process favouring a sexual preference of the group in which they live. Individuals in each region most readily attracted to or excited by mates of the type there favoured, in contrast to possible mates of the opposite type, will, in fact, be the better represented in future generations, and both the discrimination and the preference will thereby be enhanced. It appears certainly possible than an evolution of sexual preference due to this cause would establish an effective isolation between two differentiated parts of a species, even when geographical and other factors were least favourable to such separation. [Fisher, 1930:130–1]

Dobzhansky (1940:316) clearly outlined how the process of reproductive character displacement could occur, as follows: "Where hybridization jeopardizes the integrity of two or more adaptive complexes, genetic factors which would decrease the frequency or prevent the interbreeding would thereby acquire a positive selective value." He saw race formation as the development of genetic patterns which are adapted to a definite environment, and speciation as the fixation of these patterns of adaptation through reproductive isolating mechanisms. Thus: "If races are to become species, isolating mechanisms must arise when the distinct adaptive complexes are exposed to the risk of disintegration due to interbreeding" (Dobzhansky, 1940:316–17). He also considered that: "it is possible that certain forms of physiological isolation may occasionally arise as by-products of the adaptation to the environment" (Dobzhansky, 1940:319); and he was aware that:

The basic problem which remains to be settled is how frequently and to what extent can the isolating mechanisms be regarded adaptational by-products arising without the intervention of the special selective processes postulated above. Only experimental data could elucidate the situation further. [Dobzhansky, 1940:320]

He did not, however, make a distinction between those isolating factors that are directly amenable to natural selection and those that are not.

H. J. Muller (1942:103–4) considered the ways in which the "bars to crossing" (e.g., including homogamic mechanisms) might arise: "It may then be concluded that the establishment of the genes that result in the incapacitation of hybrids comes about in large measure as an automatic consequence of evolution in general. . . . By analogy, we may reason that many of the genetically conditioned bars to crossing come about in the same way." After some discussion he concludes: "Hence, in addition to arising as automatic consequences of evolution, there will also be the tendency for bars to crossing to become established as a result of selection for the isolationist advantage they confer. This can occur in the case of populations which are already fairly but not completely isolated, or which are completely isolated but by wasteful means" (H. J. Muller, 1942:104).

It is interesting to note that both Dobzhansky and Muller saw that homogamic mechanisms could arise either incidentally or through the direct action of selection favoring sexual preference; thus both writers supported the notion of reproductive character displacement, as it was later to be named. Furthermore, because Wallace (1889) was really concerned with the incapacitation of hybrids and so was dealing with incompatibility, the name "Wallace Effect" (V. Grant, 1966) may be inappropriate for reproductive character displacement. Perhaps the "Gulick Effect" is a more suitable designation for the process of reinforcement.

In his discussion on the importance of isolating mechanisms and the theories explaining their origin, Mayr (1963:548) made no clear distinction between homogamic processes and infertility and the incapacitation of hybrids. He saw natural selection acting only to improve subsidiary isolating mechanisms, and that a primary and basic mechanism must be fully efficient when contact is established. Thus: "When such cross-sterile species first come into contact, it matters little whether or not they produce (sterile!) hybrids owing to the incompleteness of the other isolating mechanisms, particularly ethological ones. . . . However, there will be strong selection in favor of the acquisition of additional isolating mechanisms to prevent such wastage of gametes" (Mayr, 1963:551). Bigelow (1964:455) also considered that: "Speciation (as defined by Mayr, . . .) must be complete before the *sympatric* divergence toward a perfection of mechanisms that inhibit interbreeding can take place under the sympatric conditions within a 'hybrid zone.' " He later noted that:

> It is important to bear in mind, however, that such mechanisms [to prevent interbreeding] cannot begin to evolve as a *direct* result of selection . . . until genetic incompatibility is sufficient to render hybrids sterile, or ill-adapted vehicles of "gamete wast-

age" – in other words, until speciation has been completed.
[Bigelow, 1965:456]

J. A. Moore (1957) was possibly the first evolutionary biologist to seriously criticize the necessity for reproductive character displacement, but it is now apparent that his arguments have several serious shortcomings. Thus, he made no clear distinction between those isolating factors that are amenable to the direct action of natural selection (e.g., homogamic mechanisms), and those that cannot be directly influenced by natural selection for their isolating effect (e.g., infertility, and hybrid inviability and sterility). The two examples used to support his case both involved only the consideration of hybrid incapacitation, and hence isolating factors upon which natural selection cannot directly act. Furthermore, both examples are more complicated than previously assumed, so that they are no longer appropriate to his argument; see the accounts by Main (1968) of the *Crinia* (now *Ranidella*) *signifera* complex, and Brown (1973) of the *Rana pipiens* complex. Moore also explained how divergence in habitat preference could result from ecological competition, so that positive assortative mating occurred as an incidental consequence of this process; but none of the proponents of reproductive character displacement have ever required that it be a universal process (Blair, 1974).

The idea that isolating mechanisms which were promoted by natural selection "should have a selective advantage in the zone of overlap and not elsewhere" was considered by J. A. Moore (1957) to be a major criticism of reproductive character displacement because the mechanisms appeared to be general characteristics of biological species, whether in sympatric or allopatric populations. The assumption that they are ad hoc mechanisms, and hence should be restricted to sympatric zones, presumably is based on the notion that all speciation occurs by way of geographic separation of large portions of species' ranges. But it is now plausible to consider other processes of speciation (Bush, 1975). Thus localized sympatric speciation, followed by expansion into allopatric areas, and coupled with the stability associated with sexual selection, could account for the uniformity of mate-selecting systems. Similarly, speciation in a small peripheral isolate, followed by a phase of complete sympatry with the parental species during which reinforcement occurred, and subsequently expansion of range of the new species into allopatric areas, again under the influence of stabilizing sexual selection, could explain the uniformity of homogamic mechanisms (Otte, 1974; Littlejohn, 1977).

Paterson (1978) argued strongly against the feasibility of reinforcement (i.e., reproductive character displacement) of homogamic mechanisms. He advanced a deterministic model based on fixation of one autosomal chromosomal translocation, which acts as a cause of hybrid sterility, and postulated that random mating occurs in an area of sympatry of unspecified area and population numbers, and closed to migration between adjacent allopatric populations. He reasoned that in such a closed system, any displacement away from an equilibrium point will lead to extinction of the rarer chromosomal arrangement (when hybrids have some fitness), or to extinction of the rarer population if the hybrids are completely sterile or inviable. He thus implied that the outcome of this hypothesis is such that the evolution of homogamic mechanisms is unlikely under the specified conditions. It seems, however, that his conclusion can only apply to the closed sympatric model, for he commented that: "The parapatric situation [in which migration and genetic input can occur] has not been dealt with here in detail as it is somewhat distinct" (Paterson 1978:370). Because the basis for reinforcement, as developed by Dobzhansky (1940), for example, rests on the assumption of complex polygenic adaptations rather than one translocation, and there is the potential for migration in most natural systems, the laboratory studies, and Crosby's later models (Crosby, 1970), it is difficult to see the general relevance of Paterson's conclusion.

Furthermore, Paterson (1978) did not discuss those cases of reproductive character displacement that have been analyzed in detail (e.g., Blair, 1955; Littlejohn, 1965; Fouquette, 1975; Ralin, 1977). Nor did he consider the influence of differing reproductive environments on the allopatric divergence of mate-selecting mechanisms (Littlejohn, 1959, 1969). Adequate variation exists within the pair-forming acoustic signals of anurans, both within and between populations (e.g., Littlejohn, 1959, 1964, 1965), to provide the basis for divergent evolution in males. Although variation in the responses by females to the calls of males has not yet been measured, comparable amounts may be expected, particularly to allow for differences in temperature between potential mates; for this environmental factor can markedly influence the key recognition features of the acoustic signals (Littlejohn, 1977).

Walker (1974) examined 254 potential cases of reproductive character displacement in species pairs of gryllids and tettigoniids, and found only five possible cases, none of which was particularly convincing. He then presented several reasons why cases of reproductive character displacement might appear scarce in these groups of singing insects. The following listing is derived from his discussion (Walker 1974:1147–9), and has been condensed and rephrased to have more general application to biparental systems.

1. Sympatric enhancement of the character under study may not occur because: (a) it is not as critical to conspecific pair formation (homogamy) as was supposed; or (b) the characters may have diverged sufficiently in allopatry so that when the two populations became sympatric no additional divergence was required.
2. Sympatric enhancement has occurred but is difficult to detect because: (a) the pattern of geographic variation typical of character displacement was not produced; or (b) the typical pattern of geographic variation did occur, but sample sizes were too small to enable its detection, the features analyzed were not the critical ones used in mate selection, or the diagnostic pattern of geographic variation was lost because differences arising in the zone of sympatry spread into the zones of allopatry.

The relative rarity of reinforcement could also be due to the restricted range of conditions under which it may be expected to occur. The following tabulation includes the more obvious of these; the listing should by no means be considered exhaustive.

1. Genetic incompatibility or ecological divergence must already exist so that cross-mating results in relatively reduced reproductive success for the individuals concerned, compared with those breeding with an individual of the same compatible or adaptive system.
2. Only those aspects of reproductive biology affecting mate preference and selection, gamete release and transfer, and the timing of reproductive events leading to fertilization, are amenable to reinforcement.
3. The potentially interactive systems must have sufficient similarity in one or more components of their prefertilization reproductive biology that mistakes of mate selection could occur, or reproductive efficiency is reduced because of the presence of the other breeding system.
4. Rates of migration, and of other inputs of genetic information, into the zone of interaction must be low enough, or the cost of mistakes high enough, to allow divergence to proceed sufficiently for the changes to be detected.
5. The extent and frequency of interaction between individuals of the two reproducing systems must be sufficient to provide an adequate selective pressure to result in divergence.
6. The components of reproductive biology must contain sufficient

between-individual variability, or suitable mutations must arise
with sufficient frequency, to provide the essential underlying
genetic changes to allow the divergence.

Waage lists four basic stages in the documentation of reproductive
character displacement, namely:

1) The characters involved should be shown to play a significant role
in species discrimination or in other aspects of premating isolation; and any displacement detected should be shown to be
perceptible to the species involved.

2) Except in situations of presumed character release (P. R. Grant,
1972), the allopatric character state(s) should be shown to represent the precontact condition; or at least it should be shown that
sympatry has arisen from previous allopatry.

3) The displacement in character states should be shown to be
unique to sympatry for one or both species and not merely a
predictable extension of trends already established for one or both
species in allopatry.

4) Character divergence in sympatry should be shown to have
resulted from the interaction of the two species and not from some
interaction with an aspect of their environment also unique to the
area of sympatry. [Waage 1979:104]

Narrow overlaps of geographic range and narrow hybrid zones are
potentially appealing areas in which evidence for reinforcement might be
sought. Three such categories of zonal interaction may be recognized:

1. Narrow hybrid zone along an ecotone in which there is hybrid
advantage. In this case the band of hybrids separates parental
individuals from direct interaction, and thus is a true hybrid zone
(sensu Short, 1969). The ecotone avoidance effect, discussed
earlier, will also mean that fewer parental individuals may
approach the hybrid zone. Hence such a situation is directly
noninteractive for pure parental genotypes, although there may be
indirect interactions at the boundaries of the ecotone where pure
parental individuals may contact the recombination products.
Even so, reproductive character displacement would not be
expected under these conditions.

2. Narrow hybrid zone in which the frequency of parental individuals exceeds 5%; that is, an overlap with hybridization (sensu
Short, 1969). Here parental individuals are in direct contact, and
hybrids are of lower fitness than either parental type. Such
conditions could arise in a relatively uniform environment where
two ecologically similar systems are interacting, or they may also
occur at an interface between two contrasting environments. This
situation is suitable for the development of reproductive character

displacement. However, the continuous immigration of naive individuals into the zone from the adjacent extensive areas of allopatry may swamp any tendency toward divergence (Watson, 1972), so that an evolutionary stalemate could be the result (Littlejohn, 1969).

3. Narrow overlap or contact (parapatry) between two taxa that are genetically incompatible but of similar ecological requirements. This situation would be similar to (2) above, but without hybrids; however, any mistakes in mate selection would be just as important in promoting reinforcing selection, but again may be subject to swamping by immigration.

Gartside (1980) has suggested a process whereby reproductive character displacement could occur in a narrow zone of overlap between two genetically compatible systems. In the southeastern United States the chorus frogs, *Pseudacris nigrita* and *P. triseriata feriarum,* overlap geographically in the east of their distributions with little or no hybridization, and their mating calls are reinforced (Fouquette, 1975). In the west, however, their calls are apparently very similar, and they interact in a narrow hybrid zone in which more than 50% of individuals sampled are putative hybrids; in spite of this hybridization, the two systems are genetically distinct outside the zone (Gartside, 1980). On the basis of the geographical variation in the nature of the interaction, Gartside (1980) suggested the following mechanism:

> The existence of a patchy environment in the east . . . may have allowed initial macro-geographic overlap . . . with a minimum of reproductive and ecological interaction. Sharp boundaries (ecotones) between the patches would provide little or no intermediate environment in which hybrids might be favored over parental types (i.e., hybrid habitats . . .), and hence should provide a selective mechanism against hybrids which could lead to reproductive character displacement or to strengthening of habitat preferences. These environmental factors might be an important precondition for geographic sympatry in taxa which are otherwise unable to co-exist at the same localities. . . . The hybrid zone in the west, then, might be a consequence of relatively uniform (but slightly different) environments separated by a strip of intermediate or novel habitat, in which hybrids are not at an absolute disadvantage to parental types. [Gartside, 1980:64–5]

The scheme advanced by Gartside (1980) may be further developed as follows: If the distributions of contrasting habitats in a zone of interaction are such that there is patchiness with sharp ecotones at the edges, then the border populations of each system would have the characteristics of small peripheral isolates, and there could be a large number of them along and within a zone of interaction. In such small peripheral isolates, with relatively long perimeters compared with their areas, most, or perhaps all, breeding individuals could interact with those of the other system under conditions favoring the development of positive assortative mating (Otte, 1974; Littlejohn, 1977). Furthermore, a large number of trials would be possible; so there is a reasonable chance that homogamic mechanisms could develop in one of the isolates. Once established, the successful carriers could increase their reproductive success and so expand their distribution along the zone.

It is evident, from the arguments presented earlier, that homogamic mechanisms will evolve, or be maintained, through the direct action of natural selection only under a rather restricted range of conditions. Thus two distinct situations may be recognized, namely, those in which there is a high probability of mating with an individual (1) of another adaptive set in the absence of intermediate or open environments, or (2) with an incompatible haploid genome. However, such a conclusion should not diminish the significance of the origin, evolution, and operation of homogamic mechanisms in their own right, particularly when they are viewed as biocommunication systems, and coupled with other evolutionary processes such as sexual selection; for these areas of reproductive biology can provide some of the most interesting aspects of the study of diversity in biparental living systems. Also, because reproductive isolation has been given so much emphasis in the development of the biological species concept, it is essential that homogamic mechanisms be viewed in their correct perspective.

An alternative classification of factors and mechanisms associated with the maintenance of genetic discontinuity in coexistent biparental systems

If the concept of reproductive isolation, which is based on the maintenance of the genetic integrity of supraindividual units of biological organization, and includes a heterogeneous selection of mechanisms and factors, is no longer accepted as operational, can a more meaningful and logical alternative scheme, in which individuals are emphasized, be advanced? It is evident from the heading to this section that no convenient and concise term can be suggested to describe such a set of processes; nor is it necessarily desirable that there be such an all-inclusive term. The following compilation represents an attempt at a provisional classification.

Directly acting homogamic mechanisms

Included here are the set-specific biocommunication systems associated with the mutual convergence and contact of potential mates which could result from:

1. Maximizing efficiency of mate selection against a background of signals emitted by other coexistent reproductive systems.
2. Sexual selection by the responding sex (receiver), based on the effectiveness of advertising by signalers (transmitters) of their fitness and suitability as mates in competition with other potential mates which are also signaling at the same time.
3. Second-order effects arising between cognate genetic systems as a consequence of: (a) an earlier phase of exposure of each system to different sets of signaling reproductive systems during geographic separation, that is, different reproductive environments (Little-john, 1959, 1969); or (b) differential effects of predators (McPhail, 1969), and other illegitimate exploiters of such signaling systems (Otte, 1974).

Also included is the temporal displacement of breeding seasons or times arising as a direct consequence of reproductive interactions (McNeilly and Antonovics, 1968).

Indirectly acting homogamic mechanisms

These mechanisms reduce the probability of communication or contact between reproductive individuals belonging to different adaptive systems, and may arise as a consequence of ecological divergence. It seems unlikely that such divergence would occur directly as a means of reducing the probability of heterogamy. Included are:

1. Host shifts as sympatric speciational events (Bush, 1975).
2. Habitat displacements because of: (a) density-dependent interactions for limiting resources, or (b) disruptive selection for background matching in response to predation (Tauber and Tauber, 1977a,b).
3. Asynchrony of breeding: (a) to avoid ecological interactions arising from the same causes listed in (2) above (these changes may be seasonal or diurnal, and result in reproductive individuals of each system being active or present at different times); and (b) as a consequence of reproductive stages getting out of phase because of the same selective agent acting on different sensitive stages in a complex life cycle (Alexander, 1968).

Gametic incompatibility

Here there is lack of mutual recognition between gametes (Monroy and Rosati, 1979), so that fusion does not occur. Such a mechanism is fundamental to sexual reproduction and is particularly important where there is external fertilization.

Recipient haploid rejection

Two categories may be recognized:

1. Gametic: incompatibility of the female genital tract to sperm in animals with internal fertilization.
2. Gametophytic: failure of pollen grains to germinate on the stigma, or of pollen tubes to reach the ovules, in flowering plants (equivalent to the incompatibility barriers of V. Grant, 1966).

Hybrid rejection syndrome

Restricted to viviparous forms in which the developing hybrid embryo is recognized as foreign and is aborted at as early a stage as possible (Coyne, 1974). This process might develop as an extension of an intrinsic mechanism for the assessment of normality in embryos resulting from homogamic matings.

Hybrid inviability

The loss of fitness of recombination products could result from three processes:

1. Intrinsic genetic incompatibility. Discordancy of haploid genomes following fusion means that normal ontogeny cannot proceed under any realized conditions.
2. Extrinsic ecological incompatibility: the loss of fitness resulting from the intermediacy of adaptation to a nonexistent environment.
3. Parental rejection syndrome: a delayed form of the hybrid rejection syndrome, in which the parents identify the progeny as "abnormal" and reject them by ceasing to administer further parental care.

F_1 hybrid sterility

This defect could arise in three ways:

1. Intrinsic genetic breakdown so that functional gametes capable of fusion are not produced, even where all other aspects of reproductive biology may be "normal."
2. Defective mating patterns or genital morphology, so that no mate can be secured, or attempts at fertilization are unsuccessful.
3. Defective parental behavior so that the progeny are not successfully reared (McPhail, 1969).

F_2 hybrid breakdown

The expressions of genetic incompatibility are delayed until this generation.

Second-order genetic incompatibility

If there are ecological interactions, reproductive interactions, or a combination of these factors, in sympatry, then the increased genetic divergence arising as a consequence could result in a higher level of genetic incompatibility between sympatric individuals of different adaptive systems relative to that between allopatric individuals (Watson and Martin, 1968), thus leading to an increased cost of mismating, so that a positive feedback develops.

References

Alexander, R. D. 1968. Life cycle origins, speciation, and related phenomena in crickets. *Q. Rev. Biol. 43:*1–41.

Ayala, F. J., and Campbell, C. A. 1974. Frequency-dependent selection. *Annu. Rev. Ecol. Syst. 5:*115–38.

Baker, H. G. 1952. The ecospecies – prelude to discussion. *Evolution 6:*61–8.

Bigelow, R. S. 1965. Hybrid zones and reproductive isolation. *Evolution 19:*449–58.

Blair, W. F. 1955. Mating call and stage of speciation in the *Microhyla olivacea–M. carolinensis* complex. *Evolution 9:*469–80.

– 1964. Isolating mechanisms and interspecies interactions in anuran amphibians. *Q. Rev. Biol. 39:*334–44.

– 1974. Character displacement in frogs. *Am. Zool. 14:*1119–25.

Brown, L. E. 1973. Speciation in the *Rana pipiens* complex. *Am. Zool. 13:*73–9.

Brown, W. L., and Wilson, E. O. 1956. Character displacement. *Syst. Zool. 5:*49–64.

Burma, B. H. 1949. The species concept: a semantic review. *Evolution 3:*369–70, 372–3.

Bush, G. L. 1975. Modes of animal speciation. *Annu. Rev. Ecol. Syst. 6:*339–64.

Cain, A. J. 1953. Geography, ecology and coexistence in relation to the biological definition of the species. *Evolution 7:*76–83.

Clausen, J., Keck, D. D., and Hiesey, W. M. 1939. The concept of species based on experiment. *Am. J. Bot. 26:*103–6.

Coyne, J. A. 1974. The evolutionary origin of hybrid inviability. *Evolution 28:*505–6.

Crosby, J. L. 1970. The evolution of genetic discontinuity: computer models of the selection of barriers to interbreeding between subspecies. *Heredity 25:*253–97.

Darwin, C. 1859. *The Origin of Species by Means of Natural Selection.* London: John Murray.

de Bach, P. 1966. The competitive displacement and coexistence principles. *Annu. Rev. Entomol. 11:*183–212.

Dobzhansky, Th. 1937. Genetic nature of species differences. *Am. Nat. 71:*404–20.

– 1940. Speciation as a stage in evolutionary divergence. *Am. Nat. 74:*312–21.

– 1951. *Genetics and the Origin of Species,* 3rd ed., rev. New York: Columbia University Press.

– 1970. *Genetics of the Evolutionary Process.* New York: Columbia University Press.

– 1975. Analysis of incipient reproductive isolation within a species of *Drosophila. Proc. Natl. Acad. Sci. U.S.A. 72:*3638–41.

Dobzhansky, Th., Ayala, F. J., Stebbins, G. L., and Valentine, J. W. 1977. *Evolution*. San Francisco: Freeman.

Ehrlich, P. R., and Raven, P. H. 1969. Differentiation of populations. *Science 165:*1228–32.

Fisher, R. A. 1930. *The Genetical Theory of Natural Selection.* Oxford: Clarendon Press.

Fouquette, M. J. 1975. Speciation in chorus frogs. I. Reproductive character displacement in the *Pseudacris nigrita* complex. *Syst. Zool. 24:*16–23.

Gartside, D. F. 1980. Analysis of a hybrid zone between chorus frogs of the *Pseudacris nigrita* complex in the southern United States. *Copeia 1980:*56–66.

Grant, P. R. 1972. Convergent and divergent character displacement. *Biol. J. Linn. Soc. 4:* 39–68.

Grant, V. 1966. The selective origin of incompatibility barriers in the plant genus *Gilia. Am. Nat. 100:*99–118.

– 1971. *Plant Speciation.* New York: Columbia University Press.

Gregor, J. W. 1939. Experimental taxonomy. IV. Population differentiation in North American and European sea plantains allied to *Plantago maritima* L. *New Phytol. 38:* 309–22.

– 1944. The ecotype. *Biol. Rev. 19:*20–30.

Gregor, J. W., Davey, V. M., and Lang, J. M. S. 1936. Experimental taxonomy. I. Experimental garden technique in relation to the recognition of the small taxonomic units. *New Phytol. 35:*323–50.

Gulick, J. T. 1890. Divergent evolution through cumulative segregation. *J. Linn. Soc. Zool. 20:*189–274.

– 1891. Intensive segregation, or divergence through independent transformation. *J. Linn. Soc. Zool. 23:*312–80.

Hagen, D. W. 1967. Isolating mechanisms in threespine sticklebacks (*Gasterosteus*). *J. Fish. Res. Bd. Can. 24:*1637–92.

Hutchinson, G. E. 1957. Concluding remarks. *Cold Spring Harbor Symp. Quant. Biol. 22:* 415–27.

– 1968. When are species necessary? In *Population Biology and Evolution* (ed. Lewontin, R. C.), pp. 177–86. Syracuse: Syracuse University Press.

Huxley, J. (ed.). 1940. *The New Systematics.* London: Oxford University Press.

Jones, D. F. 1928. *Selective Fertilization.* Chicago: University of Chicago Press.

Levin, D. A. 1971. The origin of reproductive isolating mechanisms in flowering plants. *Taxon 20:*91–113.

– 1978. The origin of isolating mechanisms in flowering plants. *Evol. Biol. 11:*185–317.

– 1979. The nature of plant species. *Science 204:*381–4.

Littlejohn, M. J. 1959. Call differentiation in a complex of seven species of *Crinia* (Anura, Leptodactylidae). *Evolution 13:*452–68.

– 1964. Geographic isolation and mating call differentiation in *Crinia signifera. Evolution 18:*262–6.

– 1965. Premating isolation in the *Hyla ewingi* complex (Anura: Hylidae). *Evolution 19:* 234–43.

– 1969. The systematic significance of isolating mechanisms. In *Systematic Biology, Proceedings of an International Conference,* pp. 459–82. Washington, D.C.: National Academy of Sciences.

– 1977. Long-range acoustic communication in anurans: an integrated and evolutionary approach. In *The Reproductive Biology of Amphibians* (ed. Taylor, D. H., and Guttman, S. I.), pp. 263–94. New York: Plenum Press.

Littlejohn, M. J., and Watson, G. F. 1973. Mating-call variation across a narrow hybrid zone between *Crinia laevis* and *C. victoriana* (Anura: Leptodactylidae). *Aust. J. Zool. 21:* 277–84.

– 1976. Effectiveness of a hybrid mating call in eliciting phonotaxis by females of the *Geocrinia laevis* complex (Anura: Leptodactylidae). *Copeia 1976:*76–9.

McNeilly, T., and Antonovics, J. 1968. Evolution in closely adjacent plant populations. IV. Barriers to gene flow. *Heredity 23:*205–18.

McPhail, J. D. 1969. Predation and the evolution of a stickleback (*Gasterosteus*). *J. Fish. Res. Bd. Can. 26:*3183–208.

Main, A. R. 1968. Ecology, systematics and evolution of Australian frogs. *Adv. Ecol. Res. 5:* 37–86.

Mather, K. 1955. Polymorphism as an outcome of disruptive selection. *Evolution 9:*52–61.

Maynard Smith, J. 1978. *The Evolution of Sex.* Cambridge: Cambridge University Press.

Mayr, E. 1942. *Systematics and the Origin of Species.* New York: Columbia University Press.

– 1957. Difficulties and importance of the biological species concept. In *The Species Problem* (ed. Mayr, E.), pp. 371–88. Washington, D.C.: American Association for the Advancement of Science.

– 1963. *Animal Species and Evolution.* Cambridge: Belknap Press of Harvard University Press.

– 1969. The biological meaning of species. *Biol. J. Linn. Soc. 1:*311–20.

– 1970. *Populations, Species, and Evolution.* Cambridge: Belknap Press of Harvard University Press.

Monroy, A., and Rosati, F. 1979. The evolution of the cell–cell recognition system. *Nature (Lond.) 278:*165–6.

Moore, J. A. 1957. An embryologist's view of the species concept. In *The Species Problem* (ed. Mayr, E.), pp. 325–38. Washington, D.C.: American Association for the Advancement of Science.

Moore, W. S. 1977. An evaluation of narrow hybrid zones in vertebrates. *Q. Rev. Biol. 52:* 263–77.

Muller, C. H. 1952. Ecological control of hybridization in *Quercus:* a factor in the mechanism of evolution. *Evolution 6:*147–61.

Muller, H. J. 1940. Bearings of the "*Drosophila*" work on systematics. In *The New Systematics* (ed. Huxley, J.), pp. 185–268. London: Oxford University Press.

– 1942. Isolating mechanisms, evolution and temperature. *Biol. Symp. 6:*71–125.

Murray, J. 1972. *Genetic Diversity and Natural Selection.* Edinburgh: Oliver and Boyd.

Otte, D. 1974. Effects and functions in the evolution of signaling systems. *Annu. Rev. Ecol. Syst. 5:*385–417.

Paterson, H. E. H. 1978. More evidence against speciation by reinforcement. *S. Afr. J. Sci. 74:*369–71.

Ralin, D. B. 1977. Evolutionary aspects of mating call variation in a diploid-tetraploid species complex of treefrogs (Anura). *Evolution 31:*721–36.

Riley, H. P. 1952. Ecological barriers. *Am. Nat. 86:*23–32.

Rivas, L. R. 1964. A reinterpretation of the concepts "sympatric" and "allopatric" with proposal of the additional terms "syntopic" and "allotopic." *Syst. Zool. 13:*42–3.

Robson, G. C. 1928. *The Species Problem. An Introduction to the Study of Evolutionary Divergence in Natural Populations.* Edinburgh: Oliver and Boyd.

Robson, G. C., and Richards, O. W. 1936. *The Variation of Animals in Nature.* London: Longmans, Green.

Romanes, G. J. 1886. Physiological selection; an additional suggestion on the origin of species. *J. Linn. Soc. Zool. 19:*337–411.

– 1897. *Darwin, and After Darwin,* vol. 3, *Post-Darwinian Questions: Isolation and Physiological Selection.* London: Longmans, Green.

Rosenzweig, M. L. 1978. Competitive speciation. *Biol. J. Linn. Soc. 10:*275–89.

Short, L. L. 1969. Taxonomic aspects of avian hybridization. *Auk 86:*84–105.

Simpson, G. G. 1951. The species concept. *Evolution 5:*285–98.

– 1953. *The Major Features of Evolution.* New York: Columbia University Press.

– 1961. *Principles of Animal Taxonomy.* New York: Columbia University Press.

Straw, R. M. 1955. Hybridization, homogamy, and sympatric speciation. *Evolution 9:*441–4.

Tauber, C. A., and Tauber, M. J. 1977a. A genetic model for sympatric speciation through habitat diversification and seasonal isolation. *Nature (Lond.) 268:*702–5.

– 1977b. Sympatric speciation based on allelic changes at three loci: evidence from natural populations in two habitats. *Science 197:*1298–9.

Turesson, G. 1922a. The species and the variety as ecological units. *Hereditas 3:*100–13.

– 1922b. The genotypical response of the plant species to the habitat. *Hereditas 3:*211–350.

– 1923. The scope and import of genecology. *Hereditas 4:*171–6.

– 1929. Zur Natur und Begrenzung der Arteinheiten. *Hereditas 12:*323–34.

Turrill, W. B. 1946. The ecotype concept. A consideration with appreciation and criticism, especially of recent trends. *New Phytol. 45:*34–43.

Valentine, D. H. 1948. Studies in British primulas. II. Ecology and taxonomy of primrose and oxlip (*Primula vulgaris* Huds. and *P. elatior* Schreb.) *New Phytol. 47:*111–30.

– 1949. The units of experimental taxonomy. *Acta Biotheor. 9:*75–88.

Van Valen, L. 1976. Ecological species, multispecies, and oaks. *Taxon 25:*233–9.

Waage, J. K. 1979. Reproductive character displacement in *Calopteryx* (Odonata: Calopterygidae). *Evolution 33:*104–16.

Walker, T. J. 1974. Character displacement and acoustic insects. *Am. Zool. 14:*1137–50.

Wallace, A. R. 1889. *Darwinism. An Exposition of the Theory of Natural Selection with Some of its Applications.* London: Macmillan.

Watson, G. F. 1972. The *Litoria ewingi* complex (Anura: Hylidae) in south-eastern Australia. II. Genetic incompatibility and delimitation of a narrow hybrid zone between *L. ewingi* and *L. paraewingi. Aust. J. Zool. 20:*423–33.

Watson, G. F., and Martin, A. A. 1968. Postmating isolation in the *Hyla ewingi* complex (Anura: Hylidae). *Evolution 22:*664–6.

White, M. J. D. 1978. *Modes of Speciation.* San Francisco: Freeman.

Williams, G. C. 1966. *Adaptation and Natural Selection.* Princeton: Princeton University Press.

– 1975. *Sex and Evolution.* Princeton: Princeton University Press.

Wilson, A. C., Bush, G. L., Case, S. M., and King, M.–C. 1975. Social structuring of mammalian populations and rate of chromosomal evolution. *Proc. Natl. Acad. Sci. U.S.A. 72:*5061–5.

Wright, S. 1932. The roles of mutation, inbreeding, crossbreeding and selection in evolution. *Proc. 6th Int. Congr. Genetics 1:*356–66.

16   The maintenance of B chromosomes in *Brachycome dichromosomatica*

S.SMITH-WHITE AND C.R.CARTER

Accessory or B chromosomes occur in perhaps 10% of diploid short-lived higher plants. Often, or even usually, they are subject to accumulation or "drive" mechanisms. Most frequently such mechanisms involve nondisjunction and preferential segregation into the generative nucleus during the first pollen grain division (Muntzing, 1974), but in maize it occurs in the second pollen grain division (Rhoades and Dempsey, 1972). Kimura and Kayano (1961) have shown that accumulation occurs in the tetrasporic embryo sacs in *Lilium callosum*.

Accumulation mechanisms result in a drive akin to meiotic drive (Hartl, 1977), which in the absence of selection should lead to the permanent establishment of B chromosomes in a species. That this does not happen requires that the Bs be subject to adverse selection. There is some evidence that a single B may be selectively neutral or even advantageous, but multiple B chromosomes appear to be generally disadvantageous.

Kimura and Kayano (1961) have presented a detailed model for the situation in *L. callosum*, which is a herbaceous perennial with both sexual and vegetative reproduction. No other equivalent treatment of B-chromosome population genetics is known to the present authors. Hartl suggests that the issue involving B chromosomes has never been resolved, but has simply gone out of fashion.

Brachycome dichromosomatica is an ephemeral annual with two pairs of chromosomes. It is essentially self-incompatible and apparently has no means of asexual reproduction. On the basis of data presented by Carter (1978) and Carter and Smith-White (1972) it would seem to offer favorable material for the analysis of B-chromosome population genetics, differing in detail from the case of *L. callosum*.

Method of study

The system

In *B. dichromosomatica* the number of Bs varies from 0 to 3. The Bs are mitotically stable, are not heterochromatic, and have no obvious phenotypic effect on the plants that possess them.

At first anaphase of pollen mother cell (P.M.C.) meiosis in plants that have one B there is a small proportion of loss of this univalent. It is found to lag on the anaphase spindle, and thus fails to be included in either of the daughter nuclei. For the data presented by Carter and Carter and Smith-White, this loss amounts to about 14%, or 0.14 proportion.

In plants with two Bs, these pair with one another in a normal way and with complete regularity. Thus P.M.C. meiosis in 2B plants yields four microspores per mother cell, each with one B.

No quantitative data on the meiotic behavior of 3B plants are available. For simplicity it will be assumed that the third B is regularly lost during meiosis. Although this assumption is probably not strictly true, it is permitted by the failure to find plants with four or more Bs. It allows 3B plants to be treated like 2B plants, yielding microspores each with one B.

On the basis of very little microscopic evidence and some breeding data (Carter, 1978) it is assumed that conditions of meiosis in embryo sac mother cells (E.S.M.C.s) are similar to those in P.M.C.s.

The mature pollen is trinucleate. In the first pollen grain (P.G.) division, yielding the generative and vegetative nuclei, there is a high frequency of nondisjunction and directed segregation of the daughter Bs when one is present in the microspore. The result is that a microspore with one B yields a generative nucleus with two Bs. This generative nucleus undergoes a second division to yield the male gamete nuclei. Two lines of evidence, one the examination of the second P.G. mitosis, and the other the pollination data provided by Carter (1978), suggest that this directed nondisjunction may be 100% (mitosis data) or only 85% (pollination data).

In the embryo sacs, which undergo three haploid mitoses, there is no nondisjunction or preferential segregation of the B chromosomes. Megaspores with one B yield egg nuclei with one B.

The species is highly self-incompatible and outbreeding. Random mating, limited only by distance, can be inferred.

The symbols

For convenience, the symbols used in this paper are listed here.

N The size of a sample. NF_i and Nf_i are numbers in a B chromosome class.

$[f_i]$ Frequencies of classes before selection ($i = 0, 1, 2, 3$). These are the frequencies immediately following fertilization.

$[F_i]$ Frequencies of classes after selection. If selection operated during growth and before maturity F_i represents frequencies at maturity and at sampling. In the absence of selection, $F_i = f_i$.

p, q Frequencies of meiospores (microspores and megaspores) with 0 or 1 B chromosome. $q = 1 - p$. Two Bs in a meiospore are not allowed in the system.

a The proportion of B chromosome loss at meiosis in 1B plants. Estimated from data at $a = 0.14$.

k The proportion of 0B meiospores yielded by 1B plants. $k = \frac{1}{2}(1 + a)$, $\frac{1}{2} \le k \le 1$. From the data $k = 0.57$.

c The frequency of occurrence of normal or regular disjunction of the B chromosome in the first P.G. division. From the data $c = 0$ or $c = 0.15$.

$\bar{B}$ The weighted frequency of B chromosomes. $\bar{B} = iF_i$.

$[b_i]$ The fitnesses of the corresponding B chromosome classes where selection operates during growth before maturity ($i = 0, 1, 2, 3$).

$[d_i]$ The fitnesses of the corresponding B chromosome classes where selection operates after maturity (and hence after sampling). These fitnesses could involve differences in the number of seeds produced by the several classes. No data on this point are available.

$[g_i]$ With gametic selection g_i ($i = 0, 1$) represents the relative viabilities of 0B and 1B embryo sacs.

$[j_i]$ With gametic selection, the relative viabilities of 0B, 1B, and 2B male gametes.

$[s_{ij}]$ Where selection involves pollen-style interaction, s_{ij} represents the compatibility of pollen with i B chromosomes with styles carrying j Bs ($i = 0, 1, 2; j = 0, 1, 2, 3$).

$\bar{W}$ Mean fitness. For the case of selection before maturity, $\bar{W} = \Sigma\, b_iF_i$. For the case of gametic selection, $\bar{W}_{es}$ and $\bar{W}_{pg}$ represent mean fitnesses of embryo sacs and pollen.

The gamete vectors and the generation matrix

Note that the gametes are, respectively, the embryo sac egg nuclei and the pollen gamete nuclei derived from the generative nuclei. On the model presented the vectors and matrix are:

$$
\female \begin{bmatrix} p \\ q \\ 0 \end{bmatrix} \overset{\male}{\begin{bmatrix} p & cq & (1-c)q \end{bmatrix}} = \begin{bmatrix} p^2 & cpq & (1-c)pq \\ pq & cq^2 & (1-c)q^2 \\ 0 & 0 & 0 \end{bmatrix}
$$

This gives the f_i set:

$$\{f_0 \quad f_1 \qquad f_2 \qquad f_3\}$$
$$\{p^2 \quad (1+c)pq \quad (1-c)pq + cq^2 \quad (1-c)q^2\} \tag{1}$$

In the special case, with $c = 0$, this reduces to

$$\{p^2 \quad pq \quad pq \quad q^2\} \tag{2}$$

In the following gamete generation p becomes $F_0 + kF_1$. In the absence of selection this is $f_0 + kf_1$.

The consequence of no selection

In the absence of selection, p', the derived value of p for the next generation, is given by

$$p' = f_0 + kf_1 = p^2 + k(1+c)pq$$

$\Delta p = p' - p$, where Δp is the change in p. Substituting $(1-p)$ for q,

$$\Delta p = [1 - (1+c)k]\,(p^2 - p)$$

Equilibrium, with $\Delta p = 0$, is only possible if $(1+c)k = 1$, and the equilibrium is then static at any value of p. If $(1+c)k$ is less than unity, Δp is negative, and p must decrease to zero. If $(1+c)k$ is greater than unity, Δp is positive, and p must increase to unity. In the special case, with

Table 16.1. *Estimation of* p *from observed data*

Population	p	F_0	F_1	F_2	F_3	N	$\overline{B}$
A. Central							
Observed nos.	0.9448	706	36	25	2	769	
Frequencies		0.9181	0.0468	0.0325	0.0026		0.1196
Theoretical							
$c = 0.15$	0.9448[a]	686.5	46.1	34.4	2.0		0.1572
	0.9582[b]	706.1	35.4	26.4	1.1		0.1190
$c = 0$	0.9448[a]	686.5	40.1	40.1	2.3		0.1654
	0.9582[b]	706.1	30.8	30.8	1.3		0.1252
B. Marginal							
Observed nos.	0.7846	111	20	22	3	156	
Frequencies		0.7115	0.1282	0.1410	0.0192		0.4679
Theoretical							
$c = 0.15$	0.7846[a]	96.1	27.3	25.6	6.9		0.6359
	0.8435[b]	111.1	23.6	18.0	3.2		0.4436
$c = 0$	0.7846[a]	96.1	26.3	26.3	7.2		0.6442
	0.8435[b]	111.1	20.5	20.5	3.8		0.4673

[a] $p = F_0 + kF_1$.
[b] $p = \sqrt{F_0}$.

$c = 0$, if $k \neq 1.0$, p must decrease continuously with the passage of generations.

The estimation of p from observed values. In the absence of selection, and with $(1 + c)k \neq 1.0$, the value of p must change with each generation. It is not correct to estimate p from an observed set $[F_i]$ $(= f_i)$ as $F_0 + kF_1$. The best estimate is $p = \sqrt{F_0}$. To illustrate this point, the data from Carter (1978) have been summed in two groups, first the central populations of demes 1, 2, and 4 of *B. dichromosomatica* var. *dichromosomatica* together with the normal seasons data for *B. dichromosomatica* var. *alba,* and second, the marginal data for *B. dichromosomatica* var. *dichromosomatica* with the abnormal season data for var. *alba.* These summations are justified by the heterogeneity values. For both sets p values have been computed by both methods, using the two cases of $c = 0.15$ and $c = 0$, and the consequent $[F_i]$ $(= f_i)$ values have been calculated. These calculations are presented in Table 16.1, and clearly show a much better fit of $[F_i]$ and $\overline{B}$ obtained using $p = \sqrt{F_0}$.

In Table 16.2 the changing values of p, $[F_i]$, and $\overline{B}$ have been computed over 20 generations from the expression $p_{n+1} = kp_n + (1 - k)p_n^2$, using $c =$

Table 16.2 *Change in B chromosome frequency*[a]

Generation	p	q	F_0	F_1	F_2	F_3	$\overline{B}$
0	0.958	0.042	0.9178	0.0402	0.0402	0.0018	0.1260
1	0.9407	0.0593	0.8849	0.0558	0.0558	0.0035	0.1779
2	0.9167	0.0833	0.8403	0.0764	0.0764	0.0069	0.2499
3	0.8839	0.1161	0.7813	0.1026	0.1026	0.0135	0.3483
4	0.8397	0.1603	0.7051	0.1346	0.1346	0.0257	0.4809
5	0.7819	0.2181	0.6114	0.1705	0.1705	0.0476	0.6543
6	0.7085	0.2915	0.5020	0.2065	0.2065	0.0850	0.8745
7	0.6197	0.3803	0.3840	0.2357	0.2357	0.1446	1.1409
8	0.5184	0.4816	0.2687	0.2497	0.2497	0.2319	1.4448
9	0.4111	0.5889	0.1690	0.2421	0.2421	0.3468	1.7667
10	0.3070	0.6930	0.0942	0.2128	0.2128	0.4802	2.0790
11	0.2155	0.7845	0.0464	0.1691	0.1691	0.6154	2.3535
12	0.1428	0.8572	0.0204	0.1224	0.1224	0.7348	2.5716
13	0.0902	0.9098	0.0081	0.0821	0.0821	0.8277	2.7294
14	0.0549	0.9451	0.0030	0.0519	0.0519	0.8932	2.8353
15	0.0326	0.9674	0.0011	0.0315	0.0315	0.9359	2.9022
20	0.0021	0.9979	0.0000	0.0021	0.0021	0.9958	2.9937

[a]No selection; $k = 0.57$, $c = 0$.

0 and $k = 0.57$. Starting with the observed value for the central populations, $p = 0.958$, it is seen that, in the absence of selection, there would be effective fixation, all plants having three Bs, within 20 generations. The introduction of $c \neq 0$ gives the end point $F_2 = c$, $F_3 = (1 - c)$.

Zygotic selection

As Carter (1978) has noted, it must be assumed that selection operates to prevent this fixation of B chromosomes, and that it maintains values of p and $[F_i]$ in stable equilibria. Unstable equilibria can only be transient, and the evidence from the several demes shows that this cannot be the case.

Selection may operate in many ways. It may be zygotic or gametic, and in the former case it may operate during vegetative growth, or after maturity as by affecting the mean seed production of the several B chromosome classes. An additional method could involve differential compatibility relationships between pollens and styles of different B chromosome constitutions.

Zygotic selection during vegetative growth (before sampling)

Different fitnesses of the B chromosome classes are involved, and these fitnesses are represented by the set $[b_i]$ ($i = 0, 1, 2, 3$). The $[F_i]$ values no longer equal the $[f_i]$ values, but

$$[F_i] = \left[\frac{b_i f_i}{\overline{W}}\right]$$

The $[F_i]$ set becomes:

$$\{b_0 p^2 \quad b_1(1 + c)pq \quad b_2[(1 - c)pq + cq^2] \quad b_3 (1 - c)q^2\} \times \frac{1}{\overline{W}}$$

Mean fitness becomes:

$$\overline{W} = [b_0 - (1 + c)b_1 - (1 - 2c)b_2 + (1 - c)b_3] p^2$$
$$+ [(1 + c)b_1 + (1 - 3c)b_2 - 2(1 - c)b_3] p + [cb_2 + (1 - c)b_3]$$

In the next generation

$$p' = F_0 + kF_1 = \frac{b_0 p^2 + kb_1(1 + c)pq}{\overline{W}}$$

and at equilibrium

$$\Delta p = p' - p = 0 = \frac{b_0 p^2 + kb_1(1 + c)pq - p\overline{W}}{\overline{W}}$$

which, for $p \neq 0$, yields the quadratic equation

$$[(1 + c)b_1 + (1 - 2c)b_2 - (1 - c)b_3 - b_0]p^2$$
$$+ [b_0 - (1 + c)(1 + k)b_1 - (1 - 3c)b_2 + 2(1 - c)b_3]p$$
$$+ [k(1 + c)b_1 - cb_2 - (1 - c)b_3] = 0 \quad (3)$$

from which the equilibrium value $\hat{p}$ can be computed.

In the special case, with $c = 0$, this equation reduces to

$$[b_1 + b_2 - b_3 - b_0]p^2$$
$$+ [b_0 - (1 + k)b_1 - b_2 + 2b_3] p + [kb_1 - b_3] = 0 \quad (4)$$

Equations 3 and 4 define the conditions for equilibrium. Both are of the form

$$Ap^2 - (A + C)p + C = 0$$

where $A = -\frac{1}{2} d^2\overline{W}/dp^2$ and $C = k(1 + c)b_1 - cb_2 - (1 - c)b_3$. This equation factors to

$$(p - 1)(Ap - C) = 0$$

with the roots $\hat{p} = 1$ and $\hat{p} = C/A$.

If A and C have the same sign, and if C is numerically smaller than A, the root C/A will be biologically real. Because A is of opposite sign to $d^2\overline{W}/dp^2$, if A and C are both positive, the equilibrium will be stable. If both are negative, the equilibrium will be unstable.

The following conditions preclude a real biological root other than the trivial $p = 0$ or $p = 1$:

$$C = A, \quad C > A, \quad A = 0, \quad C = 0, \quad A \text{ and } C \text{ of opposite sign}$$

Any given set of k, c, and $[b_i]$ that meets the defined conditions defines a precise $\hat{p}$. However, a given $\hat{p}$ does not define a unique set of k, c, and $[b_i]$. It is to be noted that linearity in the effects of one, two, and three Bs gives $A = 0$.

Computing the fitness set $[b_i]$ *from data.* For any observed set of data the fitness of each class is computed as $[b_i] = [F_i/f_i]$. The b_i values thus obtained are relative, and it is convenient to transform them by dividing each by the raw value b_0; then b_0 takes the value 1.0, and the other b_is generally take values less than unity. The value of p is taken as the value before selection (i.e., $p = F_0 + kF_1$). Computations for the central and marginal populations, using $k = 0.57$ and $c = 0$, are presented in Table 16.3.

The estimates of the $[b_i]$ sets presented in this table have a low order of reliability. In particular, the very high value of the fitness b_3 obtained for the central data is probably due to sampling error, because only two plants in 769 possessed three B chromosomes. This high value of b_3 is the cause of the unstable equilibrium indicated by the set.

Although Jones and Rees (1969) have presented evidence that, in rye, plants with two Bs may be superior to plants with one B, and Carter (1978) believed that the same situation held for *B. dichromosomatica* under marginal (i.e., stress) conditions, the analysis presented here gives no support to this view. Even under marginal conditions, the fitness of 1B and 2B plants is less than that of 0B plants, and the fitness values b_1 and b_2 are hardly significantly different. Certainly, in the comparison of marginal and central populations there is a substantial change in relative fitness values, with b_2 increasing relative to b_1 under stress conditions.

The characteristics of the formulation given in this section are examined in detail in Appendixes 1, 2, and 3 of this chapter, and reference to these appendixes will be made in the general discussion.

Zygotic selection after maturity

It is envisaged that such selection might operate by differential seed production from the several B chromosome classes. The class frequencies in the actual samples would be unaffected by this selection. They are

Table 16.3. *Computing the fitnesses* b_i *from data*

Class		0B	1B	2B	3B
A. Central					
populations:	F_i^a	0.91808	0.04681	0.03251	0.00260
$p = 0.94476$	f_i^b	0.89257	0.05219	0.05219	0.00305
$\overline{W} = 0.97222$	b_i^c	1.02857	0.89702	0.62293	0.85230
	b_i^d	1.00000	0.87210	0.60562	0.82862
unstable	$b_i f_i$	0.89257	0.04551	0.03161	0.00253
	F_i^e	0.91808	0.04681	0.03251	0.00260
B. Marginal					
populations:	F_i^a	0.71154	0.12821	0.14103	0.01923
$p = 0.78462$	f_i^b	0.61562	0.16899	0.16899	0.04639
$\overline{W} = 0.8652$	b_i^c	1.15581	0.75864	0.83450	0.41454
	b_i^d	1.00000	0.65637	0.72201	0.35862
stable	$b_i f_i$	0.61562	0.11092	0.12201	0.01664
	F_i^e	0.71154	0.12821	0.14103	0.01923

[a]Observed. [b]Computed. [c]Crude. [d]Transformed. [e]F_i (computed) $= b_i f_i / \overline{W}$.

therefore representative of the $[f_i]$ set and not of the $[F_i]$ set. It follows that if $c = 0$, f_1 and f_2 must be equal, and any difference in the observed values must be due either to sampling error, or to c having a value not equal to zero.

In matching theoretical and observed values of p, the value giving rise to the f_i set must be used (i.e., $p = \sqrt{f_0}$). The $[F_i]$ set still represents the frequency distribution after selection, and the value p' in the next generation is again $F_0 + kF_1$.

To avoid confusion, the set $[d_i]$ is here used to represent the fitness values of the B chromosome classes. Because the $[F_i]$ values are dependent on the $[d_i]$ set, the latter cannot be computed from the available observational data.

Theoretically, the equations for the present case are identical with the case in which selection operates before sampling. At equilibrium

$$\Delta_p = \frac{d_0 p^2 + kd_1(1 + c)pq - p\overline{W}}{\overline{W}} = 0$$

which yields the quadratic equation

$$[(1 + c)d_1 + (1 - 2c)d_2 - (1 - c)d_3 - d_0]p^2$$
$$+ [d_0 - (1 + c)(1 + k)d_1 - (1 - 3c)d_2 + 2(1 - c)d_3]p$$
$$+ [k(1 + c)d_1 - cd_2 - (1 - c)d_3] = 0$$

which is again of the form $Ap^2 - (A + C)p + C = 0$, with the biological root $\hat{p} = C/A$.

Gametic selection

This model introduces different fitnesses or viabilities for embryo sacs and male gametes. Let $[g_i]$ $(i = 0, 1)$ represent the relative viabilities of 0B and 1B embryo sacs (or egg nuclei), and let $[j_i]$ $(i = 0, 1, 2)$ represent the relative viabilities of male gamete nuclei with i Bs. Because the fitness values are relative, g_0 and j_0 may both be made unity. Then with k in the domain $\frac{1}{2} \le k \le 1.0$, and c in the domain $0 \le c \le 1.0$, the gamete vectors and generation matrix become:

$$\begin{bmatrix} p \\ g_1 q \\ 0 \end{bmatrix}^{1/\overline{W}_{es}} \quad [p \quad j_1 cq \quad j_2(1 - c)q]^{1/\overline{W}_{pg}} = \begin{bmatrix} p^2 & j_1 cpq & j_2(1 - c)pq \\ g_1 pq & g_1 j_1 cq^2 & g_1 j_2(1 - c)q^2 \\ 0 & 0 & 0 \end{bmatrix}^{1/\overline{W}}$$

$$\overline{W} = \{1 - g_1 - cj_1 - (1 - c)j_2 + cg_1 j_1 + (1 - c)g_1 j_2\}p^2$$
$$+ \{g_1 + cj_1 + (1 - c)j_2 - 2[cg_1 j_1 + (1 - c)g_1 j_2]\}p$$
$$+ \{cg_1 j_1 + (1 - c)g_1 j_2\}$$

The class frequencies become:

$$[F_i] = \{p^2 \quad (g_1 + j_1 c)pq \quad j_2(1 - c)pq + g_1 j_1 cq^2 \quad g_1 j_2(1 - c)q^2\}\frac{1}{\overline{W}}$$

At equilibrium

$$\Delta p = \frac{p^2 + k(g_1 + cj_1)pq - p\overline{W}}{\overline{W}} = 0$$

giving the quadratic equation

$$p + k(g_1 + cj_1)q - \overline{W} = 0$$

or, substituting $(1 - p) = q$,

$$[g_1 + cj_1 + (1 - c)j_2 - cg_1 j_1 - (1 - c)g_1 j_2 - 1]p^2$$
$$+ [1 - (1 + k)(g_1 + cj_1) - (1 - c)j_2 + 2cg_1 j_1 + 2(1 - c)g_1 j_2]p$$
$$+ [k(g_1 + cj_1) - cg_1 j_1 - (1 - c)g_1 j_2] = 0$$

which is again of the form

$$Ap^2 - (A + C)p + C = 0$$

and the only real biological root is C/A, provided A and C are of the same sign and C is numerically smaller than A.

A is again equal to half the second differential $d^2\overline{W}/dp^2$ but of opposite sign, and a stable equilibrium requires that A and C both be positive.

In the special case of $c = 0$ the quadratic equation reduces to

$$[g_1 + j_2 - g_1 j_2 - 1]p^2$$
$$+ [1 - (1 + k)g_1 - j_2 + 2g_1 j_2]p + [kg_1 - g_1 j_2] = 0$$

This formulation matches that for zygotic selection, and it is evident that, for the case where $c = 0$, there is the following equivalence:

$$g_0 \sim b_0 = 1.0; \quad g_1 \sim b_1; \quad j_2 \sim b_2; \quad g_1 j_2 \sim b_3$$

but with the restriction that $g_1 j_2$ is precisely determined by the values g_1 and j_2, whereas b_3 may be independent of the other b_i parameters.

This restriction determines more precise limits to the values g_1 and j_2 capable of giving either stable or unstable equilibria. For any given value g_1, real biological equilibria require that j_2 lie within the limits defined by k and by $1 - (1 - k)g_1$. The equilibrium is then stable if g_1 is greater than unity, and unstable if g_1 is less than unity. For $g_1 = 1.0$ and $j_2 = k$, there is a static equilibrium at any value of p.

It is of interest that with this model of gametic selection stable biological equilibria require that the viability of 1B embryo sacs must exceed that of 0B embryo sacs, and the viability of 2B pollen gametes must be, within strict limits, less than that of 0B pollen gametes.

Fitting of the gametic model to the observational data, for both stable and unstable equilibria, is demonstrated in Appendix 4.

Selection involving pollen-style interactions

It is possible that pollen of different B chromosome constitutions might behave differently on styles of the same or different constitution. If such interactions were negative, with B chromosome pollen more compatible with 0B styles than with styles carrying one, two, or three B chromosomes, frequency-dependent selection would be involved, which even in the presence of pollen drive might be capable of yielding stable balanced equilibria.

Pollen-style interactions could follow one of two general systems. Style behavior in both is sporophytic. Pollen behavior might be determined by its own gametophytic constitution, such as occurs in incompatibility systems of the *Nicotiana* type, or it might be determined by the constitution of its sporophytic parent, as in the *Primula* pin-thrum system. The latter becomes extremely complex, and only the former is examined in the present paper.

If c is taken as zero (i.e., there is complete preferential nondisjunction in the pollen grain mitosis), pollen gamete nuclei will have either no Bs or two Bs. Eight pollen-style relationships are then involved. Each is given its own compatibility function, s_{ij}, where i represents the pollen constitution ($i = 0$, 2) and j the style constitution ($j = 0, 1, 2, 3$). The formulation is tabulated in Table 16.4.

At equilibrium, if such is possible, this formulation yields a set of

Table 16.4. *Formulation of the pollen-style interaction model*

Pollen	Style	Compatability	Pollen frequency	Style frequency	Product
0B	0B	s_{00}	p	F_0	$s_{00}F_0p$
	1B	s_{01}		F_1	$s_{01}F_1p$
	2B	s_{02}		F_2	$s_{02}F_2p$
	3B	s_{03}		F_3	$s_{03}F_3p$
2B	0B	s_{20}	q	F_0	$s_{20}F_0q$
	1B	s_{21}		F_1	$s_{21}F_1q$
	2B	s_{22}		F_2	$s_{22}F_2q$
	3B	s_{23}		F_3	$s_{23}F_3q$

equations

$$[F'_i] = \begin{bmatrix} F'_0 \\ F'_1 \\ F'_2 \\ F'_3 \end{bmatrix} = \begin{bmatrix} [s_{00}F_0 + ks_{01}F_1]p \\ [(1 - k)s_{01}F_1 + s_{02}F_2 + s_{03}F_3]p \\ [s_{20}F_0 + ks_{21}F_1]q \\ [(1 - k)s_{21}F_1 + s_{22}F_2 + s_{23}F_3]q \end{bmatrix}^{1/\overline{W}}$$

where $\overline{W}$ is the sum of frequencies by fitnesses and the F'_i are the values of F_i in the next generation. At equilibrium $[F'_i] = [F_i]$.

No formal solution for $\hat{p}$ has been found for this system. However, it is possible, using chosen values for s_{ij}, to demonstrate that stable equilibria can be achieved. The method is simply the tedious one of iteration through successive generations, an example of which is presented in Appendix 6. It seems unlikely that a set of $[s_{ij}]$ values could be found that would give a reasonable fit to the observed data.

Discussion

The genetic system offered by *B. dichromosomatica* is in some respects more simple than that of *Lilium callosum*, which was studied by Kimura and Kayano. The species is annual and outbreeding, has no means of vegetative reproduction, and is apparently incapable of apomixis. Preferential nondisjunction of a B chromosome in the first pollen grain mitosis is either complete or nearly so, and in meiosis in 2B plants the Bs pair and segregate regularly.

The approach made in the present study differs from that made by Kimura and Kayano. Our approach has been to examine separately the consequences of three different model systems – zygotic selection, gametic selection, and pollen-style interaction – and to attempt to fit each of these to the available observational data. For *B. dichromosomatica* it would seem to be unprofitable to attempt any formulation involving a combination of these systems, even though such is very likely to occur in nature. The number of parameters becomes too formidable.

It is demonstrated that, in the absence of selection, there can be no equilibria for B chromosome frequency other than the trivial ones. With strong pollen drive, if Bs are present at all, fixation would be achieved within 20 generations when all plants would possess the maximum permissible number of Bs. Selection must operate to prevent this fixation, and B chromosomes must be generally deleterious.

The formulation of the model for zygotic selection reduces to a very simple form. If the fitness set $[b_i]$ (or d_i) is determined, equilibrium is attained when a quadratic equation of the form $Ap^2 - (A + C)p + C = 0$ is satisfied, yielding $\hat{p} = C/A$; and this equilibrium is biologically real and stable if A and C are both positive and C is numerically smaller than A.

For zygotic selection, and provided the selection operates before sampling, fitness sets $[b_i]$ can be derived from the available data, but the values thus computed (see Table 16.3), and especially the value b_3, are unreliable.

The computations given in Appendix 1 show that it is possible to modify the fitnesses determined from one set of data (the marginal data) so as to match another set (the central data) by small changes in one only of several of the parameters, such as the values c and k, and the fitness values b_0, b_1, and b_2. The changes required are quite small, and are of a magnitude that could well have environmental causation, but are unfortunately at least an order of magnitude less than the reliability of the estimates of the parameters. In this degree, the estimates are ill-conditioned.

In Appendix 2 it is shown that for stable equilibria with any given relationship of the fitnesses of 0B and 1B plants, restrictions are imposed on the permissible values of b_2 and b_3. These restrictions enable possible arbitrary fitness sets to be suggested, and the set chosen for examination is:

$$k = 0.57, c = 0, [b_i] = \{1.0 \quad 0.75 \quad 0.75 \quad 0.375\}$$

This set gives a stable equilibrium at $\hat{p} = 0.42$.

In Appendix 3 this arbitrary set is modified in one of four ways, by changing the parameters c, k, b_2, and b_0. All four methods of modification allow the attainment of good fits to both the central and marginal data, with probability values lying between $P = 0.4$ and $P = 0.99$. Modification of b_1 does not permit an acceptable fit.

The model of gametic selection also proves to be simple, and is of the same form as that for zygotic selection. However, it is even more restricted. Stable equilibria require that the fitness of 1B embryo sacs exceed that of 0B embryo sacs. In Appendix 4 it is shown that the model gives stable equilibria that are rather poor fits to the observational data, and unstable equilibria that fit well. The latter, however, are considered unacceptable. The model of gametic selection, at least operating alone, is not favored.

Systems of pollen-style interaction do not permit simple formulation, and the equations for the model examined appear to be insoluble. In Appendix 5 it is shown that the model is capable of yielding stable equilibria against the pressure of pollen drive. However, it seems likely that the model is only capable of yielding stable equilibria for B-chromosome frequencies that are remote from the observational data, and it is concluded that such systems are not likely to be significant in the species.

Conclusion and summary

B chromosomes occur in all the populations of *B. dichromosomatica* that have been studied. The frequency of B chromosomes tends to be higher in populations in marginal environmental situations than in central localities of the species distribution. Carter infers that the fitness of 1B and especially of 2B plants increases relative to that of 0B plants under water stress in dry conditions. On the basis of the models presented, this inference appears to be justified, although 1B and 2B fitnesses remain lower than 0B fitness.

There is a strong pollen drive or distortional load due to nondisjunction and preferential segregation in the first P.G. mitosis. In the absence of selection this drive must lead to a fixation of Bs at the maximum possible frequency.

Because this does not happen, B chromosomes must be generally deleterious in the species. Selection must balance the drive to permit stable equilibria, but these equilibria must be capable of environmental modification.

Three model systems are examined. Zygotic selection is capable of giving close fits to the observational data, and only small changes in one of several parameters are required to permit the environmental effects observed by Carter. Quite small changes in the relative fitness values of 0B, 1B, and 2B plants, well within the scope of environmental causation, yield substantial changes in the equilibrium frequencies of the B chromosome classes. Unfortunately, the magnitude of these changes is less than the reliability of the estimates of the parameters.

The model of gametic selection appears to be less satisfactory. Pollen-style interactions are capable of giving stable equilibria against pollen drive provided the interactions are negative. However, they seem to be capable only of yielding equilibria that are remote from the observational data.

It is possible that both zygotic and gametic selection act in combination. With the data available at present, the examination of such combination systems is not profitable.

The models presented are simplistic. A complete model would need to be stochastic, and would need to take into account annual variation in weather conditions and population size within localities. In all desert environments such variations may be extreme.

Appendixes

In these appendixes examples are worked out to demonstrate the consequences of change in the values of c, k, and items of the fitness sets, and the magnitude of such changes required to fit the observed B-chromosome distributions of the central and marginal populations. Computations have been made to seven or nine decimal places, and are presented rounded off to five places. This rounding off is responsible for slight apparent arithmetical errors. Chi-square values have been computed after lumping the 3B class with the 2B class, leaving two degrees of freedom.

Appendix 1

This exercise commences with $c = 0$, $k = 0.57$, and the $[b_i]$ set determined from the marginal-populations data (see Table 16.3), namely,

$$[b_i] = \{1.00000 \quad 0.65637 \quad 0.72201 \quad 0.35862\}$$

with $\hat{p} = C/A = 0.01547/0.01976 = 0.78465$.

In order to fit the set to the central-populations data, with $\hat{p} = 0.94472$, these statistics are modified successively in one of five ways: (1) by introducing $c \neq 0$, (2) by changing the value of k, and (3, 4, and 5) by changing the values of b_2, b_1, and b_0.

Method 1 (the introduction of $c \neq 0$). The $[b_i]$ set and k remain unchanged.

The coefficients of the equation $Ap^2 - (A + C)p + C = 0$ become:

$$\underset{(c \neq 0)}{A} = \underset{(c-0)}{A} + c(b_1 + b_3 - 2b_2)$$

$$\underset{(c \neq 0)}{C} = \underset{(c-0)}{C} + c(kb_1 + b_3 - b_2)$$

and the value for $c \neq 0$ is determined by iteration to give, as near as practicable, $C/A = 0.94476$.

Method 2 (by modifying k). The coefficient A remains unaltered.

Because $C/A = p$, $c \, (= kb_1 - b_3) = pA$ and $k = (pA + b_3)/b_1$.

Method 3 (by modifying b_2). C is independent of b_2 and remains unchanged.

Because $A = C/p$, the new value of b_2 becomes:

$$b_2 = A + b_0 + b_3 - b_1$$

Method 4 (by modifying b_1). Both A and C are changed. Let b_1 be increased by the quantity e. Then:

$$A_{(\text{mod})} = A_{(\text{orig})} + e, \text{ and } C_{(\text{mod})} = C_{(\text{orig})} + ke$$

and the new value for b_1 is found by iteration from

$$(C + ke)/(A + e) = p = 0.94476$$

Method 5 (by modifying b_0). The values of the other b_is and of C are not *immediately* altered.

Again, let b_0 be increased by a quantity e. Then:

$$A_{(b_0-1+e)} = C/p = C/0.94476 \text{ and } e = A_{(b_0-1)} - A_{(b_0-1+e)}$$

The set is then retransformed by dividing all b_i values by the value $b_0 = (1 + e)$.

These five methods give the following results:

1. $c = 0.00759$; $C/A = 0.0155545/0.0164627 = 0.94483$
2. $k = 0.57481$; $C/A = 0.018628/0.01972 = 0.94462$
3. $b_2 = 0.718628$; $C/A = 0.01547/0.016378 = 0.94456$
4. $e = -0.008435$; $b_1 = 0.647935$; $C/A = 0.010663/0.011285 = 0.94488$
5. $e = +0.00335$ and the b_i set, after retransforming is
   $$[b_i] = \{1.00000 \quad 0.65418 \quad 0.71960 \quad 0.35746\}$$
   with $C/A = 0.01542/0.01632 = 0.94485$

These modifications of the marginal-populations statistics give fits to the central-populations data as set out in Table 16.5.

Appendix 2. Deriving an arbitrary fitness set
Using $c = 0$, the conditions for stable equilibria are:

$$(b_1 + b_2) > (b_0 + b_3); b_3 < kb_1; (b_1 + b_2 - b_3 - b_0) > (kb_1 - b_3)$$

Table 16.5. *Goodness of fit*[a]

Class	0B	1B	2B	3B	$\chi^2_{(2)}$	P
Observed NF_i	706	36	25	2		
Predicted NF_i						
method 1	710.9	27.5	29.8	0.9	3.11	0.21
method 2	710.8	27.3	30.0	0.9	3.30	0.19
method 3	711.0	27.3	29.9	0.9	3.25	0.20
method 4	711.4	26.8	29.9	0.9	3.67	0.17
method 5	711.0	27.2	29.9	0.8	3.35	0.19

[a]Marginal statistics modified to central data.

Because $b_0 = 1.0$ by definition, the last inequality can be rearranged to give

$$b_2 > 1 - (1 - k)b_1$$

Thus a minimum value is set for b_2, and b_3 must lie within the range $0 \leq b_3 \leq kb_1$. With $b_2 = [1 - (1 - k)b_1]$ and $b_3 = kb_1$, there is static equilibrium at any value of p. With values outside these limits, the $[b_i]$ set is incapable of giving equilibria other than the trivial ones. If b_2 exceeds its minimum, then $\hat{p}$ has a maximum value less than unity when b_3 is minimum (i.e., $b_3 = 0$).

It is clear that, on the basis of the model, Carter's inference (Carter, 1978:119, last line et seq.) cannot be sustained. If b_2 is equal to b_0 ($= 1.0$), and if $b_3 = 0$, then the equilibrium value of $\hat{p}$ is equal to k (0.57) for any value of b_1. Even if $b_2 = 0.9$, an equilibrium value of about $p = 0.95$ would require $b_3 = 0$ and $b_1 = 0.25$.

Such a discrepancy between b_1 and b_2 could not fit the data. From general considerations, b_2 could not greatly exceed b_1. For the purposes of this examination, an arbitrary initial $[b_i]$ set has been chosen, using $c = 0$ and $K = 0.57$:

$$[b_i] = \{1.0 \quad 0.75 \quad 0.75 \quad 0.375\}$$

where $b_2 = b_1$ and $b_3 = \frac{1}{2}b_1$. These values conform to the necessary constrictions and yield

$$\hat{p} = C/A = 0.0525/0.125 = 0.420$$

Appendix 3. Fitting the arbitrary set

This arbitrary fitness set is matched to the observed data for the central (*) and marginal (†) populations by modifying it in four of the ways set out in Appendix 1. Method 4 of that appendix, involving change of b_1, is not capable of matching either set of data.

Method 1 (the introduction of $c \neq 0$). By iteration the values are found:

    * $c = 0.16125.$ $C/A = 0.060966/0.064531 = 0.94475$

    † $c = 0.13145.$ $C/A = 0.059401/0.075706 = 0.78462$

Method 2 (changing the value of k)

    * $k = 0.65746.$ $C/A = 0.118087/0.125 = 0.94470$

    † $k = 0.63077.$ $C/A = 0.098077/0.125 = 0.78462$

Method 3 (by modifying b_2)
* b_2 becomes 0.680570. $C/A = 0.0525/0.055570 = 0.94475$
† b_2 becomes 0.691911. $C/A = 0.0525/0.066911 = 0.78462$

Method 5 (by changing b_0)
* $e = 0.069430$
† $e = 0.058089$

Retransformation gives the b_i sets:
* $[b_i] = \{1.0 \quad 0.701308 \quad 0.701308 \quad 0.350654\} \quad p = 0.94476$
† $[b_i] = \{1.0 \quad 0.708825 \quad 0.708825 \quad 0.354413\} \quad p = 0.78462$

The goodnesses of fit achieved by the four methods to the central and marginal data are presented in Table 16.6. The chi-square values are surprisingly low, with probabilities ranging from $P = 0.4$ to $P = 0.99$.

Appendix 4. Gametic selection

It has been shown that, for stable equilibria, g_1 must exceed 1.0, and j_2 must lie in the range $[1 - (1 - k) g_1] \le j_2 \le k$. Conversely, for unstable equilibria g_1 must be less than unity, and j_2 must lie between k and $[1 - (1 - k)g_1]$.

For any specified observational or theoretical value of $\hat{p}$, if g_1 is fixed,

Table 16.6. *Goodness of fit comparisons*[a]

Class	0B	1B	2B	3B	$\chi^2_{(2)}$	P
A. Central data:						
Observed NF_i	706	36	25	2		
Predicted NF_i						
method 1	706.0	35.9	26.3	0.8	0.049	0.99
method 2	706.2	30.9	30.9	0.9	1.57	0.43
method 3	708.8	31.1	28.2	0.9	0.94	0.60
method 5	709.9	29.1	29.1	0.9	1.94	0.40
B. Marginal data:						
Observed NF_i	111	20	22	3		
Predicted NF_i						
method 1	108.1	25.2	20.1	2.6	1.36	0.50
method 2	108.3	22.3	22.3	3.1	0.31	0.86
method 3	109.5	22.6	20.8	3.1	0.36	0.84
method 5	110.2	21.4	21.4	2.9	0.12	0.93

[a]Modified arbitrary set (Appendix 2) to the central and marginal populations data.

then j_2 and $g_1 j_2$ are also precisely determined. Further, the ratio of the class values F_2/F_1 is given by the ratio j_2/g_1. As g_1 falls to 1.0, j_2 rises to k, and the F_2F_1 ratio approaches $k/1.0$. Then as g_1 falls below 1.0, j_2 rises above k. Clearly the model is not capable of yielding acceptable stable equilibria.

Using $c = 0$ and $k = 0.57$, and g_1 values from 1.2 down to 0.8, computations have been made for values of j_2 required to fit the $\hat{p}$ values 0.94476 and 0.78462. These, and the goodness-of-fit comparisons with the central and marginal data, are presented in Table 16.7.

Appendix 5. Pollen-style interaction

With the formulation given above in the section on "Selection involving pollen-style interactions" no general solution for $\hat{p}$ seems possible. However, using arbitrary values of $[F_i]$ and suitable values of $[s_{ij}]$, it can be demonstrated that stable equilibria are possible.

In the example presented, $c = 0$ and $k = 0.57$. The initial F_i values are set equal, $F_0 = F_1 = F_2 = F_3 = 0.25$, which gives $p = F_0 + kF_1 = 0.3925$. The chosen s_{ij} values are symmetrical, to avoid any suggestion of the

Table 16.7. *Goodness of fit comparisons, gametic selection*

Data	F_0	F_1	F_2	F_3	$\chi^2_{(2)}$	P
Central populations						
Observed NF_i	706	36	25	2		
Theoretical NF_i:						
g_1 j_2						
1.2 0.48964[a]	698.6	49.0	20.0	1.4	5.01	0.08
1.1 0.52960[a]	700.8	45.1	21.7	1.4	2.52	0.29
0.9 0.61085[b]	705.4	37.1	25.2	1.3	0.04	0.98
0.8 0.65215[b]	707.5	33.2	27.0	1.3	0.31	0.85
Marginal populations						
Observed NF_i	111	20	22	3		
Theoretical NF_i:						
g_1 j_2						
1.2 0.50531[a]	103.0	34.0	14.3	4.7	8.24	0.017
1.1 0.53697[a]	104.4	31.5	15.4	4.7	5.86	0.05
0.9 0.60448[b]	107.3	26.5	17.8	4.4	2.08	0.34
0.8 0.64052[b]	108.8	23.9	19.1	4.2	0.41	0.82

[a]Stable equilibria.
[b]Unstable equilibria.

Table 16.8. *Equilibrium with a pollen-style interaction system*

Generation	F_0	F_1	F_2	F_3	p'^a
0	0.25	0.25	0.25	0.25	0.3925
1	0.05852	0.33398	0.38631	0.22119	0.24889
2	0.04007	0.26832	0.28283	0.40878	0.19301
3	0.02816	0.28205	0.26574	0.42404	0.18893
4	0.02878	0.28234	0.25895	0.42993	0.18971
5	0.02899	0.28437	0.26037	0.42628	0.19108
6	0.02930	0.28437	0.26084	0.42550	0.19139
7	0.02933	0.28439	0.26109	0.42518	0.19144
8	0.02934	0.28435	0.26110	0.42520	0.19142
9	0.02933	0.28435	0.26110	0.42522	0.19141
10	0.29332	0.28434	0.26109	0.42523	0.19141
15	0.02933	0.28434	0.26109	0.42523	0.19141

^{a}Note that p' refers to the p value *resulting from* the F_i values given for the generation (line) *not* the p value giving rise to that line.

involvement of gametic selection. These values are:

$$s_{00} = 0.1 \quad s_{02} = 0.7 \quad s_{20} = 1.0 \quad s_{22} = 0.4$$
$$s_{01} = 0.4 \quad s_{03} = 1.0 \quad s_{21} = 0.7 \quad s_{23} = 0.1$$

Changes in F_i and in p have been computed, and are presented in Table 16.8. Equilibrium is approached very close to $\hat{p} = 0.19141$ after 10 generations, but it is noted that Δp fluctuates from negative to positive in sequences of three or four generations. Change in Δp is rapid at first, but after 17 generations is only of the order 10^{-1}.

Stable equilibrium is thus achieved against the pressure of pollen drive. It seems unlikely that any set of s_{ij} values can be found that would yield stable equilibrium values as high as those found in the observational data. The system is probably not involved in the maintenance of B chromosome equilibria in *B. dichromosomatica*.

References

Carter, C. R. 1978. The cytology of *Brachycome*. 8. The inheritance, frequency and distribution of B chromosomes in *B. dichromosomatica* (n = 2), formerly included in *B. lineariloba*. *Chromosoma (Berl.)* 67:109–21.

Carter, C. R., and Smith-White, S. 1972. The cytology of *Brachycome lineariloba*. 3. Accessory chromosomes. *Chromosoma (Berl.)* 39:361–79.

Hartl, D. L. 1977. Applications of meiotic drive in animal breeding and population control. *Proc. Int. Conf. on Quantitative Genetics* (ed. Pollak, E., Kemthorne, O., and Bailey, T. B.), pp. 63–88. Ames: Iowa State University Press.

Jones, R. N., and Rees, H. 1969. An anomalous variation due to B chromosomes in rye. *Heredity 24:*265–71.

Kimura, M., and Kayano, H. 1961. The maintenance of supernumerary chromosomes in wild populations of *Lilium callosum* by preferential segregation. *Genetics 16:*1699–712.

Muntzing, A. 1974. Accessory chromosomes. *Annu. Rev. Genet. 8:*243–66.

Rhoades, M. M., and Dempsey, E. 1972. On the mechanism of chromosome loss induced by the B chromosome in maize. *Genetics 71:*73–96.

17 Genetic aspects of ant evolution.

ROSS H.CROZIER

Genetic system

Ants are male-haploid insects and comprise a large group of animals. Brown and Taylor (1970) estimate that the number of living species is probably about 15,000. Ants thus form a group with more species than there are species of birds (Mayr, 1969:11), for example, and they are extremely important components of temperate and tropical ecosystems (Wilson, 1971:1). The group is thus worthy of study in its own right. However, the additional feature of sociality increases the significance of evolutionary studies on this group.

Because the males are haploid, various geneticists have expected that ants and other Hymenoptera should have lower levels of genetic variation than animals in which both sexes are diploid. However, the extent of this depression of gene diversity level will depend on the extent of fitness correlation between the sexes (Crozier, 1970b, 1979; Pamilo and Crozier, 1978, and in prep.). In agreement with theoretical expectations, the prevailing consensus has been that ants and other Hymenoptera do, in fact, have unusually low levels of gene diversity for insects (e.g., Crozier, 1977a; Ayala, 1978; Pamilo et al., 1978). However, Nevo's (1978) recent review shows that it is probably not the case that Hymenoptera have unusually low levels of gene diversity, but that *Drosophila* have unusually high levels (averaging 0.140) and have inflated the overall estimate for insects. Thus, the average level for eight orthopteran species (0.038) is close to the average levels in ant species of the *Iridomyrmex purpureus* group (0.038, Halliday, 1978), the *Rhytidoponera impressa* group (0.036, Ward, 1978), and in the genus *Formica* (0.043, Pamilo et al., 1978). The chief lesson of this shift in appraisal is not, however, that theory has now to be readjusted in light of the data, but rather that the diversity of life is still inadequately surveyed in terms of gene diversity, in large part owing to a disproportionate attention having been given to *Drosophila* (Nevo lists 43 *Drosophila* but only 23 other insect species in his survey). The lability of the data is underscored by the fact that the data sets of Nevo (1978) and Powell (1975) overlap but are not coextensive – each includes some "personal communications" omitted by the other.

Ants are karyotypically a highly diverse group, with haploid numbers now known to range from 3 to about 46 (Crozier, 1975:45–50; Imai et al., 1977; Bishop and Crozier, in prep.). The lowest number known was found in heterozygous condition in the Japanese species *Ponera scabra* (Imai and Kubota, 1972), whereas the highest number of 46 or possibly 47 (it is hard to count 90-odd chromosomes!) occurs in the enigmatic "living fossil" *Nothomyrmecia macrops* (Bishop and Crozier, in prep.), which has been recently rediscovered (Taylor, 1978). Changes in chromosome number during evolution seem to have involved Robertsonian rearrangements rather than other mechanisms; although the diversity of rearrangements now known in ants is large (Imai et al., 1977).

There is no experimental evidence bearing on the genic-level mechanism of sex determination in ants, but the occasional discovery of diploid males suggests that the most likely mechanism is the multiple-locus one that is plausible for most Hymenoptera (Crozier, 1977a). Under the multiple-locus model, sex is determined by a number of loci: Individuals heterozygous at one or more loci become female, but individuals homozygous or hemizygous for all of these loci develop as males.

Chromosomal and allozymic differences between species
Prevalence of sibling species in ants

It has been known for a long time that in numerous ant groups only slight morphological differences serve to distinguish many species (Wilson, 1955; Wing, 1968). Traditional taxonomists have shown no hesitation in describing such species, but have so far refrained from nomenclatural recognition of sibling species only detected through chromosomal or allozymic characters, and it may be necessary for geneticists to step in and supply such descriptions (as is becoming common for *Drosophila* taxa – e.g., Ayala, 1973; Ayala and Dobzhansky, 1974). Apparently "good" species dissolve into clusters of siblings upon genetic analysis so frequently (Crozier, 1977a) that I think it would be big news if any widely distributed ant species were shown *not* to consist of a cluster of siblings!

Allozyme differences between closely related ant species are generally slight, and most genetically characterized sibling species have been discovered on the basis of chromosomal differences. Thus, there is strong chromosomal evidence that morphologically uniform species are in fact clusters of siblings in the cases of *Aphaenogaster "rudis"* (Crozier, 1977b, and this chapter), *Myrmecia pilosula* (Imai et al., 1977), *M. fulvipes* (Imai et al., 1977), and *Rhytidoponera "metallica"* (Crozier, 1969; Imai et al., 1977; this chapter), and strongly suggestive evidence for *Camponotus "compressus"* (see Crozier, 1975:55), and *Myrmica "sulcinodis"* (Hauschteck-Jungen, in Crozier, 1975:50). In addition, chromosomal evidence indicates that *Amblyopone fortis* should probably be resurrected from synonymy with *A. australis* (Imai et al., 1977) and that *Myrmecia "brevinoda"* includes more than one biological species (A. D. Bishop, pers. comm.). *Conomyrma bicolor* (Arizona) and *C. ?thoracicus* (Peru) have morphologically indistinguishable workers but divergent karyotypes (Crozier, 1970a).

Although allozyme differences between closely related ant species are generally slight, they have nevertheless contributed to the recognition of sibling species even in the absence of chromosomal information when a reasonable number of loci have been surveyed. Thus, Halliday (1978) concluded that the Australian "meat-ants" placed in the species *Iridomyrmex "purpureus"* actually consist of eight sibling species. Halliday notes that his allozyme data are not definitive in themselves, but are concordant with body color and nest form differences. Ward (1978, 1980), in his study of the *Rhytidoponera impressa* group, detected two new species, *enigmatica* and *confusa*. Although the most reliable characters separating *enigmatica* from the very similar *confusa* and *chalybaea* seem, from Ward's description, to be electrophoretic, there are moderately consistent morphological differences as well. However, Ward explicitly notes that *confusa* and *chalybaea* are "morphologically almost indistinguishable," but are genetically distinct (especially at the *Est-4* locus, with both *confusa* allozymes having less anodal migration than both *chalybaea* allozymes). Ward thus seems to have begun the era in which genetically characterized new ant species are given nomenclatural recognition, although it is unfortunate that he does not detail the crucial allozyme differences between *confusa* and *chalybaea* in his descriptions of the species, or in his taxonomic key to the *impressa* group.

Sibling species in Rhytidoponera "metallica"

The "greenhead" ant *Rhytidoponera "metallica"* is one of the most common Australian insects. As with most *Rhytidoponera* species, normal queens are unnecessary for the maintenance of the colonies, their place being taken by a corps of mated workers (Haskins and Whelden, 1965).

Although there is considerable variation in morphology and in degree of irridescence throughout the range of *"metallica"* (Brown, 1958), the variation is not consistent enough to have enabled systematists to discern natural subdivisions. Furthermore, in some areas, much of this variety can be found in single colonies. The chromosomal evidence, however, is very strong that there are at least two, and possibly three, sibling species currently subsumed under *"metallica"* (Crozier, 1969; Imai et al., 1977). Briefly, there is an eastern sibling with a Robertsonian polymorphism (n = 17–23), but in parts of western New South Wales, northwestern Victoria, and South Australia, *"metallica"* populations have quite a different karyotype, with haploid numbers ranging from 10 to 12. As detailed further by Imai et al. (1977), these western populations might also represent two distinct species, as shown by some complex chromosomal changes uncovered between the more completely analyzed karyotypes from this region.

The known distribution of the two *"metallica"* chromosome number groups is shown in Figure 17.1, except that a further "eastern" colony has been collected near York, in Western Australia. This last collection suggests that the "western" sibling(s) may, in fact, be somewhat restricted in distribution, rather than occupying the western half of mainland Australia. The "western" ants seem to be largely restricted to mallee sites, although some collections were made in other associations (recently disturbed?), and "eastern" *"metallica"* also occur in mallee. The discovery of "eastern" *"metallica"* in Adelaide is noteworthy, and leads to an eventual nomenclatural problem or, looked at another way, to a possible solution to a nomenclatural problem! The problem is: Which entity should retain the name *"metallica"*? The species was described from specimens collected in the Adelaide region (Smith, 1858:94); "eastern" ants occur in Adelaide itself, but "western" ones occur on the other side of the Adelaide hills. Thus, unless consistent morphological differences are found between the forms, the sibling species to which the holotype corresponds is likely to be uncertain. However, this situation also provides the solution to much of the inconvenience that would result from changing the name of such a common and widespread insect, because the name *"metallica"* can now in reasonably good conscience be retained for the widely distributed "eastern" sibling!

The on-going *"metallica"* work is part of a larger study testing the application of phenetic and cladistic methods to ants and using various

Figure 17.1. Collections of *Rhytidoponera "metallica"* eastern species (triangles) and western species (circles). As discussed in the text, the western species may itself comprise two entities. A further collection was made of the eastern species, from York, Western Australia.

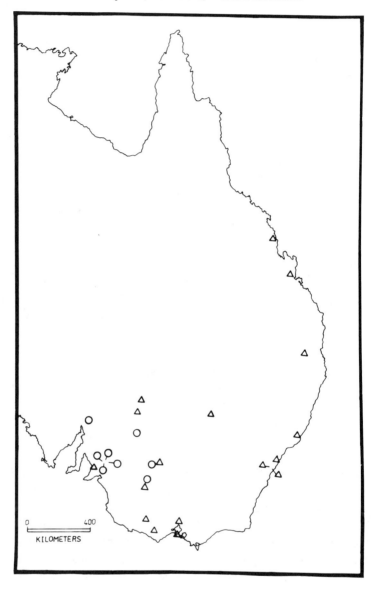

Rhytidoponera species. The allozyme evidence so far indicates that the *"metallica"* siblings are extremely similar, but it appears likely that they do possess different acid phosphatase alleles. If this distinction is upheld by further analysis, it should enable a more rapid mapping of these species than is possible at present (for chromosomal analysis, brood have to be present – and found).

The occurrence of the "western" sibling(s) in South Australia is interesting in view of the great diversity of meat ant sibling species (Halliday, 1978), *and* the extravagant speciation in morabine grasshoppers of the genus *Vandiemenella* (White, 1978:178–86) in the same area.

Sibling species in the Aphaenogaster "rudis" group

Aphaenogaster species are among the most common ants encountered in the hardwood forests of eastern North America, occurring in small colonies averaging a few hundred workers in size (Wilson, 1971:436–7). The most frequently found of the nominal species is *"rudis,"* in which, however, a number of different karyotypes have been found (Crozier, 1975:55, 1977b).

I estimated allozyme frequencies for four *"rudis"* cytotypes and two related species from various localities in Georgia and one in Florida (Crozier, 1977b). These are the 18-, 20-, and 22-chromosome (haploid numbers) Georgian *"rudis,"* a *"rudis"* entity from Alachua County, Florida (with another, different, 22-chromosome karyotype), *fulva,* and *treatae.*

The various karyotypes are sufficiently different to make a confident assessment of their affinities impossible with present knowledge, but a diagrammatic distillation of the relationships as outlined previously (Crozier, 1977b)* is given in Figure 17.2. A key point is that *fulva* and 18-chromosome Georgian *"rudis"* have indistinguishable karyotypes. The 18-chromosome *"rudis"* also possess in weak degree some character states traditionally used to separate *"rudis"* from *fulva* in taxonomic keys.

How well does similarity at the genic level accord with that perceived at the chromosomal? According to the information from four loci, the answer is "not very well." Figure 17.3 shows a phenogram derived by WPGMA clustering (Sneath and Sokal, 1973:234) of Nei's minimum genetic distance (Nei, 1972) values between all populations for which information for all four loci was collected. The populations are numbered according to the scheme given previously (Crozier, 1977b). Heavy lines lead to popula-

---

* Crozier's (1977b) Figure 18C should be read as Figure 18A and vice versa.

Figure 17.2. Relationships among the karyotypes of 20-chromosome *"rudis"* (triangle), 22-chromosome Georgian *"rudis"* (circle), 22-chromosome *"rudis"* from Alachua County, Florida (A), *fulva* (F), 18-chromosome Georgian *"rudis"* (square), and *treatae* (T). The basis for this scheme is discussed further in the text.

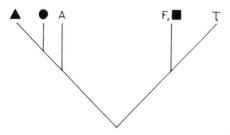

Figure 17.3. WPGMA clustering of *Aphaenogaster* populations listed by Crozier (1977b) on the basis of Nei's minimum genetic distance values for four loci. Heavy lines lead to populations with five or more colonies sampled (three parental genomes per colony), and the scale is in genetic distance units (×100). Distances between OTUs are obtained by adding the vertical lines connecting them. Symbols are as for Figure 17.2 except that solid symbols denote those Georgian *"rudis"* populations for which all colonies were karyotyped, and open ones those whose karyotype was not determined but inferred from worker morphology. Population numbers are those listed by Crozier (1977b). The population 8 sample is basically 20-chromosome species in morphology, but somewhat intermediate to Georgian 22-chromosome *"rudis"* morphology.

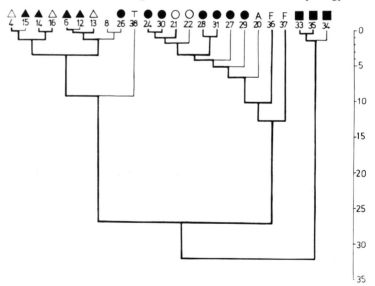

tions sampled for at least five colonies, and give some idea of the reliability that can be assumed for different sections of the phenogram.

The most anomalous placement is perhaps that of population 26, which is a 22-chromosome Georgian population, yet is closely associated in the phenogram with 20-chromosome populations. However, not only was the sample size small for population 26, but also this population occurred in the transition zone between 20- and 22-chromosome populations. The presence of the "wrong" alleles in colonies karyotyped as one cytotype or the other strongly indicates that, although there is no evidence for introgression from the chromosomal evidence (no intermediate karyotypes were found), some gene exchange does occur across this transition zone. In this connection, the grouping of population 26 with the 20-chromosome populations is understandable, and its close association with population 8 is especially so because population 8 (which was not karyotyped) perhaps also involved introgression, as it contained some colonies classed as "intermediate" in morphology (color).

The most significant differences between the phenogram and the scheme based on karyotypes concerns the relative placement of 18-chromosome Georgian *"rudis"* and *fulva:* Although these entities are extremely similar with respect to karyotype, they occur far apart on the basis of the genetic distance clustering. The separation of *treatae* and *fulva* is also interesting, but less reliable because of the small sample size for the *treatae* population and the relatively tentative nature of the assessment of their karyotypic affinities. The differing placements of the other entities are also interesting, but subject to the same reservations.

The essential points made by the phenogram are also reflected in another cluster analysis of the allozyme data. Figure 17.4 shows a Wagner tree derived from the allele frequencies (Crozier, 1977b:Table 2), using the program CLADIST. The root to the tree is here that point equalizing the median distances to the operational taxonomic units (OTUs) on either side. The tree shown is the equal shortest of 215 obtained by varying the input order of the OTUs, and was chosen because it is somewhat more similar in branching sequence to the WPGMA phenogram than is the other. In Figure 17.4, as in Figure 17.3, *fulva* clusters with the 22-chromosome populations, and not with 18-chromosome *"rudis,"* and various other major similarities, as well as some differences, can be readily seen between the two schemes.

The contrasting allozyme and chromosomal data suggest at least two conclusions: first, that changes at the genic and chromosomal levels are imperfectly correlated, so that chromosomally similar forms can be genically distant, and chromosomally different forms can be relatively similar with respect to allozymes; and second, that karyotypes can retain high levels of integrity (in the features of gross chromosome morphology detectable by cytogeneticists) despite evidence from allozymes of gene flow. This situation, pertaining to the transition zone between the 20- and 22-chromosome Georgian *"rudis"* forms, is reminiscent of that reported for grasshoppers of the *Vandiemenella* taxa *viatica*$_{17}$ and P24 (XY) on Kangaroo Island, where biometrical evidence indicates introgression from each entity into the other despite the transition zone's being very narrow (White, 1978:185–6). Finally, I conclude that all of the *Aphaenogaster* forms discussed here are in fact sibling species, and that the chief nomenclatural problems remaining in this case relate to the relationships between the 18- and 22-chromosome species in Georgia and the karyotypically different forms farther north (Crozier, 1975:55), as the 20-chromosome species ranges from the Georgian coastal plains to the New Jersey pine barrens with little apparent change (Crozier, 1977b).

Further significance of ant sibling species

The study of speciation or its consequences in ants may be interesting to speciationists, but it also has significance for other evolution-

Figure 17.4. Rooted Wagner tree for the populations referred to in Figure 17.3, based on the allele frequencies given by Crozier (1977b). Symbols are as for Figure 17.3. The root was chosen to equalize the median distances to the OTUs on each side. The tree shown was one of the two equal-shortest out of 215; it is the more similar one to the phenogram in Figure 17.3.

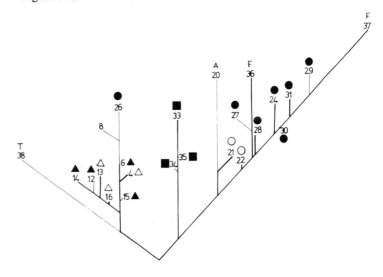

ary studies, especially ecological genetics. Thus, a major field of endeavor for evolutionists today is the attempt to relate allozyme variation to selective factors, such as climatic variables. However, if a small number of loci have been used, the presence of sibling species would probably go unrecognized. This would have been the case with the *Aphaenogaster* siblings, and would probably prove to be so for the *Rhytidoponera* *"metallica"* case as well, but for the chromosomal evidence. Thus, an apparent strong and exciting correlation between allozyme frequencies and environmental variables may simply reflect differing allozyme frequencies in two species that vary in relative abundance geographically. I am *not* trying to discourage ecological studies on ants, which are superb subjects for such investigation because of their abundance and low individual mobility, but rather I stress that such studies must take into account the likelihood that many, if not most, widely distributed ant "species" are in fact clusters of siblings.

Chromosomal evolution and life patterns

The entire range of chromosome numbers of the order Hymenoptera is contained within the single family Formicidae (n = 3–46). But this does not indicate that all ant groups are karyotypically variable; in fact there is a spectrum of ant genera ranging from ones that are almost invariant for chromosome number (such as *Formica*) to highly variable ones (especially *Myrmecia,* with n = 5–44). Of course, even in an ant genus with little change in chromosome number many structural changes can occur (Imai et al., 1977), but chromosome number provides the most readily quantified indicator for an attempt to uncover correlations between differing rates of chromosomal evolution between ant genera and the factors that influence these rates.

A highly preliminary attempt to establish a framework for such studies is given, perhaps rashly, in Table 17.1. "Perhaps rashly," because of the very low reliability of many estimates (especially those "guesstimates" of mine). The data shown pertain to the 11 ant genera for which five or more species have been karyotyped. For those species with chromosome number polymorphisms (species of *Aphaenogaster, Myrmecia, Pheidole,* and *Rhytidoponera*), the number used has been the midpoint of the range; this convention allows each species to be weighted equally.

The karyotype data, the geographic spread estimates, and the fossil information are all explicitly given in the literature, as are most of the estimates of colony size. However, the estimates of colony density and total size of genus are mostly based on my rough impressions gained from the literature.

Table 17.1. *Karyotype variability, in terms of variation in haploid number,* n, *in ant genera together with other selected attributes*

Subfamily and genus	No. spp. examined	$\bar{n}^a$	Median n	s_n	CV_n	Geog. spread[b]	Colony size[c]	Colony density[d]	Fossils[a]	Genus size[c]
Dolichoderinae										
Iridomyrmex (O. W.)	18	9.11	9	1.98	21.59	3	M-B	D	Yes	M
Iridomyrmex (O. W.)	*14	8.79	9	0.43	4.85	3	M-B	D	Yes	M
Tapinoma	6	7.67	8	1.52	19.64	11	M-B	S-D	—	S
Formicinae										
Camponotus	33	17.00	17	5.04	29.67	13	S-B	S	Yes	B
Formica	30	26.37	26	0.49	1.86	4	M-B	D	Yes	M
Polyrhachis	6	20.83	21	0.41	1.96	7	S-M	S	—	M
Myrmeciinae										
Myrmecia	11	24.68	25.5	14.36	58.17	1	S-M	S	—	M
Myrmicinae										
Leptothorax	10	11.20	11.50	2.30	20.53	11	S	S	—	S
Monomorium	7	13.29	11	4.07	30.64	10	M-B	S-D	—	M

Pheidole	20	10.37	10	1.95	18.82	12	M	D	—	B
Pheidole	*19	9.95	10	0.40	4.07	12	M	D	—	B
Solenopsis	7	13.86	16	2.67	19.29	12	M-B	S-D	—	M
Ponerinae										
Rhytidoponera	9	20.89	21	4.19	20.05	2	S-M	S-D	—	M
Rhytidoponera	*8	22.06	21.25	2.43	11.00	2	S-M	S-D	—	M

[a]The chromosome data come from Crozier (1975:46–50, 1977b), Hauschteck-Jungen and Jungen (1976), and Imai et al. (1977), plus the count of n = 44 for *Myrmecia* nr. *brevinoda* (A. D. Bishop, pers. comm.).

[b]The estimates of geographic spread were made by counting the occurrences of a genus in the regions tabulated by Brown (1973).

[c]Average completed colony sizes in terms of adults present, with S = < 100, M = 100–1000, B = > 1000, were taken from Wilson (1971:436–8) and Brian (1965:5–6) or, in some cases, represent rough estimates by me.

[e]Information about the fossil record comes from Wilson (1971:28–9).

[f]Rough estimates of genus size (in terms of number of species in the world) are also mine and rest on general impressions recorded in the literature with S = < 50 spp., M = 50–500 spp., B = > 500 spp.

[g]The characterization of the density of colonies as S (sparse) or D (dense), which categories are also meant to mirror the relative population sizes of the breeding population, rest on subjective assessments by me.

*Denotes estimates adjusted by omitting apparently aberrant species. Further explanation in text.

In *Iridomyrmex,* taxonomic problems render the estimation of variation difficult. Thus, only the Old World species are considered here because it is widely suspected among myrmecological systematists that the New World species are a separate stock (which would therefore have to be convergent cytologically (Crozier, 1975:51–2)). However, among the remaining Old World species, the *glaber* and *darwinianus* groups are chromosomally and morphologically distinct from the rest, and their placement together may only reflect that they are all members of the tribe Tapinomini lacking special distinguishing features (spines, etc.) (Crozier, 1975:51–2). The estimate denoted * in the table is for Old World *Iridomyrmex* excluding the *glaber* and *darwinianus* groups.

Species have also been excluded from *Pheidole* and *Rhytidoponera* to yield secondary estimates. *Pheidole nodus* has a range of numbers (n = 17–20) that are all much higher than those for all other congeners, which all have haploid numbers of 9 or 10, or, in one case, 12. For *Rhytidoponera,* as has been shown above, the western form (or forms) of *"metallica"* is quite divergent from the other members of the genus karyotyped so far. *Pheidole nodus* and *Rhytidoponera "metallica"* (western sibling) have therefore been omitted in making the secondary estimates denoted * in the table. It is these secondary estimates that are used from here on.

The most useful karyotypic quantities that should be distinguished in assessing the results shown in Table 17.1 are the mean chromosome number, and the extent of chromosome number variability within genera. Although the correlation between $\bar{n}$ (mean haploid number in a genus) and s_n (its standard deviation) is not statistically significant ($r = 0.40$), it is much higher than between $\bar{n}$ and the coefficient of variation (CV_n) ($r = 0.16$). Using $\bar{n}$ and CV_n thus seems the best way to examine variation in number and variation in variability separately.

The table shows striking differences between genera, not only in mean number, but also in variability. Sometimes these differences occur between closely related genera, as witnessed by the differences in CV_n between *Camponotus* and *Polyrhachis.* There is no apparent taxonomic relationship explaining this variability between genera – no subfamily has a monopoly on either variable or invariant genera, or on genera with high or low numbers (the low numbers in Dolichoderinae and high ones in Myrmeciinae probably reflect the small numbers of genera examined in these subfamilies).

There is a statistically significant negative correlation between $\bar{n}$ and geographic spread ($r = -0.60$, $0.05 > p > 0.01$), but all other correlations for n̄ are both low and nonsignificant. Why there is a correlation between n̄ and geographic spread is unclear, although a number of unsatisfactory hypotheses can be framed.

For CV_n, there are no statistically significant correlations with any of the quantities in the table, but the correlation with colony density is high ($r = -0.56$) and close to being significant at the 0.05 level (for the analysis, quantities such as colony size and density were coded on a numerical scale, S = 1, M = 2, B = 3, etc.). A negative correlation between chromosome number variability, if this reflects rate of intrageneric chromosomal evolution, and population size (reflected as colony density) is expected; chromosome rearrangements should become established more easily in species with small rather than large populations.

Thus, the evidence now is overall more suggestive than compelling for correlations between the course of karyotype evolution and other attributes. I hope that other myrmecologists will provide firmer estimates of the various necessary quantities, if only in shock at seeing mine, so that this approach can be placed on a firmer basis. Of course, the strong possibility remains that internal genetic factors (such as the rate of evolutionary duplication of heterochromatic and centromeric material) may also affect the rate of evolution of ant chromosome numbers and form, leading to different kinds of "karyotype orthoselection" (White, 1973:450–4) in different ant groups.

I thank A. D. Bishop, M. Archer, and P. I. Dixon for critical comments on the manuscript, and S. Hand and C. Stahel for drawing the diagrams. This work has been supported by grants from the United States National Science Foundation, the Australian Research Grants Committee, and the Ian Potter Foundation, and by funds from the University of Georgia and the University of New South Wales.

References

Ayala, F. J. 1973. Two new subspecies of the *Drosophila willistoni* group (Diptera: Drosophilidae). *Pan-Pacific Entomol. 49:*273–9.
- 1978. The mechanisms of evolution. *Sci. Am. 239(1):*48–61.
Ayala, F. J., and Dobzhansky, T. 1974. A new subspecies of *Drosophila pseudoobscura* (Diptera: Drosophilidae). *Pan-Pacific Entomol. 50:*211–19.
Brian, M. V. 1965. *Social Insect Populations.* New York: Academic Press.
Brown, W. L. 1958. Contributions towards a reclassification of the Formicidae. II. Tribe Ectatommini (Hymenoptera). *Bull. Mus. Comp. Zool. Harv. Univ. 118:*175–362.
- 1973. A comparison of the Hylean and Congo–West African rain forest ant faunas. In *Tropical Forest Ecosystems in Africa and South America: A Comparative Review* (ed. Meggers, B. J., Ayensu, E. S., and Duckworth, W. D.), pp. 161–85. Washington, D.C.: Smithsonian Institution Press.
Brown, W. L., and Taylor, R. W. 1970. Superfamily Formicoidea. In *The Insects of Australia* (ed. Waterhouse, D. F.), pp. 951–9. Melbourne: Melbourne University Press.

Crozier, R. H. 1969. Chromsome number polymorphism in an Australian ponerine ant. *Can. J. Genet. Cytol. 11:*333–9.

— 1970a. Karyotypes of twenty-one ant species (Hymenoptera: Formicidae), with reviews of the known ant karyotypes. *Can. J. Genet. Cytol. 12:*109–28.

— 1970b. On the potential for genetic variability in haplo-diploidy. *Genetica 41:*551–6.

— 1975. *Animal Cytogenetics 3 Insecta 7 Hymenoptera.* Berlin: Borntraeger.

— 1977a. Evolutionary genetics of the Hymenoptera. *Annu. Rev. Entomol. 22:*263–88.

— 1977b. Genetic differentiation between populations of the ant *Aphaenogaster "rudis"* in the southeastern United States. *Genetica 47:*17–36.

— 1979. Genetics of sociality. In *Social Insects,* vol. 1 (ed. Hermann, H. R.), pp. 223–86. New York: Academic Press.

Halliday, R. B. 1978. Genetic studies of meat ants, *Iridomyrmex purpureus.* Unpublished Ph.D. thesis, University of Adelaide.

Haskins, C. P., and Whelden, R. M. 1965. "Queenlessness," worker sibship, and colony versus population structure in the formicid genus *Rhytidoponera. Psyche 72:*87–112.

Hauschteck-Jungen, E., and Jungen, H. 1976. Ant chromosomes I. – The genus *Formica. Insectes Soc. 23:*513–24.

Imai, H. T., and Kubota, M. 1972. Karyological studies of Japanese ants (Hymenoptera, Formicidae). III. Karyotypes of nine species in Ponerinae, Formicinae, and Myrmicinae. *Chromosoma (Berl.) 37:*193–200.

Imai, H. T., Crozier, R. H., and Taylor, R. W. 1977. Karyotype evolution in Australian ants. *Chromosoma (Berl.) 59:*341–93.

Mayr, E. 1969. *The Principles of Systematic Zoology.* New York: McGraw-Hill.

Nei, M. 1972. Genetic distance between populations. *Am. Nat. 106:*283–92.

Nevo, E. 1978. Genetic variation in natural populations: patterns and theory. *Theor. Pop. Biol. 13:*121–77.

Pamilo, P., and Crozier, R. H. 1978. Effects of haplo-diploidy on genic variation under some selection models. *XIV Int. Congr. Genet., Moscow, Contributed Paper Sessions Abstracts Sections 21–32:*6.

Pamilo, P., Rosengren, R., Vepsäläinen, K., Varvio-Aho, S. L., and Pisarski, B. 1978. Population genetics of *Formica* ants I. Patterns of enzyme gene variation. *Hereditas 89:*233–48.

Powell, J. R. 1975. Protein variation in natural populations of animals. *Evol. Biol. 8:*79–119.

Smith, F. 1858. *Catalogue of Hymenoptera,* part 6. London: British Museum.

Sneath, P. H. A., and Sokal, R. R. 1973. *Numerical Taxonomy.* San Francisco: Freeman.

Taylor, R. W. 1978. *Nothomyrmecia macrops:* A living-fossil ant rediscovered. *Science 201:*979–85.

Ward, P. S. 1978. Genetic variation, colony structure, and social behavior in the *Rhytidoponera impressa* group, a species complex of ponerine ants. Unpublished Ph.D. thesis, University of Sydney.

— 1980. A systematic revision of the *Rhytidoponera impressa* group (Hymenoptera: Formicidae) in Australia and New Guinea. *Aust. J. Zool.* (in press).

White, M. J. D. 1973. *Animal Cytology and Evolution.* Cambridge: Cambridge University Press.

— 1978. *Modes of Speciation.* San Francisco: Freeman.

Wilson, E. O. 1955. A monographic revision of the ant genus *Lasius. Bull. Mus. Comp. Zool. Harv. Univ. 113:*1–201.

— 1971. *The Insect Societies.* Cambridge, Massachusetts: Harvard University Press.

Wing, M. W. 1968. Taxonomic revision of the Nearctic genus *Acanthomyops. Mem. Cornell Univ. Agric. Exp. Sta. 405:*1–173.

18 Chromosomal evolution and morphometric variability in the thelytokous insect *Warramaba virgo* (Key)

WILLIAM R.ATCHLEY

A discussion of the genetic consequences of thelytoky still rests largely on inference from general theoretical considerations rather than direct factual evidence.

M. J. D. White (1973)

For many years biologists have debated the evolutionary consequences of asexual reproduction. As a result, considerable theory exists about the consequences of genetic recombination, the biological use of sex, the long-term advantages of sexual reproduction, and the resultant evolutionary potential of parthenogenetic species. This has been particularly true for *thelytoky,* a form of parthenogenesis in which virgin females produce exclusively female offspring by a process not involving fertilization.

Although theoretical developments have outpaced experimentation (reviews in Williams, 1975; Maynard Smith, 1978; White, 1977), significant progress has been made recently toward providing direct evidence on the amount of genic, chromosomal, and morphometric variability in thelytokous organisms and the possible role of this variation in evolution. Many of the critical data on cytogenetic variability in thelytokous organisms and closely related sexually reproducing species have been provided by the elegant studies of M. J. D. White and his co-workers. White has been critically examining the morabine grasshopper genus *Warramaba* (formerly *Moraba*) (Orthoptera, Eumastacidae) which contains both thelytokous and sexual species and is rather widely distributed in Australia.

White (1973) has noted that "there have been very few studies on the biometry of thelytokous populations." Therefore, in this essay dedicated to Michael White I propose to quantify some of the cytogenetic variability and explore further the morphometric variation between clones of the thelytokous species *Warramaba virgo* Key. The principal question to be examined is that of whether there is congruence in the patterns of cytogenetic and morphological variability.

The genus *Warramaba*

The genus *Warramaba* is comprised of five morabine grasshopper species occurring in southern Australia. Included are *W. virgo* (Key), a widespread parthenogenetic species, and four sexually reproducing taxa including *W. picta* Key and three undescribed species, P196, P169, and P125 (P125 includes the chromosomal race P188). "P" stands for "provisional species" and these are the code numbers used in the Australian National Insect Collection. The sexual species are confined to Western Australia, whereas *W. virgo* occurs in both eastern and Western Australia. The distributions of the various taxa are mapped by Webb et al. (1978).

W. virgo is a diploid, all-female species reproducing by obligate thelytoky. It is a wingless species of low vagility that is widely distributed throughout the sandy regions of much of southern Australia. This species feeds on shrubs of various species of *Acacia*. *W. virgo* has a highly disjunct distribution in that it occurs at many localities in Western Australia as well as in New South Wales and Victoria in eastern Australia; however, a gap of approximately 1600 km exists between the eastern and western localities. Extensive collecting efforts have failed to find *W. virgo* in the intervening gap.

Since its discovery by White et al. (1963), much has been learned about *W. virgo* from classical cytogenetic methods (White et al., 1963; White, 1966; White and Webb, 1968; White et al., 1973; Hewitt, 1975), by hybridization studies (White et al., 1973; White and Contreras, 1979), by chromosome banding studies (Webb et al., 1978), and by biometrical analyses of morphological, ecological, and physiological variation (Atchley, 1977a,b, 1978).

Reproduction in *W. virgo* is clonal in that the progeny of any female are isogenic. The meiotic mechanism in *W. virgo* and its genetic consequences have been well studied, and White et al. (1963) have shown that the chromosome set in the oocyte undergoes a premeiotic replication to the tetraploid number prior to a two-division meiosis. The final product of meiosis is a diploid egg. Synapsis is completely restricted to sister chromosomes and never occurs between what would otherwise be termed nonsister homologues. Meiotic bivalents exhibit chiasmata, but crossing-over is without genetic consequence because it occurs between chromosomes that are molecular copies of each other. Such a meiotic system has maintained the high level of heterozygosity derived from the original hybrid constitution and augments it with mutations that only become homozygous by recurring equivalent base substitutions.

There is now little doubt that *W. virgo* arose by hybridization between two sexual species of *Warramaba* whose modern descendants are the species P169 and P196 (Hewitt, 1975; White et al., 1977; White and Contreras, 1979). Hybrids between P169 and P196 have the same karyotype as the "standard" karyotype of *M. virgo,* and a few of the "synthetic virgo" females produced by hybridization in the laboratory were parthenogenetic. Although the hybrids have the same karyotype as their mother, it is not known if the mechanism of parthenogenesis in the hybrids is the same as in *W. virgo.*

All available evidence suggests the point of origin of *W. virgo* was in Western Australia at a point where the periphery of the range of the two parapatric sexual species P169 and P196 intersected (White and Contreras, 1979; White, in prep.).

The cytogenetic diversity in *W. virgo,* the highly disjunct distribution of its clones, the correlation of distribution to known geological and historical events, the very low vagility of morabines, and so on, led White (in prep.) to hypothesize that the hybridization (or hybridizations) giving rise to *W. virgo* occurred at least a half million years ago. If this hypothesis is substantiated, it would make *virgo* at least an order of magnitude older than the currently estimated ages of various parthenogenetic vertebrates that have arisen by hybridization (White, 1979).

Cytogenetic variability in *W. virgo*

Webb et al. (1978) examined C- and G-banding patterns in 22 local populations of *W. virgo* and concluded that 15 cytogenetically unique clones could be recognized [at least 9 additional cytogenetically distinguishable clones are now recognized (White, pers. comm.), giving a total of 24 clones, but complete cytogenetic data were not available for their inclusion in this chapter]. To quantify cytogenetic variability in *W. virgo,* I have analyzed 42 cytogenetic characters taken from Webb et al. (1978) and White (pers. comm.). Webb et al. (1978) indicate that about 50 identifiable bands occur in *W. virgo,* but only 34 are polymorphic between clones, and only these were used in these analyses. The remaining eight characters are chromosomal rearrangements, including fusions, fissions, and inversions (Table 18.1). Each character is coded as a binary (present or absent) variable.

Matrices of euclidean distance coefficients and simple matching coefficients (Sneath and Sokal, 1973) were computed for the 15 clones. The simple matching coefficient (S_m) is a useful measure of similarity for binary data and is defined as:

$$S_m = \frac{a + b}{a + b + c + d}$$

where, in a 2×2 table, a = the number of positive matches, b = the number of negative matches, and c and d are mismatches. A UPGMA cluster analysis was carried out on each matrix (Sneath and Sokal, 1973). The cluster analysis results based on the simple matching coefficient were very similar to those obtained from euclidean distances. Therefore, only

Table 18.1. *Cytogenetic characters used in multivariate analyses of clonal divergence*

C-bands: Presence or absence of band unless otherwise noted

1. Band 4	18. Band 69
2. Band 4 enlarged	19. Band 71
3. Band 4 decreased	20. Band 72.5 doubled
4. Band 8 equally divided	21. Band 73
5. Band 17	22. Band 74
6. Band 17 doubled	23. Band 82.5
7. Band 23	24. Band 87
8. Band 32	25. Band 87.3
9. Band 33.5 doubled	26. Band 88
10. Band 36	27. Band 89.8
11. Band 37	28. Band 90
12. Band 44	29. Band 90.5
13. Band 49	30. Band 92
14. Band 49 enlarged	31. Band 92.5
15. Band 49 decreased	32. Band 93.
16. Band 60.5	33. Band 94
17. Band 64.5	34. Band 97

Chromosomal rearrangements: Presence or absence
   of rearrangement
35. Standard karyotype
36. 5_{196}–6_{169} translocation
37. 3_{196}–4_{169} translocation
38. 1_{169}–6_{169} translocation
39. X chromosome dissociated
40. X chromosome paracentric inversion
41. AB–3_{169} translocation
42. 1_{169}–4_{169} translocation

the phenograms based on the simple matching coefficients will be given here. Attempts to provide an interpretable ordination of the cytogenetic data by nonmetric scaling or principal coordinates analyses were unsuccessful, the reason being that the very large morphometric distances between a few of the clones relative to small distances between many others led to unsatisfactory ordinations. Thus only the phenograms will be provided.

The sexually reproducing progenitors of *W. virgo,* (P196 and P169) cannot be included in these analyses because the homologies of the various C-bands are not now known with certainty. The karyotype expected from natural hybridization between P196 and P169 can be regarded as the "primitive" condition and designated "standard."

Based on the total chromosomal data, there are three cytologically defined groups in *W. virgo.* Group 1 clones are those possessing the standard karyotype and differing from one another only in the details of the C-banding. These clones are widespread in both eastern and Western Australia, and in these analyses include the clones from Kalgoorlie, W. A., Coolgardie, W. A., Ponton Creek, W. A., Kitchener, W. A., Broken Hill, N. S. W., Little Topar, N. S. W., and Cobar, N. S. W.

Group 2 clones are those characterized by having the standard karyotype modified by various structural rearrangements as well as possessing some differences in C-banding patterns. This latter group is exemplified by clones known from single localities, which are probably derived evolutionarily from group 1 clones. Group 2 includes clones from Coonaanna Hill, W. A., South Ita, N. S. W., Hayes Hill, W. A., and Spargoville, W. A.

Group 3 clones are those with the "Boulder" karyotype, which includes a shorter X limb, a different type of X_1 chromosome, and a rearranged chromosome 5_{196}. These clones have two C-bands deleted and six added. Two cytogenetically distinct clones make up the group 3 category: Boulder (one locality) and Zanthus (two localities). This group is restricted to Western Australia.

Figure 18.1 provides a UPGMA phenogram based on simple matching coefficients from the 42 chromosomal banding and rearrangement cytogenetic characters for 15 clones of *W. virgo.* These results indicate the existence of three rather obvious cytogenetically defined groups. These groups are: (1) Boulder and Zanthus, (2) Spargoville, and (3) the remaining 12 clones. Boulder and Zanthus are group 3 clones and obviously cytologically distinct. Spargoville probably belongs in group 2 but exhibits many differences from typical group 2 clones. There is little obvious separation between group 1 and group 2 clones.

When the cluster analysis is repeated on just the 34 banding characters, the pattern of similarities between clones is altered considerably (Figure 18.2). The relationships between the three divergent clones (Spargoville, Boulder, and Zanthus) and the remaining 12 are little changed. The Boulder and Zanthus clones differ from all of the group 1 and group 2 clones by at least 18 bands, whereas the Spargoville clone differs from groups 1 and 2 by a minimum of 10 bands. Spargoville differs from the Boulder and Zanthus clones by 18 bands. The relationships between C-banding similarity and geographic distance within the group 1 and 2 clones in Figure 18.2 provide a complex picture. In New South Wales, the Cobar clone shares only 88% similarity in C-banding pattern (a difference in three bands) with both the Broken Hill and Little Topar clones but 91% similarity with Kalgoorlie and Coolgardie in Western Australia. On the other hand, there is over 97% similarity (difference in only one band) between Little Topar, N. S. W. and both Coonaanna Hill and Coolgardie in Western Australia. Further, Coolgardie exhibits a 97% similarity in banding patterns to the South Ita and Monia Gap, N. S. W. clones. The latter are group 2 clones. As noted, 97% similarity translates into a difference of a single C-band between clones that are separated from each other by at least 1600 km and that probably diverged from each other many thousands of years ago.

A matrix of average simple matching coefficients (= average percent similarities) for the major cytogenetic groups was computed to assess the extent of similarity (or divergence) within and between the chromosomal

Figure 18.1. UPGMA cluster analysis of a matrix of simple matching coefficients for the 42 cytogenetic characters given in Table 18.1 Cytogenetic groups were defined a priori and are described in the text.

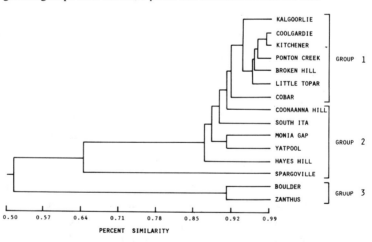

categories described above (Table 18.2). This table was generated by computing all pairwise similarities between clones within and between groups and then calculating the desired means and standard deviations (e.g., Kalgoorlie and Coolgardie, Kalgoorlie and Kitchener, etc., for the average group 1 similarity and Kalgoorlie and Coonaanna Hill, Kalgoorlie and South Ita, etc., for the group 1 vs. group 2 values). The elements of Table 18.2 give average percent similarities for the 34 banding characters alone (nonitalic values) and for the 42 banding and chromosomal characters (italic values). The standard deviation of each estimate is also provided, together with the sample sizes in parentheses.

Greatest similarity is found within the clones comprising group 1, where there is an average similarity of 93% for both the banding and the banding + chromosomal rearrangements data sets. Similarity within group 2 clones falls off to 80% for banding data and 78% for all cytogenetic variables. The two group 3 clones exhibit 88% similarity in banding traits with each other and about 91% for all cytogenetic characters.

Groups 1 and 2 have a cytogenetic similarity of 87% in banding patterns and 83% in banding + chromosomal data, whereas groups 1 and 3 have similarities of 41% and 53% respectively for the two types of data. Groups 2 and 3 have similarities of 42% and 48% for banding and banding + chromosomal data.

The preceding data provide a quantitative estimate of the cytogenetic divergence that has occurred between clones of *W. virgo*. These results indicate the geographically widespread nature of the standard karyotype

Figure 18.2. UPGMA cluster analysis of a matrix of simple matching coefficients for the 34 C-banding characters given in Table 18.1.

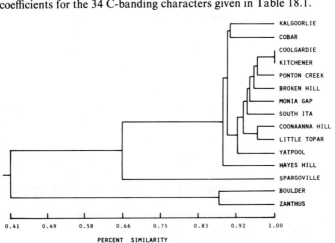

as well as something of the geographic distribution of group 1 and group 2 clones. Group 1 clones are always found in the northern portion of the species range in both eastern and Western Australia. Group 2 clones are always more southern in distribution in both the east and the west. Group 3 is currently comprised of only three clones (only two examined here) and limited to Western Australia.

These various analyses summarize the existing cytogenetic picture for *W. virgo,* and provide an overview of the cytogenetic diversity and major cytological groupings. Some caution about interpretation must be interjected at this point because it cannot be assumed that the phenogram of the cytogenetic data is a completely faithful reflection of the detailed phylogeny of the various clones, particularly in group 1. These results need to be supplemented with a phylogenetic analysis.

One strategy in producing a phylogenetic tree from these cytogenetic data is to describe the descendent and ancestral karyotypes with as few changes as possible (i.e., find the most parsimonious tree). Description of the most parsimonious tree is independent of defining the true ultimate ancestor (Fitch, 1977).

In defining the best (= most parsimonious) tree for the *W. virgo* data, three assumptions were made about the data: (1) there was no a priori

Table 18.2. *Percent similarity within and between major cytogenetic groupings of* Warramaba virgo[a]

	Group 1[b,c]	Group 2[b,c]	Group 3[b,c]
Group 1	0.936 ± .034 *0.942 ± .024* (21)		
Group 2	0.880 ± .100 *0.848 ± .095* (42)	0.823 ± .123 *0.791 ± .112* (15)	
Group 3	0.412 ± .037 *0.532 ± .027* (14)	0.417 ± .045 *0.480 ± .025* (12)	0.882 *0.907* (1)

[a]Diagonal elements give percent similarity within cytogenetic groupings, and off-diagonal elements give similarity between cytogenetic groupings.
[b]Standard deviations are given for each value, and sample sizes are given in parentheses.
[c]Nonitalic values relate to 34 banding characters alone, and italic values to 42 banding and chromosomal characters.

determination of ancestral and derived chromosomal character states; (2) cytogenetic changes were not irreversible; and (3) gains and losses of C-bands and rearrangements were equi-probable events. At the present time, it is not clear what relative probabilities can be assigned to the gain and loss of a C-band.

In constructing the best tree, we started with an initial estimate of the tree, which in this case was the phenogram given in Figure 18.1. Then all possible interchanges between ancestor neighbor nodes were examined with the goal of making a better tree. The final tree was the one from which no better tree could be derived. In all, 1.8×10^{11} trees are possible for these 15 clones.

Figure 18.3 gives the most parsimonious tree resulting from an analysis of the 42 banding and rearrangement characters. Two additional and equally parsimonious trees would have Spargoville arising from node B and Coonaanna Hill arising from node I. A minimum of 53 cytogenetic changes are needed to produce this tree, and the numbers on the legs indicate the number of cytogenetic changes (banding as well as chromosomal rearrangements) between nodes.

Several conclusions are evident. The major cytogenetic groupings described earlier persist in the phylogenetic analyses. Boulder and Zanthus (group 3 clones) are quite distinct, as is the Spargoville clone. In eastern

Figure 18.3. Phylogenetic tree based on 42 cytogenetic characters for 15 clones. Numbers on the legs give the number of cytogenetic changes between nodes. Double lines on legs indicate independent invasions of eastern Australia from Western Australia.

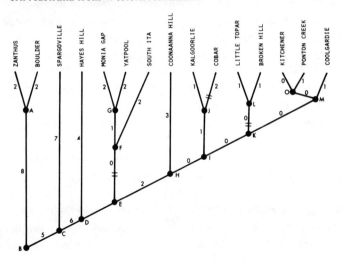

Australia, Monia Gap, Yatpool, and South Ita reflect a single phylogenetic lineage, as do Little Topar and Broken Hill. Further, the Kitchener, Ponton Creek, and Coolgardie assemblage of clones seems to reflect a close-knit cluster. Kitchener seems to possess the primitive karyotype for the group 1 cytogenetic clones as evidenced by the absence of cytogenetic changes between Kitchener's position and node H, which defines group 1. In general, there is reasonably close agreement between the phenetic and phylogenetic analyses.

The tree given in Figure 18.3 is the most parsimonious tree in a cytogenetic sense. However, it may not be the most parsimonious tree in a geographic sense because it requires three independent invasions of eastern Australia. (An invasion in this case is equated to crossing the 1600-km gap that now exists between the eastern and Western Australian clones.) These three invasions are demarcated in Figure 18.3 by double lines on legs arising from nodes E, J, and K. A tree that is most parsimonious in a geographic sense is one that would involve only a single invasion of eastern Australia. Such a tree would require at least 55 chromosomal changes to arrive at the best tree. This is an increase of two chromosomal band changes over that shown in Figure 18.3.

The greatest incongruity in the phylogenetic analyses as compared to geographic distribution is the relationship between Kalgoorlie and Cobar. They occur at opposite ends of the species range in eastern and Western Australia but are placed together in Figure 18.3 by the mutual sharing of C-band 87 as described by Webb et al. (1978).

The phylogenetic analyses are tentative and will be the subject of another paper. The number of cytogenetic characters useful in delimiting the phylogeny of *W. virgo* is limited. Although we have employed 42 characters, 18 of these characters are the same in all taxa except one. These 18 characters are useful in distinguishing particular clones, but they are useless in a phylogenetic analysis. Further, 17 of the remaining characters serve to distinguish the Boulder, Zanthus, and Spargoville clones. Only the remaining eight characters are useful in a phylogenetic sense. Thus, understanding the phylogeny is going to require more characters.

Based on these analyses of the cytogenetic data alone, one would conclude either that (1) the group 1 clones have preserved the primitive karyotypic configuration, and very little change has occurred in the C-banding patterns in spite of their long-term geographic and historical separation; or (2) the banding patterns are polyphyletic in origin and, as a result, do not provide an accurate description of evolutionary divergence among the clones in *W. virgo*.

The quantitative data emphasize a pronounced (and possibly historically early) divergence between the group 3 clones (Boulder and Zanthus) and the group 1 and 2 clones. The highly divergent group 3 clones are located geographically only a few kilometers from Kalgoorlie, W. A., where several clones occur that exhibit the standard karyotype. The most plausible explanation for this marked divergence is that it reflects a multiple origin of *W. virgo;* however, it might also be that these are clones surviving from a very early origin. In addition, the Spargoville clone is distinct and reflects a considerable cytological divergence from the remaining clones. Clearly, if Spargoville is to be considered in group 2, it must have diverged very early unless there is a selectionist basis for this marked divergence. These aspects of the overall problem require considerable further study.

Morphometric variability in *W. virgo*

Superimposed upon the cytogenetic diversity in *W. virgo* is a considerable amount of ecological, physiological, and morphological variability. Previously (Atchley, 1977a,b, 1978), I have examined the extent of differentiation and variability between samples of *W. virgo* from eastern or Western Australia, and determined the proportion of variance in individual morphometric characters occurring between chromosomal races, food plant races, clones, localities within clones, years within localities, and so on, and the morphometric consequences of contact zones between sexual and parthenogenetic species.

Important aspects of morphometric variability still to be examined include geographic variation in morphometric shape among samples within *W. virgo* and the relationship of morphometric to cytogenetic diversity. The latter is particularly timely because Webb et al. (1978) have cytogenetically characterized a large number of distinct clones of *W. virgo,* thus facilitating the quantification of their interrelationships in this chapter.

To assess overall morphometric similarity, and facilitate comparisons with the chromosomal data, a Mahalanobis distance matrix based on 10 morphometric characters and 22 samples of *W. virgo* (Table 18.3) was analyzed by a UPGMA cluster analysis (Figure 18.4). The following 10 morphometric characters (together with a character code) were used in these analyses: (1) dorsal head length measured from the posterior edge of the postocciput (HL); (2) head width at the level of the frontoclypeal suture (HW); (3) lateral head length measured from the bottom of the eye to the tip of the gena (LHL); (4) interocular distance in facial aspect (ID); (5) greatest eye length measured diagonally (EL); (6) pronotum length

Table 18.3. *Summary of locality, cytogenetic data, and foodplant data for 22 samples of* Warramaba virgo

Numeric code	Locality	Cytogenetic clone	Chromosomal karyotype	Major clonal group	Food plant
1. (COBB)	2.6 km W Cobar, N.S.W.	—	Standard	—	Acacia aneura
2. (COBC)	6.4 km W Cobar, N.S.W.	Cobar	Standard	1	A. aneura
3. (COBD)	1.6 km SE Cobar, N.S.W.	—	Standard	—	
4. (COBE)	6.4 km NW Cobar, N.S.W.	—	Standard	—	A. aneura
5. (COON)	2.1 km SW Coonavittra Tank, N.S.W.	—	Standard	—	A. loderi
6. (WILC)	1.6 km E jct. to Ivanhoe & Wilcania, N.S.W.	—	Standard	—	A. loderi
7. (TOP)	4.5 km E Little Topar Hotel, N.S.W.	Little Topar	Standard	1	A. loderi
8. (SEBH)	4.8 km SE Broken Hill, N.S.W.	Broken Hill	Standard	1	A. loderi
9. (SVC)	Sterling Vale Creek, SW of Broken Hill, N.S.W.	Broken Hill	Standard	1	A. loderi
10. (BH)	16 km S Broken Hill, N.S.W.	—	Standard	—	A. loderi
11. (NET)	16 km E Netley, H.S., N.S.W.	—	Standard	—	A. loderi
12. (KUDG)	9.7 km N Kudgee, H.S., N.S.W.	—	Standard	—	A. loderi
13. (ITA)	South Ita Sandhills, 159 km N Wentworth, Victoria	South Ita	South Ita	2	A. wilhelmiana
14. (NOW)	Nowingi, Victoria	—	Standard	—	A. wilhelmiana
15. (POON)	40 km S Pooncarie, N.S.W.	Pooncarie	Pooncarie	2	A. wilhelmiana
16. (MGAP)	14.5 km WNW Monia Gap, N.S.W.	Monia Gap	Monia Gap	2	A. wilhelmiana
17. (SHUT)	11.3 km S Shuttleton, N.S.W.	—	Standard	—	A. wilhelmiana
18. (KALA)	3 km NNW Kalgoorlie, W.A.	Kalgoorlie	Standard	1	A. graffiana
19. (KALB)	9 km NNW Kalgoorlie, W.A.	Kalgoorlie	Standard	1	A. graffiana
20. (KALC)	4 km SW Kalgoorlie, W.A.	Kalgoorlie	Standard	1	A. graffiana
21. (BOUL)	21 km S Boulder, W.A.	Boulder	Boulder	3	A. graffiana
22. (HAYS)	6 km W Hayes Hill, N of Norseman, W.A.	Hayes Hill	Hayes Hill	2	A. acuminata

(PL); (7) pronotum width (PW); (8) fore femur length (FFL); (9) hind femur length (HFL); (10) hind femur width (HFW).

Samples from the same cytogenetic clone generally clustered out together (e.g., Cobar B–C reflecting the Cobar clone; SE Broken Hill, Sterling Vale Creek, and Broken Hill from the Broken Hill clone; and Kalgoorlie A–C representing the Kalgoorlie clone). The eight samples from eastern Australia known to be included in cytogenetic group 1 (i.e., Cobar, Little Topar, and Broken Hill clone samples) fall out in the same major cluster although they are separated considerably from the cytogenetic group 1 clones from Western Australia (i.e., Kalgoorlie). Samples known to be from cytogenetic group 2 (i.e., South Ita, Monia Gap, and Hayes Hill) exhibit greater divergence from each other and from the group 1 clones. Finally, there is considerable morphometric divergence between the three Kalgoorlie samples and the remainder (i.e., Boulder, which is a group 3 clone, and Pooncarie and Shuttleton, whose cytological affinities are not known at present).

The phenogram given in Figure 18.4 provides evidence about the major divisions in the data (e.g., the distinctiveness of the Boulder, Shuttleton, and Pooncarie samples) and the overall groupings of the data for compari-

Figure 18.4. UPGMA cluster analysis of a Mahalanobis distance matrix for 10 morphometric characters.

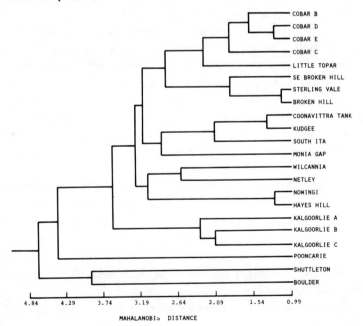

son with cytological results. However, other statistical procedures, such as canonical variate analysis, will give a more refined analysis of morphometric variability. A canonical variate analysis will provide answers to at least three pertinent questions about morphometric variability in *W. virgo*. First, canonical analysis will answer the question of whether there are significant differences between samples when all characters are considered simultaneously. Second, this type of analysis provides information about the directions in which the sample means differ, and the characters that contribute the most to these differences. Another way of looking at this aspect is to ascertain the spatial configuration of the groups in the reduced subspace of a canonical variate analysis. Third, a canonical variate analysis provides data on the relative morphometric distances between the groups.

Each axis in a canonical variate analysis reflects a complex pattern of morphometric variation in which the various characters contribute differentially. Each sample can be projected onto the relevant canonical axes, resulting in a series of canonical variate "scores" that position the samples along various axes of multivariate character variation. These axes are the principal discriminating patterns of variability.

With the analysis of variance and canonical variate analysis approach to geographic variation, two methods exist for clarifying character variation between the samples. (1) Variation between the various samples can be described for those individual characters that make the largest contribution to each canonical vector (i.e., those characters having the largest coefficients). However, divergence between samples based on variation in individual characters does not always coincide with the patterns produced by the simultaneous covariation of several characters. (2) The sample means projected onto each canonical axis can be ranked, and multiple comparisons testing carried out to elucidate multivariate character variation on each canonical axis. Use of multiple comparisons procedures is an improvement over the common practice of drawing confidence ellipsoids on ordination diagrams.

In these analyses, Fisher's protected least significant difference (LSD) procedure was used for multiple comparisons testing. This method was shown by Carmer and Swanson (1973) to be the most powerful multiple comparisons procedure in widespread use; that is if population means were indeed different for two groups in the analysis, the LSD procedure was more likely than other procedures to declare that the two groups were significantly different.

Thus, the univariate analyses document the variation patterns for each individual character, and each canonical axis describes the contribution of every character to a major axis of morphometric variation as well as the relative position of every sample in each axis of multivariate variation. As a result, we can define the major patterns of morphometric variation, elucidate the relative contribution of each character to the individual pattern, and determine how samples differ along each major axis of variation.

Figure 18.5. Multiple comparisons testing of 10 univariate morphometric characters. Character codes are explained in the text, and locality codes refer to Table 18.3. Critical value is 0.01.

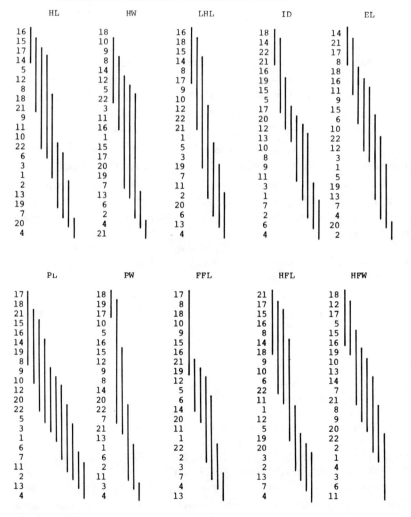

To determine the directionality of the morphometric variability, the first three canonical axes are plotted pairwise. A Prim network (Sneath and Sokal, 1973) based on a Mahalanobis distance matrix was used to link samples exhibiting greatest similarity for the characters examined. The Prim network is superimposed on the plot of the first two axes only.

The univariate analyses of variance for the 10 morphometric characters indicated highly significant differences ($P < 0.001$) between means for every character. Multiple comparisons results for the 22 samples are given in Figure 18.5. Samples with the smallest values for a character are always given first, and each line defines the maximum homogeneous subset of samples for the critical value of $\alpha = 0.01$. A within-groups correlation matrix is given in Table 18.4 to document the intercorrelations for the 10 characters. All pairwise correlations are positive and significantly different from zero at the $P < 0.01$ level.

A general ecological and cytogenetic pattern is evident in the univariate analyses. In terms of ecology, those grasshoppers collected from the *Acacia aneura* food plant at Cobar, N. S. W. are consistently the largest grasshoppers, whereas those from the *A. wilhelmiana* food plant are consistently the smallest. One major exception to this pattern is the sample from South Ita Sandhills, which was collected from *A. wilhelmiana* but is represented by large grasshoppers. I have commented elsewhere (Atchley, 1977b) on the morphometric peculiarities of this sample and its possible mode of origin. The grasshoppers collected from the *A. loderi* and *A. graffiana* food plants are intermediate in size between the previously described groups.

Table 18.4. *Within-groups correlation matrix from 22* Warramaba virgo clones[a]

	HL	HW	LHL	ID	EL	PL	PW	FFL	HFL	HFW
HL	1.000									
HW	0.732	1.000								
LHL	0.873	0.786	1.000							
ID	0.645	0.652	0.647	1.000						
EL	0.674	0.601	0.680	0.502	1.000					
PL	0.789	0.736	0.825	0.602	0.623	1.000				
PW	0.700	0.723	0.704	0.637	0.485	0.724	1.000			
FFL	0.681	0.639	0.716	0.524	0.558	0.714	0.663	1.000		
HFL	0.793	0.677	0.827	0.541	0.637	0.791	0.669	0.781	1.000	
HFW	0.433	0.460	0.419	0.440	0.227	0.477	0.551	0.444	0.374	1.000

[a]The degrees of freedom are 568, and all correlation coefficients are significant at $P < 0.01$.

Figure 18.6. Canonical variate analysis of 10 morphometric characters. Upper figure gives the projection of sample means onto the first two canonical variates. Samples are connected by a minimum spanning tree derived from a Mahalanobis distance matrix. Sample codes refer to Table 18.3. Lower figure provides a vector diagram of the canonical variate coefficients. The longer the vector, the greater the contribution of the character; the angle of the vector indicates the relative importance on each variate. Character codes are given in the text.

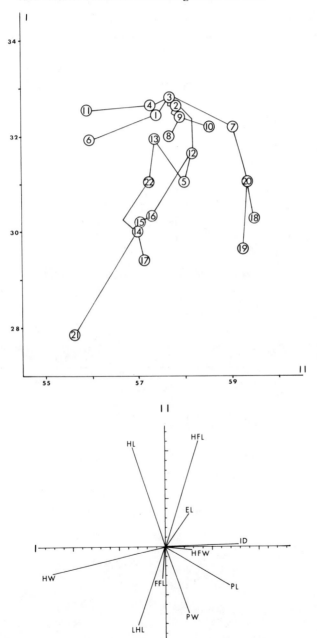

In terms of cytogenetic patterns, the grasshoppers from Cobar, Little Topar, and South Ita, which are group 1 clones from eastern Australia, are consistently largest in size, whereas those group 2 clones from the Pooncarie, Monia Gap, and Shuttleton clones are smallest. The Kalgoorlie clone is

Figure 18.7. Canonical variate analysis of 10 morphometric characters with projection of sample means onto variates I and II. See Figure 18.6 for further explanation.

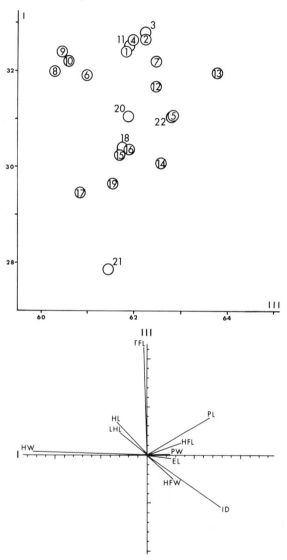

heterogeneous in that two samples have large values for several characters, whereas a third sample has small values. The grasshoppers from the Boulder clone are generally small in size, and those from the Hayes Hill clone are intermediate.

Turning to the canonical variate analysis, we find that the first canonical vector does not have a large canonical root, as is often found in canonical analyses of sexually reproducing species. The first canonical vector only accounts for 27% of the intersample variability. In fact, the first root is only about three times as large as the fifth root (1.35 vs. 0.43). This finding suggests that the amounts of variability associated with each major axis of intersample dispersion are not disproportionately different, and provides an interesting contrast to the results previously reported on sexually reproducing morabine species (Atchley, 1974; Atchley, 1977b; Atchley and Cheney, 1974).

The projection of sample means onto canonical axes I–III is given pairwise in Figures 18.6–18.8. Canonical axis I reflects an inverse relationship between head width on the one hand and pronotum length and interocular distance on the other (Table 18.5). The remaining seven characters contribute much less to this variate. Those samples with the highest scores on CV I (e.g., samples 2, 3, 4, 6, 7, and 11) have a wide head, long pronotum, and large interocular distance. Those samples with low scores on axis I (e.g., 14–19 and 21) generally have a narrow head and pronotum and a small interocular distance. The major exception to this

Table 18.5. *Standardized canonical variate coefficients for 10 morphometric characters for 22 samples of* Warramaba virgo

	I	II	III	IV	V
Head length (HL)	−0.336	1.030	0.330	1.328	−0.124
Head width (HW)	−1.230	−0.267	0.059	0.275	0.831
Lateral head length (LHL)	−0.297	−0.819	0.232	0.193	−1.526
Interocular distance (ID)	0.780	−0.010	−0.545	−0.425	0.139
Eye length (EL)	0.245	0.342	−0.057	−0.206	0.300
Pronotum length (PL)	0.684	−0.395	0.391	−0.873	−0.733
Pronotum width (PW)	0.241	−0.690	0.000	0.315	0.655
Fore femur length (FFL)	−0.057	−0.332	1.105	0.393	−0.230
Hind femur length (HFL)	0.346	1.100	0.126	−0.444	1.029
Hind femur width (HFW)	0.280	−0.061	−0.343	0.509	−0.080
Canonical root	1.352	1.096	0.805	0.641	0.427
Percent variance	0.270	0.219	0.161	0.128	0.052

Figure 18.8. Canonical variate analysis of 10 morphometric characters with projection of sample means onto variates II and III. See Figure 18.6 for further explanation.

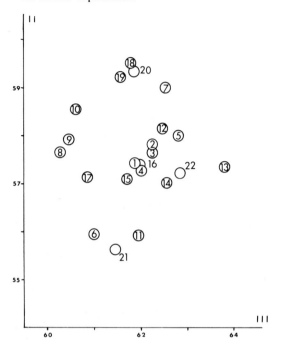

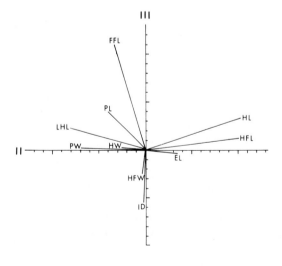

pattern is sample 21 (Boulder, W. A.), which has a very wide head but a short pronotum and small interocular distance.

The pattern of homogeneous subsets of samples reflected by the LSD results for CV I serves to isolate the Boulder clone, which is significantly different from all other samples on CV I. There is no obvious demarcation of homogeneous subsets into eastern and Western Australian samples (Figure 18.9). However, all samples having large scores on CV I except sample 5 (Coonavittra Tank, N. S. W.) and sample 17 (Shuttleton, N. S. W.) are from the northern part of the range in eastern Australia. Samples from the southern part of the range in eastern Australia or from Western Australia have small scores on CV I. In addition to this geographic pattern for the samples on CV I, there is an ecological relationship in that samples 1–4 generally have, the largest scores on CV I. Samples 1–4 were collected from the *Acacia aneura* food plant. Of the next eight eastern Australian samples displayed on CV I, seven of them (samples 5–12) were collected from the *A. loderi* food plant. The remaining five samples from eastern Australia (13–18) have small scores for CV I and were collected from *A. wilhelmiana*. Results such as these prompted me (Atchley, 1977a,b) to designate these various groups of samples the *A. aneura, A. loderi,* and *A. wilhelmiana* food plant races. Thus, CV I contains a strong geographic as well as ecological element.

CV II reflects an inverse relationship between hind femur length and head length with pronotum width and lateral head length. Samples 4, 7, 13, 2, 3, and 20 generally have large values for hind femur length and head

Figure 18.9. Multiple comparisons testing of canonical variate scores for 22 samples of *W. virgo.* Critical value is 0.01. Locality codes refer to Table 18.3.

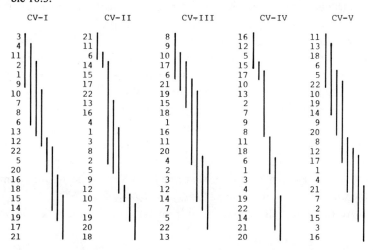

length. Samples from Cobar (samples 1–4) and samples 6 and 11 have the largest values for pronotum width, whereas samples 10 and 17–19 have the smallest values. Samples 4, 6, 11, and 13 have large values for lateral head length, whereas 14–16 and 18 have small values. The eastern Australian samples from Cobar (samples 1–4 collected from *A. aneura*) fall out together into a homogeneous subset of canonical variate scores along with two other samples. Further, all of the samples from *A. loderi* except sample 6 occur together on CV II. For the samples from *A. wilhelmiana,* only sample 12 is misplaced from the others on CV II. The Kalgoorlie samples (18–20) occur together, but are well separated from the remaining sample from Western Australia (cytogenetically distinct sample 21 from Boulder). Thus, again there is a geographic and ecological element to the morphometric variation on this canonical variate.

CV III reflects an inverse relationship between fore femur length and interocular distance. Samples 13, 4, 7, 3, 2, and 1 have long fore femora, whereas samples 17, 8, 10, 15, 19, and 9 have short ones. For interocular distance, samples 4, 6, 2, 7, 1, 3, and 13 have large values, and samples 18, 14, 22, 21, and 16 have small ones.

CV IV concerns an inverse relationship between head length–hind femur width and pronotum length–hind femur length, whereas CV V involves a dipolar relationship between lateral head length–pronotum length and hind femur length–head width–pronotum width.

Congruence of cytogenetic and morphological patterns of variation

With the results of Webb et al. (1978) and Atchley (1977a,b, 1978) and those depicted here, we should be able to elucidate the congruence of cytogenetic and morphometric patterns of variation within *W. virgo*. A major question to be resolved is that of whether a classification of samples of *W. virgo* based on cytogenetic data would be the same as that based on morphometric data. Does one type of data provide information not evident in the other type? Is there a biological interpretation for instances of incongruity in such classifications?

It is evident from the morphometric results depicted by Figures 18.4–18.9 that the samples from the cytogenetically defined clones cluster out together, namely, Cobar (samples 1–4), Broken Hill (8–10), and Kalgoorlie (18–20). Further, the Cobar samples are most similar phenotypically to those from Broken Hill, whereas those from Kalgoorlie are most similar to the sample from Little Topar (7).

All of the samples known to be from cytogenetic group 1 clones (Figure

18.1, Table 18.1) from eastern Australia (localities 1–4, 7–10) cluster together in Figure 18.6, whereas the group 1 clones in Western Australia cluster together and are linked by the smallest distance to the group 1 clones in the east. The samples known to be from group 2 in eastern Australia (13, 16) cluster out together in Figure 18.6 with the exception of Pooncarie, which is discussed below.

The only group 2 clone from Western Australia examined morphometrically was Hayes Hill (22). It is morphometrically distant from the other known group 2 clones, but, when all taxa are considered, the Hayes Hill sample is most similar to the South Ita clone, which is also group 2.

The only group 3 clone examined morphometrically was Boulder, W. A. (21). This sample is morphometrically distinct throughout the canonical variate analysis and appears to have a distinct shape for the 10 characters examined. The Boulder sample is also quite distinct cytologically, as evidenced by its inclusion in group 3, and may be a member of a small group of clones that arose independently of the group 1 and 2 clones.

After completing the morphometric analysis described in Figures 18.4–18.9, we were informed by White (pers. comm.) of unpublished information showing that the samples from Pooncarie and Netley each constitute cytologically distinct clones. They apparently fall into the group 2 category in that they have a rearranged standard karyotype, but the C-banding studies are incomplete.

At this point we might inquire if parallel changes have occurred between clones in cytogenetic and morphometric characters; that is, are cytogenetic divergence and morphometric divergence correlated? Comparable cytogenetic and morphometric data are available from the previous analyses for eight cytogenetically defined clones, including Cobar, Little Topar, South Ita, Monia Gap, Kalgoorlie, Boulder, and Hayes Hill. To ascertain if parallel cytogenetic and morphometric divergence has occurred, the pairwise distance values from a Mahalanobis distance matrix for the 10 morphometric characters have been plotted against the pairwise euclidean distances for the 34 banding characters alone and the 42 banding + chromosomal rearrangement characters (Figure 18.10).

It is obvious from Figure 18.10 that two groups of points exist in each plot. Those points at the top of each plot (given as open circles for contrast) are the various distance comparisons with the Boulder clone. Thus, there is a much larger degree of cytogenetic divergence as opposed to morphometric divergence between the Boulder clone and the remaining clones.

For the banding data alone, there is a positive linear relationship betwen cytogenetic divergence and morphometric divergence for the non-Boulder

Figure 18.10. Bivariate plot of pairwise euclidean distances (D_c) for cytogenetic characters and Mahalanobis distances (D_m) for morphometric characters. Plot *A* gives the relationship between 34 banding characters and 10 morphometric characters. Plot *B* gives the relationship of 42 banding and chromosomal rearrangement characters versus 10 morphometric characters. The open circles are the various comparisons with the Boulder clone. See text for further detail.

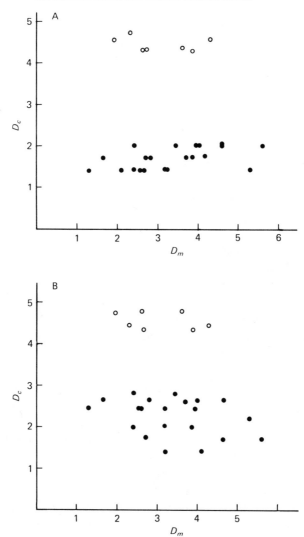

clones (correlation coefficient = 0.469, $P < 0.05$). No relationship between morphology and C-banding could be found for the Boulder clone comparisons (correlation = 0.069). Certain distributional aspects of distance coefficients inhibit a more rigorous statistical analysis of these data. However, these plots suggest an important biological relationship between morphometric and C-banding pattern divergence in cytogenetic groups 1 and 2 of *W. virgo* that does not persist in group 3. These results would seem to reinforce other data suggesting a possible independent hybrid origin of the group 3 clones.

This correlated change in external morphology and chromosomal structure does not persist when the eight chromosomal rearrangements are added to the C-banding data. Thus, unlike the C-banding patterns, changes in chromosomal rearrangements have not paralleled morphological change in *W. virgo*. It will be most interesting to superimpose the results of allozyme analyses onto these conclusions and see if the concordance between C-banding and morphological divergence persists at the genic level.

The morphometric data diverge from the cytogenetic data in at least two major respects. For the samples examined, geographic proximity as well as cytogenetic homogeneity is manifested in several important instances in the morphometric data. In flightless insects having very low vagility and very little chance of accidental transport (as in the case of morabine grasshoppers), geographic position can also be an important measure of evolutionary relationship. This is particularly true for asexual forms where new colonies can be easily founded by a single individual. The cytogenetic results described previously indicated several instances in which geographically highly distant clones (i.e., Coonaanna Hill and Little Topar; Broken Hill, Ponton Creek, and Coolgardie) were very similar. However, this pattern is not usually reflected in the morphometric data, in which clones in geographic proximity tend to cluster together. The morphometric results described here and elsewhere also faithfully reflect other cytogenetically unique clones (e.g., Pooncarie, Netley, Boulder, etc.).

Another interesting contrast between the cytological and morphometric analyses concerns levels of variability within cytogenetically defined clones that are homogeneous for chromosome karyotype and C-banding pattern while the morphometric data indicate significant differences between samples. Atchley (1977b, 1978) has shown that the three samples from the Broken Hill clone exhibit statistically significant differences in 4 of the 14 morphometric characters. The four samples from the Cobar clone have 9 of the 14 characters showing significant intersample differences, whereas the three samples from the Kalgoorlie clone have 9 of the 14 characters exhibiting significant intersample divergence.

The underlying cause of this morphological variability is still not completely understood. It is not known at present whether these rather marked differences in morphology are a result of non-genetically-defined intraclonal variability. It is hoped that studies of allozyme variability currently underway will provide independent evidence on variability in *W. virgo* within these cytogenetically defined clones.

Summary

Quantitative analyses of the cytogenetic data for *W. virgo* indicate marked divergence between the Spargoville, Boulder, and Zanthus clones as opposed to the remaining 12 clones. These three clones have diverged markedly in their C-banding patterns and morphometrics. There is a positive and linear relationship between divergence in banding patterns and morphometric differentiation in the group 1 and group 2 clones, but this relationship is absent in the group 3 clones as evidenced by the data from the Boulder sample. The data analyzed here suggest the possibility of a multiple hybridization origin of *W. virgo* with the group 3 clones arising independently of groups 1 and 2. The relationship between mode of origin and degree of divergence in the Spargoville clone is less clear at this time.

There are pronounced correlations between univariate and multivariate morphometric character divergence on the one hand and ecological and cytogenetic variation patterns on the other. The largest grasshoppers are those from the Cobar clone, and these insects feed on *Acacia aneura*. The smallest grasshoppers are those from group 2 clones from eastern Australia (i.e., Shuttleton, Monia Gap and Pooncarie), which feed on *A. wilhelmiana*.

Two important discrepancies exist between the morphometric and cytogenetic results. First, cytogenetically highly similar clones often occur at great geographic distance from each other (about 2000 km apart). The morphometric results do not indicate such discrepancies but rather indicate that morphometrically similar clones are also geographically similar. Second, there is considerable morphometric variability with cytogenetically homogeneous clones.

This research was supported by the College of Agriculture and Life Sciences of the University of Wisconsin, Madison and by research grants from the National Science foundation (GB-39788) and the National Institutes of Health (GM-07212). I am deeply indebted to M. J. D. White for his continued encouragement and support throughout this research. Walter M. Fitch very generously did the computations involved in the phylogenetic analyses. N. A. Campbell and R. A. Pimentel offered helpful suggestions on an earlier draft of this manuscript.

References

Atchley, W. R. 1974. Morphometric differentiation in chromosomally characterized parapatric races of morabine grasshoppers. *Aust. J. Zool. 22:*25–37.

– 1977a. Evolutionary consequences of parthenogenesis: evidence from the *Warramaba virgo* complex. *Proc. Natl. Acad. Sci. U.S.A. 74:*1130–4.

– 1977b. Biological variability in the parthenogenetic grasshopper *Warramaba virgo* (Key) and its sexual relatives. I. The eastern Australian populations. *Evolution 31:*782–99.

– 1978. Biological variability in the parthenogenetic grasshopper *Warramaba virgo* (Key) and its sexual relatives. II. The Western Australian taxa. *Evolution 32:*375–88.

Atchley, W. R., and Cheney, J. 1974. Morphometric differentiation in the *viatica* group of morabine grasshoppers. *Syst. Zool. 23:*400–15.

Carmer, S. G., and Swanson, M. R. 1973. An evaluation of ten pairwise multiple comparison procedures by Monte Carlo methods. *J. Am. Stat. Assoc. 68:*66–74.

Fitch, W. M. 1977. On the problem of discovering the most parsimonious tree. *Am. Nat. 111:*223–57.

Hewitt, G. M. 1975. A new hypothesis for the origin of the parthenogenetic grasshopper *Moraba virgo. Heredity 34:*117–36.

Maynard Smith, J. 1978. *The Evolution of Sex.* Cambridge: Cambridge University Press.

Sneath, P. A., and Sokal, R. R. 1973. *Numerical Taxonomy.* San Francisco: Freeman.

Webb, G. C., White, M. J. D., Contreras, N., and Cheney, J. 1978. Cytogenetics of the parthenogenetic grasshopper *Warramaba* (formerly *Moraba*) *virgo* and its bisexual relatives. IV. Chromosome banding studies. *Chromosoma (Berl.) 67:*309–39.

White, M. J. D. 1966. Further studies on the cytology and distribution of the Australian parthenogenetic grasshopper, *Moraba virgo. Rev. Suisse Zool. 73:*383–98.

– 1973. *Animal Cytology and Evolution.* Cambridge: Cambridge University Press.

– 1977. *Modes of Speciation.* San Francisco: Freeman.

– 1979. The genetic system of the parthenogenetic grasshopper *Warramaba virgo.* Paper delivered at the Symposium on Insect Cytogenetics at Royal Entomological Society of London.

White, M. J. D., and Contreras, N. 1979. Cytogenetics of the parthenogenetic grasshopper *Warramaba* (formerly *Moraba*) *virgo* and its bisexual relatives. V. Interaction of *W. virgo* and a bisexual species in geographic contact. *Evolution 33:*85–94.

White, M. J. D. and Webb, G. C. 1968. Origin and evolution of parthenogenetic reproduction in the grasshopper *Moraba virgo* (Eumastacidae: Morabinae). *Aust. J. Zool. 16:*647–71.

White, M. J. D., Cheney, J., and Key, K. H. L. 1963. A parthenogenetic species of grasshopper with complex structural heterozygosity (Orthoptera: Acridoidea). *Aust. J. Zool. 11:*1–19.

White, M. J. D., Webb, G. C., and Cheney, J. 1973. Cytogenetics of the parthenogenetic grasshopper *Moraba virgo* and its bisexual relatives. I. A new species of the *virgo* group with a unique sex chromosome. *Chromosoma (Berl.) 40:*199–212.

White, M. J. D., Contreras, N., Cheney, J., and Webb, G. C. 1977. Cytogenetics of the parthenogenetic grasshopper *Warramaba* (formerly *Moraba*) *virgo* and its bisexual relatives. II. Hybridization studies. *Chromosoma (Berl.) 61:*127–48.

Williams, G. C. 1975. *Sex and Evolution.* Princeton: Princeton University Press.

19 Speculations on the evolution of the insect storage proteins

JOHN A. THOMSON

The realization of complete metamorphosis in insects must surely be accounted a noteworthy step in the evolutionary saga. There followed the great adaptive radiations of the Coleoptera, Lepidoptera, Hymenoptera, and Diptera, now so significant in both number of species and number of individuals, and the appearance of the lesser endopterygote orders. Division of the functional life of the insect into distinctive larval and adult phases connected by a transitional pupal stage has provided unequaled flexibility of life plan. It has resulted in an enormous potential for specialized resource exploitation without prejudice to subsequent dispersal, for limited vulnerability to particular predator pressures, and for adjustment of the life cycle to environmental periodicities and fluctuations.

There is nothing inherently new about the regulation of gene expression or the involvement of programmed cell death in the development of endopterygote insects. Remodeling, reconstruction, and addition of new organs is a feature also of hemimetabolous development. It is the difference in timing and scale of changes from one molt to the next that distinguishes hemimetabolous from holometabolous patterns. What made a complete metamorphosis possible was the evolution of special food reserves supplying energy and material resources during the "pupal" phase (Tiegs, 1922). Of particular quantitative and qualitative significance in metamorphosis are the storage proteins providing nitrogen and sulfur reserves, and especially amino acids, for adult development within a system that is generally closed except for gaseous exchange. These larval storage proteins are for the developing imago what the yolk proteins synthesised by the adult fat body are for the developing embryo. In fact, the production of both the larval storage proteins and the yolk precursors in adult fat body, and comparative studies of the subsequent fate of these proteins, suggest that there might be some relationship between them. The analysis of the ancestry of the larval storage-protein genes of the holometabolous insects, of the regulatory loci controlling them, and of the developmental systems utilizing them is thus particularly intriguing.

It is true, of course, that much of the amino acid requirement for adult morphogenesis and development is derived from breakdown of cellular and extracellular proteins of larval tissues replaced during metamorphosis. Such protein stores should be clearly distinguished from those represented by the storage proteins proper. These proteins comprise specific gene products whose identifiable function is to provide a source of amino acids at key developmental stages (embryo and pupa), and which follow a pattern of synthesis, post-translational modification, storage, and mobilization that reflects this function. Just as gene products of the larval fat body are utilized to support imaginal development, gene products of the adult fat body are involved in embryonic development. Thus the yolk proteins (vitellins) and their precursors (vitellogenins) are treated here as adult storage proteins, reflecting their synthesis rather than their fate.

The larval fat body and its products

Recent reviews (including Kilby, 1963; Price, 1973; Wyatt, 1974; Thomson, 1975; Wyatt and Pan, 1978; Hagedorn and Kunkel, 1979) have stressed from different standpoints the diverse roles of insect fat body in lipid, carbohydrate, and protein metabolism. The two activities of the fat body assuming paramount significance in the present context are, first, the synthesis of at least the quantitatively predominant hemolymph proteins and, second, the synthesis, chiefly but not exclusively in female insects, of the vitellogenins. These two essential contributions of the fat body to the protein economy of the insect are typically separated in time as activities of juvenile and adult tissues respectively. This temporal separation is most clearly seen in holometabolous insects where a phase of protein uptake from hemolymph into cells follows the end of active larval feeding. The presence of protein granules (albuminoid granules, proteinaceous spheres) in the late larval fat body has long been recognized as a characteristic feature of holometabolous development (Wigglesworth, 1972; Price, 1973; Thomson, 1975).

The function of the fat body in protein synthesis is mirrored in its structure at least periodically. Structural specializations reflecting the storage role of the fat body are clearest in the holometabolous groups and fall into two patterns, depending on the degree of replacement of the larval fat body at metamorphosis. A phase of cell growth and secretion of protein into the hemolymph coincides with the feeding phase of larval development. In preparation for pupation, as the gut empties, protein granules appear in the fat body cells, and hemolymph protein levels drop correspondingly. Thus the fat body and hemolymph together form a dynamic protein storage system. Where the fat body cells are not replaced at metamorphosis, the larval storage proteins are withdrawn during pharate adult development, the protein granules become smaller and less numerous, and the fat body shrinks (e.g., in mosquitoes, Clements, 1963; in *Hyalophora,* Bhakthan and Gilbert, 1972). Later the fat body cells expand as adult functions commence, centered on vitellogenin synthesis. Where replacement of the fat body cells does occur at metamorphosis, the size and number of the protein granules diminish, and the cells finally lyze (Tiegs, 1922; Butterworth et al., 1965; Butterworth and La Tendresse, 1973), at a characteristic stage depending on the species (e.g., Ferrar, 1979) and on genotype (e.g., contrast between autogenous and anautogenous strains of *Lucilia,* Williams et al., 1977). The adult fat body is in these species a new organ constituted from relatively undifferentiated cells associated with the body wall.

The significance of fat body replacement in the evolution of the holometabolous groups presumably lies in the degrees of specialization for protein synthesis and uptake that it permits. Terminal differentiation related to larval function involves in the Diptera a complex program apparently geared to an enormous burst of protein synthesis during the larval feeding period. The case history of *Calliphora* illustrates these features well (Thomson, 1975).

At hatching, relative overreplication of the rDNA cistrons is observed. Polytenization of the fat body chromosomes follows, and then massive proliferation of nucleolar RNP. At the commencement of third instar, mRNAs for storage proteins predominate in the fat-body poly (A) RNA fraction (Kemp et al., 1978), and protein synthesis and secretion build up to maximum levels. Synthesis of storage proteins is switched off in the dipteran fat body as the larva ceases to feed actively, and the crop empties. At this stage the protein granules begin to form in the fat body as storage proteins are withdrawn from the hemolymph.

In *Calliphora,* 60% of the total extractable protein of the late larva is represented by the storage protein calliphorin (Munn and Greville, 1969; Munn et al., 1971; Thomson, 1975). At the end of the synthetic phase in larval life, calliphorin constitutes 75% of the total hemolymph protein content (20 mg/ml). Calliphorin synthesis is controlled selectively with respect to that of other fat body proteins, and uptake appears also to be preferential, judged by the relative concentrations of other hemolymph proteins. Studies on the specificity of uptake of the storage proteins by fat body, using isotopically labeled larval storage proteins from other species of varying relationship to the recipient, would be valuable. ^{14}C-calliphorin taken up into the fat body can be reextracted intact at the subunit level. During metamorphosis, some ^{14}CO$_2$ is released from animals injected with ^{14}C-calliphorin, and most of the isotope is recovered in imaginal proteins. Adult proteins are known to be synthesized de novo from amino acids (Williams and Birt, 1972) so that complete hydrolysis of calliphorin is indicated. Consistent with its postulated storage role, most of the 7 mg calliphorin produced per larva is utilized before emergence of the adult, and all traces have disappeared by the time the fly is 7 days old.

It is now clear that there exists at least in holometabolous insects a class of storage proteins of about 500,000 daltons, rich in aromatic amino acids, which are hexamers of subunits of about 80,000 daltons. Apart from calliphorin, another hexameric protein of similar molecular weight, designated protein II (Munn and Greville, 1969) or protein B (Thomson, 1975; Kinnear and Thomson, 1975), has been identified in *Calliphora*. It is also synthesized in the larval fat body, although in small quantities (8% of hemolymph in instar 3), but synthesis starts earlier and ends later than in the case of calliphorin. Protein B disappears from the hemolymph during adult development, but at a rate distinct from that shown by calliphorin, so that again a storage function is indicated.

Calliphorin and its close homologue, lucilin, from the sheep blowfly *Lucilia*, both represent a series of heteropolymers of 83,000-dalton subunits. These polypeptides are immunologically related to each other and give similar tryptic digestion products, but are electrophoretically distinguishable to a degree depending on the strain examined. Subunit diversity has not been found among the 81,000-dalton components of protein B.

Directly comparable to calliphorin and protein B are larval storage proteins LSP 1 (drosophilin) and LSP 2 of *Drosophila* (Roberts et al., 1977). Both are hexamers. LSP 1 resembles calliphorin in amino acid composition, occurs as the quantitatively predominant larval hemolymph protein, shows the same developmental pattern as calliphorin, with which it is immunologically cross-reactive, and exists as a series of heterohexamers of three immunologically related subunits of 75,000–81,000 daltons.

LSP 2 resembles protein B in pattern of occurrence during development, and is a homohexamer of a 78,000–83,000-dalton subunit. It is quantitatively subordinate to LSP 1 in larval hemolymph.

Striking similarities between calliphorin and two proteins (protein 1 and protein 2) isolated from the larval fat body in *Hyalophora cecropia* pupae have now been demonstrated (Tojo et al., 1978; Wyatt and Pan, 1978). Cecropia proteins 1 and 2 are hexamers of 85,000- and 89,000-dalton subunits respectively. Their amino acid compositions resemble that of calliphorin, but they are lower in tyrosine and phenylalanine. The pattern of synthesis, uptake, and utilization of these proteins is consistent with that shown by calliphorin. Two similar hexameric proteins of 85,000-dalton subunits have been observed as larval fat body products in *Bombyx* (see Wyatt and Pan, 1978).

The adult fat body and its products

Structural specializations of the adult fat body reflect its function in the synthesis and release of proteins, including vitellogenin in females, into the hemolymph (e.g., *Calliphora,* Thomsen and Thomsen, 1974). Protein granules of the kind seen in the larval fat body have not been observed even where protein storage might perhaps be expected, as for instance after blood meals in mosquitoes (Clements, 1963). Synthesis of the vitellogenins is generally coordinated with ovarian development, and hemolymph titers of these proteins are usually low except in certain ovariectomized females (see Wyatt and Pan, 1978; Hagedorn and Kunkel, 1979, for references).

Like the larval storage proteins, the vitellogenins and the vitellins are lipoglycoproteins, but generally they contain more conjugated lipid and carbohydrate than the larval proteins. Holoprotein molecular weights seem to be generally about 300,000 daltons in the Diptera, based on *Aedes,* and for *Calliphora* and *Drosophila* on estimates due to Harnish (cited by Wyatt and Pan, 1978). The corresponding holoproteins range between 500,000 and 600,000 daltons in the Dictyoptera, Orthoptera, and Lepidoptera.

Most vitellogenins/vitellins so far examined have at least one subunit of about 500,000 daltons. In *Drosophila melanogaster,* three subunits ranging from 44,000 to 46,000 daltons are involved (Bownes and Hames, 1977; Warren and Mahowald, 1979), whereas in other drosophilids fewer components may be recognizable (Srdic et al., 1978). *D. melanogaster* vitellogenins are encoded by monocistronic mRNA, which can be isolated from the adult fat body (Bownes and Hames, 1978; Postlethwait and

Kaschnitz, 1978). Structural relationships of the three vitellin polypeptides in this species are still open to some questions: Certainly two of them appear on serological grounds and by peptide mapping to be similar in structure (Warren and Mahowald, 1979).

Pulse-labeling experiments on the vitellogenins of *Locusta* showed that the initial product of translation is about 250,000 daltons. Intracellular processing involving cleavage then follows (Chen et al., 1978). The smaller units secreted into the hemolymph do not occur in simple molar proportions in this instance.

Two electrophoretically distinguishable vitellogenins occur in the hemolymph of certain cockroaches and in *Triatoma,* but their relationship to one another is not yet known (Wyatt and Pan, 1978).

Similarities of amino acid composition among the vitellogenins (and vitellins) of the relatively few insects studied are marked, generally reflecting a high aspartic acid/asparagine and glutamic acid/glutamine content. Unfortunately, immunological analyses which would provide evidence of homology between groups are also limited. Cross-reactivity of vitellogenins is usually limited to closely related species. Thus the evolutionary relationships of these proteins are likely to remain little known until sequence analyses become available (Hagedorn and Kunkel, 1979), or until comparative studies of tertiary structure giving an indication of similarities in molecular shape have been made.

Genetic basis of storage proteins

Genetic information of any kind, direct or indirect, bearing on the storage proteins is scarce. Evidence of the differential control of synthesis, as well as immunological and compositional distinctions, suggests that calliphorin and protein B from *Calliphora,* like LSP 1 and LSP 2 from *Drosophila,* are independent genetically (Akam et al., 1978a). Like those for the calliphorin subunits (Sekeris et al., 1977; Kemp et al., 1978), the mRNAs for LSP 1 (drosophilin) subunits are apparently monocistronic (Sekeris et al., 1977).

Genetic variation of the subunits of lucilin (homologous with calliphorin) does not affect the phenotype with respect to the protein B homologue of *Lucilia,* just as variation in LSP 2 subunits is not correlated with any change in the properties of LSP 1 in *Drosophila.* Nevertheless, a general similarity exists in amino acid composition, subunit structure and molecular weight, function, and ontogenetic pattern between these protein pairs, as between larval storage proteins 1 and 2 of *Bombyx* and of *Hyalophora,* respectively (Wyatt and Pan, 1978). These similarities open up the question of some quite recent evolutionary relationship between the genetic systems coding for the proteins of each pair.

Genetic studies of lucilin (Thomson et al., 1976; Thomson, 1975) based on electrophoretically detected differences in subunit charge, suggest that a series of closely related cistrons code for the polypeptides involved. In both *Calliphora stygia* and *Lucilia cuprina,* wild populations are highly polymorphic for subunit pattern. Pure breeding strains of *Lucilia* show up to seven different subunit bands in a stable pattern, indicating the presence of at least this number of gene products. These subunit band patterns are additive (co-dominant) in crosses between strains, and are therefore interpreted as revealing allelic variation at structural loci specifying the subunits. Allowing for overlap in electrophoretic mobility reflected by differences in band density and demonstrable for certain allele pairs, at least 12–14 loci would be required to account for the lucilin subunit patterns. Two of this series of genes were assigned to chromosome 2 by following genetic markers on all chromosomes. Recombinant phenotypes have not been observed in the crosses carried out so far, and the lucilin cistrons may therefore all be clustered on chromosome 2. Consistent with this possibility, cDNA prepared from calliphorin mRNA hybridized to only one band, or possibly two closely adjacent bands, in the polytene chromosomes of *C. vicina* (Kemp et al., 1978).

In *D. melanogaster,* the LSP 2 locus has been placed by linkage analysis based on electrophoretic variation of its subunits to 36–37 on chromosome 3 (Akam et al., 1978a) and by deletion mapping to band 68E3 or 4 on the left arm of chromosome 3 (Akam et al., 1978b).

Loci coding for the vitellogenin subunits in *Drosophila* have apparently not yet been mapped. Bownes (1979) has adduced evidence in support of the claim that the three vitellogenin/vitellin polypeptides of *D. melanogaster* are likely to be encoded by three separate structural loci. The vitellogenin mRNAs are monocistronic, and the products of translation are apparently identical or closely similar in size to the subunits of vitellins from the oocytes (Bownes and Hames, 1978; Postlethwait and Kaschnitz, 1978).

Regulatory loci affecting larval or adult storage proteins, or their subunits, have not been identified, unless the hemoglobins of chironomid larvae be included in the storage-protein category (Thomson, 1975).

Control of the larval storage proteins: synthesis and accumulation

A clear relationship between accumulation in the hemolymph of larva-specific proteins and the molt cycle can be recognized in locusts and cockroaches (reviewed by Chen, 1971; Wyatt and Pan, 1978), the molting hormones (MH) of the ecdysone group being implicated here (Kunkel, 1975).

In *Calliphora* (Thomson, 1975) the synthesis of calliphorin subunits begins late in instar 2, and in *Drosophila* that of LSP 1 starts at the commencement of instar 3 (Roberts et al., 1977). Synthesis of the subunits of calliphorin is coordinately shut down in *Calliphora* (Kinnear and Thomson, 1975) and *Lucilia* (Thomson, 1975) about one third through instar 3, too early to be related to the rise in MH level that precedes pupariation (Shaaya and Karlson, 1965) at a time of falling juvenile hormone (JH) levels (Shaaya and Levinson, 1966). Wyatt and Pan (1978) mention work by Tojo suggesting the possibility that storage-protein synthesis in *Bombyx* may be related to falling JH titers in the last larval instar. Attempts to establish experimentally a link between calliphorin synthesis and JH levels have so far failed (Thomson, 1975). Significantly, synthesis of protein B of *Calliphora* (Kinnear and Thomson, 1975), and apparently of LSP 2 in *Drosophila*, follows a different pattern (see above section on "The larval fat body and its products"). Thus the controls on storage protein synthesis are specific to certain sets of gene products, at least at the translational level. Elucidation of the controls involved is an urgent and important task.

Evidence concerning the possible hormonal control of protein uptake into the larval fat body is again apparently inconclusive. In *Calpodes*, MH clearly controls protein granule formation (Collins, 1969), whereas in *Drosophila*, MH may stimulate this process while appearing to be unnecessary for its initiation (Thomasson and Mitchell, 1972). Consistent with the *Drosophila* result, withdrawal of protein from the hemolymph into the fat body certainly starts in *Calliphora* (Martin et al., 1971) before MH is detectable, but the rate of uptake accelerates from the time near pupariation when MH titers rise. A problem in studying protein uptake is posed by the presence of a variety of inclusion types in insect fat body (Thomson, 1975): Observations of granule formation should be accompanied by measurements of the uptake of specific proteins.

Of particular interest in connection with the regulation of protein storage is the observation that substantially more protein is accumulated per cell in the fat body of autogenous compared with anautogenous strains of *Lucilia* (Williams et al., 1977) and the finding of a region-specific axial gradient in the rate of accumulation of the protein granules in *Drosophila* fat body (Tysell and Butterworth, 1978). This apparent gradient is maintained in vitro even by portions of the isolated fat body, and must therefore be embodied in some way in the developmental program of the fat body cells themselves.

Control of the adult storage proteins: synthesis and accumulation

The timing of vitellogenin synthesis generally appears to be tightly controlled, and closely related to uptake in the ovary. In most species, vitellogenin synthesis is strongly sex-limited. Small amounts of vitellogenin are sometimes found in hemolymph of males, however, as in pupae of *Hyalophora* (Telfer, 1954) or *Rhodnius* (Chalaye, 1979). The basis of the sex-specifity of vitellogenin synthesis is not known.

Recent reviews (Kambysellis, 1977; Wyatt and Pan, 1978; Hagedorn and Kunkel, 1979) emphasize that control of vitellogenin is not always directly under the control of JH, although this is certainly true of many hemimetabolous insects. In *Locusta,* JH has been shown to induce vitellogenin synthesis both in vivo and in vitro; control is here at the transcriptional level (Wyatt et al., 1976). Stimulation of vitellogenin synthesis follows administration of JH to suitable (generally allatectomized) preparations (references in Hagedorn and Kunkel, 1979) in *Periplaneta, Leucophaea,* and *Triatoma.*

Humoral influences on vitellogenin synthesis in the holometabolous orders may be complex and not necessarily consistent even within groups. Thus JH controls vitellogenin synthesis in the monarch butterfly *Danaus* (Pan and Wyatt, 1971, 1976), but not in the cecropia silkmoth *Hyalophora,* in which, like *Bombyx,* ovarian development occurs in the pharate adult. In the latter cases, MH might possibly be implicated (Hagedorn and Kunkel, 1979), as in vitellogenin synthesis in *Aedes* (Hagedorn et al., 1975), where the ovary is the source of the MH responsible. The fat body of *Aedes* produces vitellogenin in response to MH in vitro as well as in vivo (Fallon et al., 1974), but it must have previously been exposed to JH or to effectors stimulated in response to JH in order to become competent (Flanagan and Hagedorn, 1977). Hagedorn and Kunkel (1979) have reached the general conclusion that the mechanism of regulation of vitellogenin synthesis in the adult fat body is still not clear in either the higher Diptera or in the Hemiptera. Where JH has an effect in these groups, it may be an indirect one.

Uptake of vitellogenin from hemolymph into the oocytes is a selective process, apparently again under hormonal control, so that vitellogenin may be concentrated to 100 times or more the level found in hemolymph (*Periplaneta,* Bell, 1970). Recognition is also selective in the sense of being species-specific. Kunkel and Pan (1976) used isotopically labeled vitellogenins to show that cockroach and silkmoth ovaries each take up conspecific vitellogenin selectively. Experiments involving reciprocal transplantation of ovaries in *Drosphila* species suggest that electrophoretically distinct yolk proteins may be successfully incorporated in some heterospecific tests. Failure of vitellogenesis in some such transplant situations is likely to be due to the specificity of gonadotropic agents or their receptor proteins with resultant lack of competence of the ovary to take up vitellogenin (Srdic et al., 1978).

A large literature supports the idea that yolk accumulation in the ovary is influenced by JH as well as by other hormonal agents (e.g., Wigglesworth, 1972; Adams, 1974). Such a role for JH is clearly established in *Drosophila* as in other cases (Kambysellis, 1977). Here vitellin accumulation does not occur in *apterous*[4] homozygotes in spite of the presence of vitellogenins in the hemolymph, unless JH is administered (Postlethwait and Weiser, 1973).

There are clearly certain special features involved in control of synthesis of vitellogenins in the adult fat body, such as its sex limitation. Nevertheless regulation of adult storage-protein synthesis appears to involve variations in the different insect groups of the MH/JH theme fundamental to control of molting and metamorphosis at earlier stages. What is different from the larval situation is that the fat body active in protein synthesis and secretion is here separated from the site of uptake of that protein in space rather than time. The cyclical nature of vitellogenin uptake in successive oogenic cycles may also be important here, demanding interactive regulatory communication between source and sink.

Genetic models of metamorphosis

In broad overview, what goes on in the fat body during development should be typical of all tissues active in both larvae and adults. The production of specialized sets of gene products tends to be separated in time into two phases concerned respectively with growth and with reproduction. Holometaboly involves special requirements for storage of reserves, associated with a trend toward breaking the life history of the fat body into two phases. Replacement of larval cells during metamorphosis is thus seen as permitting that extreme specialization for larval life that would render reprogramming for adult functions increasingly complex and possibly inefficient.

A widely held view of insect development in general is that larval characteristics are determined by the activities of one set of genes, and adult characters by those of another set of genes (e.g., Wigglesworth,

1961). Williams and Kafatos (1971) have recently proposed an interesting model of holometabolous development that goes further and implicates differential activation of three sets of genes: larval, pupal, and adult. However, it is difficult to identify gene products that are specific to the pupal stage in a strict sense (Thomson, 1975). Pupal structures, conspicuously those related to the integument, generally appear to be of larval origin, whereas new proteins arising during pupal life are generally referable to the developing tissues of the pharate adult. Thus a two-gene-set postulate appears adequate.

Although a model based on differential activation of two partially overlapping gene sets might be held to gain support from isolated lines of evidence such as the existence of stage-specific influences on translation rates (Ilan and Ilan, 1973), none of this evidence points unequivocally to this model. On the other hand, there are many indications of the relationship of gene control with the evolutionarily ancient molt cycle. There is certainly no evidence that production of the stage-specific larval proteins in the hemimetabolous insects involves any massive changeover in patterns of gene readout. Rather, these changes in gene activity are similar in scale to those very generally involved in differentiation and morphogenesis in animals and plants. The existence of almost every conceivable gradation between replacement and carryover of larval and adult tissue also seems to argue against a model implicating the coordinate switching of banks of genes. Both tissue replacement and the patterns of gene activation are graded and overlapping processes. Whereas calliphorin and protein B are entirely larval products, or protein D and vitellogenin entirely adult products in *Calliphora,* protein A is synthesized in both larval fat body and in an adult tissue (presumably the adult fat body, Kinnear and Thomson, 1975). Where larval fat body is carried over without complete replacement, vitellogenin synthesis may be initiated in the larval fat body very early – just after spinning of the cocoon in *Hyalophora* and *Bombyx* – or much later, after adult emergence, as in *Danaus.*

We have seen that MH and JH are often, if not always, implicated, directly or indirectly, in the control of synthesis, uptake, and mobilization of both larval and adult storage proteins. Thus, right through development, variations on the same hormonal theme of JH/MH interplay may well be involved, and the lack of a discernible general pattern suggests the opportunistic specialization of previously existing mechanisms to fit the control needs of individual insect groups. Williams and Kafatos (1971) have suggested, for instance, that the gonadotropic function of JH seen in insect species with long-lived adults may be a lately evolved ancillary function substantially different from the primary role of the hormone in controlling molting. Surely this suggestion argues against the concept of gene sets. If metamorphosis was a big advance in insect evolution marked by takeover of many gene functions and their integration into coordinately controlled sets, evidence should remain in the form of systems of gene control common to many, if not all, the holometabolous groups. There is, then, much to support Wyatt and Pan (1978), whose "impression is that ... it can be questioned whether regulatory gene sets ... have yet been recognized at all."

In passing, mention may be made of the view of metamorphosis embodied in the "sequential polymorphism" concept of Wigglesworth (1961, and earlier papers). This idea was particularly developed around the distinctive differential states of epidermal cells going through metamorphosis in a series of molts. Kafatos (1976) points out that although this concept serves to emphasize the dynamic aspects of gene regulation through time rather than the static end states of the process, "temporal programming" is a preferable term. Moreover, there is a sense in which possible confusion with other usages of the term "polymorphism" might arise. In the labial gland of *Manduca* the cells are already determined for specific adult roles while exhibiting their larval differentiation. The temporal programming involved here is such that cells expressing the same larval phenotype, as duct cells, may be in a number of different states of determination (Kafatos, 1976). Complex as such programming may appear, it may be no different in principle to that implicated in, for example, the stepwise programming of hemimetabolous fat body cells, which undertake the synthesis first of larval, and then later of adult storage proteins.

Evolution of gene families coding for the storage proteins

Presently available evidence does not permit assessment of the evolutionary age of the typical larval storage proteins beyond the common ancestry of the Lepidoptera and the Diptera. Within these two orders, recognition of homologous genetic systems seems reasonably certain, and is based on correspondence of composition, subunit structure, function, and fate. But storage proteins, if that is their key role unassociated with some as yet undiscovered enzymatic or other activity, might well be expected to accumulate amino acid substitutions rather rapidly (see, for example, Hagedorn and Kunkel, 1979). In fact, the extraordinarily close similarity of calliphorin and drosophilin, and also the remarkable similarity of the dipteran and lepidopteran storage proteins points to powerful evolutionary constraints on the divergence of the larval storage proteins. It may therefore be anticipated that the evolutionary origins of these proteins can be traced into the hemimetabolous groups, if not even further back among the ametabolous insects. Comparative analytical data permitting the relationships of the vitellogenins and vitellins to be traced are not yet available (Hagedorn and Kunkel, 1979). But the vitellins, of all proteins, should surely reflect long evolutionary continuities, at least through annelid–arthropod lineages.

What conservative influences might have limited evolutionary diversification of at least the larval storage proteins in the Holometabola? The need for maintenance of particular sequences may be related to recognition of the products of translation for processing, for transport, and for uptake. Sequestration in the protein granules involves dehydration, stabilization, and packaging (often in paracrystalline arrays, e.g., Tojo et al., 1978). Each of these latter steps may also occur preferentially, depending on amino acid sequence. Certainly uptake into the oocyte is selective and requires specific recognition in the case of the vitellogenins. Similarly, differential recognition of the larval storage proteins by the enzymes involved in their mobilization and degradation is likely to be required during adult development. Selective patterns of utilization, as well as synthesis, shown by calliphorin and protein B in *Calliphora* and by LSP 1 and LSP 2 in *Drosophila,* may reflect the operation of such processes. Whatever the effective constraints of diversification of the insect storage proteins, similarly conservative patterns of evolution are seen in their close analogues, the cotyledonary storage proteins of legumes (Thomson and Doll, 1979).

Conservative influences may not necessarily have involved selection for maintenance of primary structure as such. In the presumed absence of cofactors and substrate binding sites, the amino acid sequence may be significant only locally at positions where processing or other modifications occur. In transport, sequestration, and packaging, the tertiary (and quaternary) structure may be the relevant unit rather than the amino acid sequence, the latter being conserved only as required to preserve the former. Thus, homologies between storage proteins of more distant groups may be better sought at the level of molecular shape than from compositional or even sequence data (see, e.g., Levine et al., 1978; Tang et al., 1978).

A conspicuous feature of certain insect storage proteins is the multiplicity of subunits that appear to be related structurally but are specified by separate genes. Examples are the lucilin subunits, possibly specified by at least 12–14 related structural loci, and the vitellogenin subunits of *Droso-*

phila, for which the data of Warren and Mahowald (1979) indicate a structural relationship between two of the three loci involved (Bownes, 1979). A comparable situation certainly exists in calliphorin from *C. vicina* and protein C from *C. stygia,* as well as in drosophilin (LSP 1, Wolfe et al., 1977). Only the general occurrence of duplicate genes coding for these products can provide a plausible explanation in view of the evidence that processing loci are not involved, and that translation of multiple monocistronic mRNAs is indicated. Multiplicity of charge classes among storage protein subunits would thus be a consequence of mutational divergence following gene duplication, although accumulation of mutations might be limited to particular regions of the molecule as already mentioned. Patterns of post-translational modification might also be affected by such mutational divergence, again with possible effects on net charge and perhaps chain length if cleavage sites are involved.

Small differences in the molecular weight of the products of duplicate loci, as in the case of the subunits of drosophilin, might reflect changes in post-translational modification, perhaps especially cleavage of the secreted polypeptide from a slightly larger initial product of translation (compare calliphorin, Kemp et al., 1978). Such processes, however, cannot explain the large differences in size of the transcription units coding for presumably homologous storage proteins such as the vitellogenins of *Drosophila* and *Locusta.* The finding that the transcriptional units of the eukaryotic genome consist of a mosaic of expressed (exonic) and unexpressed (intronic) sequences suggests the possibility that single nucleotide substitutions might change the "splicing" pattern of a transcribed unit (Gilbert, 1978). Addition or deletion of a whole sequence of amino acids might thus result from a single mutational event. Remembering that subunits of about 50,000 daltons are common to many vitellogenins/vitellins in a wide spectrum of insect groups, it is conceivable that the large transcription unit of *Locusta* (Chen et al., 1978) is a composite of exons transcribed separately (or in a smaller grouping) in *Drosophila.*

It is a remarkable phenomenon that so many of the storage proteins in both animals and plants show a hexameric quaternary structure. To calliphorin, its close homologues, and presumed relatives (protein B, LSP 2), it may be possible to add the vitellins of the Diptera, in which subunits of about 50,000 daltons are aggregated as holoproteins of about 300,000 daltons.

The cotyledonary storage proteins of legumes also seem to be assembled to a fundamental hexameric pattern. The subunit structure of legumin and its homologues is always $2(A + B)_3$ where A and B represent one of a limited number of possible acidic (A) and basic (B) subunits (for review see Thomson and Doll, 1979). A general structure for the proteins of the vicilin series has not yet been clearly established. One of the best-known proteins of the latter group, β-conglycinin, is a hexamer of form $2P_3$, where P represents certain of several alternative, probably related polypeptides (Thanh and Shibasaki, 1978). An underlying hexameric structure may prove universal among these seed globulins when variations of subunit origin and relationships are understood. Is it merely coincidental that the molecular organization of the insect storage proteins should be similar in this respect to that of the legume proteins, or is the resemblance based on convergence reflecting common requirements for intracellular sequestration, dehydration, packaging, or storage in the two systems?

Conclusion

Evolution of a complete metamorphosis seems unquestionably to be based on modification of the general insect molt cycle. Attention has been focused in this essay on fat body as a key organ in providing reserve materials necessary for adult development within a pupal stage, and on its roles in larval and adult life.

It has sometimes been convenient to view metamorphosis as based on successive activation of larval and adult gene sets. Gradations of organization, cell replacement, and function modify the link between the larval and adult tissues. The same hormones seem to participate in control of cell activities in the larval and adult fat body, although in a variety of perhaps opportunistic variations. On these grounds, a model of metamorphosis emphasizing temporal programming, mediated through a succession of states of determination and differentiation related to the molt cycle, may be more useful than the gene-set concept.

Crucial to distinction between those contrasting views of metamorphosis is the question of the evolutionary origin of the characteristic larval storage proteins of the holometabolous insects. The role of fat body tissue in producing both the hemolymph proteins and the vitellogenins is general in insects. In the Holometabola, at least, there are marked similarities between the hemolymph proteins and the yolk proteins in relation to synthesis, secretion, sequestration, and perhaps even subunit organization. An interesting possibility is that the two sets of proteins might have a common ancestry based on gene duplication and subsequent separation of their regulatory controls. Comparative structural and conformational studies of the hemolymph proteins of the hemi- and holometabolous groups, and of the vitellogenins, should help to solve the question of the origins of these vital insect proteins. We may then be close to elucidating the basis of metamorphosis itself.

Regardless of the ancient history of the insect storage proteins, their more recent history offers interesting and challenging opportunities for the analysis of gene evolution. Both the calliphorin/lucilin/drosophilin and the vitellogenin systems appear to involve multiple cistrons coding for the subunits of the relevant storage proteins. Recent gene duplication with subsequent limited mutational divergence appears likely to have been involved. Older cases of such duplication, with greater subsequent divergence, might be represented by proteins B and C in *Calliphora*, LSP 1 and LSP 2 in *Drosophila*, and proteins 1 and 2 in *Bombyx* and *Hyalophora*.

References

Adams, T. S. 1974. The role of juvenile hormone in housefly ovarian follicle morphogenesis. *J. Insect Physiol. 20:*263–76.

Akam, M. E., Roberts, D. B., and Wolfe, J. 1978a. *Drosophila* hemolymph proteins: Purification, characterization, and genetic mapping of larval serum protein 2 in *D. melanogaster. Biochem. Genet. 16:*101–19.

Akam, M. E., Roberts, D. B., Richards, G. P., and Ashburner, M. 1978b. *Drosophila:* the genetics of two major larval proteins. *Cell 13:*215–25.

Bell, W. J. 1970. Demonstration and characterization of two vitellogenic blood proteins in *Periplaneta americana:* an immunological analysis. *J. Insect Physiol. 16:*291–9.

Bhakthan, N. M. G., and Gilbert, L. I. 1972. Studies on the cytophysiology of the fat body of the American silkmoth. *Z. Zellforsch. 124:*433–44.

Bownes, M. 1979. Three genes for three yolk proteins in *Drosophila melanogaster. FEBS Lett. 100:*95–8.

Bownes, M., and Hames, B. D. 1977. Accumulation and degradation of three yolk proteins in *Drosophila melanogaster. J. Exp. Zool. 200:*149–56.

– 1978. Analysis of the yolk proteins in *Drosophila melanogaster*. Translation in a cell free system and peptide analysis. *FEBS Lett. 96:*327–30.

Butterworth, F. M., and La Tendresse, B. L. 1973. Quantitative studies of cytochemical and cytological changes during cell death in the larval fat body of *Drosophila melanogaster. J. Insect Physiol. 19:*1487–99.

Butterworth, F. M., Bodenstein, D., and King, R. C. 1965. Adipose tissue of *Drosophila melanogaster*. I. An experimental study of larval fat body. *J. Exp. Zool. 158:*141–53.

Chalaye, D. 1979. Étude immunochimique des proteines hémolymphatiques et ovocytaires de *Rhodnius prolixus* (Stal.). *Can. J. Zool. 57:*329–36.

Chen, P. 1971. *Biochemical Aspects of Insect Development*. Basel: Karger.

Chen, T. T., Strahlendorf, P. W., and Wyatt, G. R. 1978. Vitellin and vitellogenin from locusts (*Locusta migratoria*). Properties and post-translational modification in the fat body. *J. Biol. Chem. 253:*5325–31.

Clements, A. N. 1963. *The Physiology of Mosquitoes*. Oxford: Pergamon Press.

Collins, J. V. 1969. The hormonal control of fat body development in *Calpodes ethlius* (Lepidoptera, Hesperiidae). *J. Insect Physiol. 15:*341–52.

Fallon, A. M., Hagedorn, H. H., Wyatt, G. R., and Laufer, H. 1974. Activation of vitellogenin synthesis in the mosquito *Aedes aegypti* by ecdysone. *J. Insect Physiol. 20:*1815–23.

Ferrar, P. 1979. Absence of larval fat body in the buffalo fly, *Haematobia irritans exigua* (Diptera: Muscidae). *J. Aust. Entomol. Soc. 18:*25–6.

Flanagan, T. R., and Hagedorn, H. H. 1977. Vitellogenin synthesis in the mosquito: the role of juvenile hormone in the development of responsiveness to ecdysone. *Physiol. Entomol. 2:*173–8.

Gilbert, W. 1978. Why genes in pieces? *Nature (Lond.) 271:*501.

Hagedorn, H. H., and Kunkel, J. G. 1979. Vitellogenin and vitellin in insects. *Annu. Rev. Entomol. 24:*475–505.

Hagedorn, H. H., O'Connor, J. D., Fuchs, M. S., Sage, B., Schlaeger, D. A., and Bohm, M. K. 1975. The ovary as a source of α-ecdysone in an adult mosquito. *Proc. Natl. Acad. Sci. U.S.A. 72:*3255–9.

Ilan, J., and Ilan, J. 1973. Protein synthesis and insect morphogenesis. *Annu. Rev. Entomol. 18:*167–82.

Kafatos, F. C. 1976. Sequential cell polymorphism: A fundamental concept in developmental biology. *Adv. Insect Physiol. 12:*1–15.

Kambysellis, M. P. 1977. Genetic and hormonal regulation of vitellogenesis in *Drosophila. Am. Zool. 17:*535–49.

Kemp, D. J., Thomson, J. A., Peacock, W. J., and Higgins, T. J. V. 1978. Messenger RNA for the insect storage protein calliphorin: in vitro translation of a 20 S poly(A)-RNA fraction. *Biochem. Genet. 16:*355–71.

Kilby, B. A. 1963. The biochemistry of the insect fat body. *Adv. Insect Physiol. 1:*111–74.

Kinnear, J. F., and Thomson, J. A. 1975. Nature, origin and fate of major haemolymph proteins in *Calliphora. Insect Biochem. 5:*531–52.

Kunkel, J. G. 1975. Larval-specific protein in the order *Dictyoptera.* II. Antagonistic effects of ecdysone and regeneration on LSP concentration in the hemolymph of the oriental cockroach, *Blatta orientalis. Comp. Biochem. Physiol. 51B:*177–80.

Kunkel, J. G., and Pan, M. L. 1976. Selectivity of yolk protein uptake: comparison of vitellogenins of two insects. *J. Insect Physiol. 22:*809–18.

Levine, M., Muirhead, H., Stammers, D., and Stuart, D. I. 1978. Structure of pyruvate kinase and similarities with other enzymes: possible implications for protein taxonomy and evolution. *Nature (Lond.) 271:*626–30.

Martin, M-D., Kinnear, J. F., and Thomson, J. A. 1971. Developmental changes in the late larva of *Calliphora stygia.* IV. Uptake of plasma protein by the fat body. *Aust. J. Biol. Sci. 24:*291–9.

Munn, E. A., and Greville, G. D. 1969. The soluble proteins of developing *Calliphora erythrocephala,* particularly calliphorin, and similar proteins in other insects. *J. Insect Physiol. 15:*1935–50.

Munn, E. A., Feinstein, A., and Greville, G. D. 1971. The isolation and properties of the protein calliphorin. *Biochem. J. 124:*367–74.

Pan, M. L., and Wyatt, G. R. 1971. Juvenile hormone induced vitellogenin synthesis in the monarch butterfly. *Science 174:*503–5.

– 1976. Control of vitellogenin synthesis in the monarch butterfly by juvenile hormone. *Dev. Biol. 54:*127–34.

Postlethwait, J. H., and Kaschnitz, R. 1978. The synthesis of *Drosophila melanogaster* vitellogenins in vivo, in culture, and in a cell-free translational system. *FEBS Lett. 95:*247–51.

Postlethwait, J. H., and Weiser, K. 1973. Vitellogenesis induced by juvenile hormone in the female sterile mutant *apterous-four* in *Drosophila melanogaster. Nature New Biol. 244:*284–5.

Price, G. M. 1973. Protein and nucleic acid metabolism in insect fat body. *Biol. Rev. 48:*333–75.

Roberts, D. B., Wolfe, J., and Akam, M. E. 1977. The developmental profiles of two major haemolymph proteins from *Drosophila melanogaster*. *J. Insect Physiol. 23*:871–8.

Sekeris, C. E., Perassi, R., Arnemann, J., Ullrich, A., and Scheller, K. 1977. Translation of mRNA from *Calliphora vicina* and *Drosophila melanogaster* larvae into calliphorin-like proteins of *Drosophila*. *Insect Biochem. 7*:5–10.

Shaaya, E., and Karlson, P. 1965. Der Ecdysontiter wahrend der Insektenentwicklung. II. Die postembryonale Entwicklung der Schmeissfliege *Calliphora erythrocephala* Meig. *J. Insect Physiol. 11*:65–9.

Shaaya, E., and Levinson, H. Z. 1966. Hormonal balance in the blood of blowfly larvae. *Riv. Parassitol. 27*:211–15.

Srdic, Z., Beck, H., and Gloor, H. 1978. Yolk protein differences between species of *Drosophila*. *Experientia 34*:1572–4.

Tang, J., James, M. N. G., Hsu, I. N., Jenkins, J. A., and Blundell, T. L. 1978. Structural evidence for gene duplication in the evolution of the acid proteases. *Nature (Lond.) 251*:618–21.

Telfer, W. H. 1954. Immunological studies of insect metamorphosis. II. The role of a sex-limited blood protein in egg formation by the cecropia silkworm. *J. Gen. Physiol. 37*:539–58.

Thanh, V. H., and Shibasaki, K. 1978. Major proteins of soybean seeds. Subunit structure of β-conglycinin. *J. Agric. Food Chem. 26*:692–8.

Thomasson, W. A., and Mitchell, H. K. 1972. Hormonal control of protein granule accumulation in fat bodies. of *Drosophila melanogaster* larvae. *J. Insect Physiol. 18*:1855–99.

Thomsen, E., and Thomsen, M. 1974. Fine structure of the fat body of the female of *Calliphora erythrocephala* during the first egg-maturation cycle. *Cell Tiss. Res. 152*:193–217.

Thomson, J. A. 1975. Major patterns of gene activity during development in holometabolous insects. *Adv. Insect Physiol. 11*:321–98.

Thomson, J. A., and Doll, H. 1979. Genetics and evolution of seed storage proteins. In *Seed Protein Improvement in Cereals and Grain Legumes*, vol. 1, pp. 109–23. Vienna: International Atomic Energy Agency.

Thomson, J. A., Radok, K. R., Shaw, D. C., Whitten, M. J., Foster, G. G., and Birt, L. M. 1976. Genetics of lucilin, a storage protein from the sheep blowfly, *Lucilia cuprina* (Calliphoridae). *Biochem. Genet. 14*:145–60.

Tiegs, O. W. 1922. Researches on the insect metamorphosis. *Trans. R. Soc. S. Aust. 46*:319–527.

Tojo, S., Betchaku, T., Ziccardi, V. J., and Wyatt, G. R. 1978. Fat body protein granules and storage proteins in the silkmoth, *Hyalophora cecropia*. *J. Cell. Biol. 78*:823–38.

Tysell, B., and Butterworth, F. M. 1978. Differential rate of protein granule formation in the larval fat body of *Drosophila melanogaster*. *J. Insect Physiol. 24*:201–6.

Warren, T. G., and Mahowald, A. P. 1979. Isolation and partial characterization of the three major yolk polypeptides from *Drosophila melanogaster*. *Dev. Biol. 68*:130–9.

Wigglesworth, V. B. 1961. In *Insect Polymorphism* (ed. Kennedy, J. S.), pp. 103–13. London: Royal Entomological Society.

– 1972. *The Principles of Insect Physiology*, 7th ed. London: Methuen.

Williams, C. M., and Kafatos, F. C. 1971. Theoretical aspects of the action of juvenile hormone. *Mitt. Schweiz. Entomol. Ges. 44*:151–62.

Williams, K. L., and Birt, L. M. 1972. A study of the quantitative significance of the sheep blowfly *Lucilia*. *Insect Biochem. 2*:305–20.

Williams, K. L., Barton Browne, L., and Van Gerwen, A. C. M. 1977. Ovarian development in autogenous and anautogenous *Lucilia cuprina* in relation to protein storage in the larval fat body. *J. Insect Physiol. 23*:659–64.

Wolfe, J., Akam, M. E., and Roberts, D. B. 1977. Biochemical and immunological studies of larval serum protein 1, the major haemolymph protein of *Drosophila melanogaster* third-instar larvae. *Eur. J. Biochem. 79:*47–53.

Wyatt, G. R. 1974. Regulation of protein and carbohydrate metabolism in insect fat body. *Vehr. Dtsch. Zool. Ges. 67:*209–26.

Wyatt, G. R., and Pan, M. L. 1978. Insect plasma proteins. *Annu. Rev. Biochem. 47:*779–817.

Wyatt, G. R., Chen, T. T., and Couble, P. 1976. Juvenile hormone induced vitellogenin synthesis in locust fat body in vitro. In *Invertebrate Tissue Culture: Applications in Medicine, Biology and Agriculture* (ed. Kurstak, E., and Maramorosch, K.), pp. 195–202. New York: Academic Press.

20 Evolution in jeopardy: the role of nature reserves

O.H. FRANKEL

It is an axiom intrinsic in its concept that organic evolution is a continuing process, as stated by Darwin (1875) in the *Origin of Species:*

> It may metaphorically be said that natural selection is daily and hourly scrutinising, throughout the world, the slightest variations; rejecting those that are bad, preserving and adding up all that are good; silently and insensibly working, *whenever and wherever opportunity offers,* at the improvement of each organic being in relation to its organic and inorganic conditions of life.

Dobzhansky (1967:129) saw evolution as open-ended: "The chief characteristic . . . of progressive evolution is its open-endedness. Conquest of new environments and acquisition of new ways of life create opportunities for further evolutionary developments." And, one is left to assume, new environments would continue to arise through climatic change, human agency, accidents, or in any of the other ways in which they had arisen in the past.

In this essay I attempt to examine to what extent the premises for continuing evolution are likely to retain their validity in the future in the light of changes now in progress. To be sure, no prospects of drastic changes in the biological processes of evolutionary adaptation are discernible. But one can recognize major changes in the conditions, in Dobzhansky's word in the opportunities, for continuing evolution that cannot fail profoundly to affect the evolutionary potential of both wild and domesticated biota.

Habitat destruction and species erosion*

The first of these changes arises from the rapidly increasing destruction of relatively undisturbed habitats of natural and seminatural communities, especially in the humid tropics. Conversion of forests to agricultural or plantation crops, devastation by shifting cultivation, and drastically exploitative forest utilization increasingly contribute to habitat destruction, a process that is virtually inevitable in the face of acute population and economic pressures of developing countries. A high rate of destruction of rain forests is widespread throughout the tropical countries. One example must suffice. Indonesia has the richest forest resources in Southeast Asia. Of the 42 million hectares of lowland forests remaining in 1975, 18 million were planned to be converted to agriculture over 20 years. Shifting cultivation, having devasted 30 million hectares in the past, still destroys forests at the rate of 150,000 hectares annually. The remainder is available for forest exploitation (FAO, 1976).

It was estimated that neither the Philippines nor Malaysia would have lowland virgin forests left within a decade (Whitmore, 1973). The position is similar in most other tropical countries, where population growth and internal and external economic pressures combine to place the remaining forests at grave risk. It is widely believed that by the end of the century there will be no undisturbed tropical lowland forest left anywhere in the world, except in whatever reserves may remain (Raven, 1976). This is important in itself, considering the vulnerability of rain forest ecosystems. But it receives greatly added significance from the fact that forest exploitation now tends to be as destructive of the ecosystem as is conversion to other types of land use, as against the age-old exploitation for firewood or building timber, or even the selective tree felling of less than half a century ago. Today, forest exploitation often leads to more or less complete destruction of the habitat and, even if it is replanted, to the extinction of the majority of animals and plants it contained.

In general, the effect of widespread habitat destruction is twofold. First, it causes the loss of biota, at all taxonomic levels, specifically adapted to the habitat range affected, and the extinction of local endemic species. Second, it will greatly restrict progressive evolution in what will be either man-induced and controlled, or severely degraded ecosystems.

I have concentrated on the humid tropical forest because it is the ecosystem with by far the greatest wealth of plant and animal species, and because until recently the difficulties of exploitation had caused large

---

* Some of the issues mentioned in this and the following sections are more fully discussed in Frankel and Soulé (1981), where further references can be found.

areas to remain more or less intact. Many other terrestrial, freshwater, and marine ecosystems are similarly threatened, but the prospects for species extinction are greatest in the tropics, for reasons already discussed. Raven (1976) considers that of the 85,000 species of flowering plants in the temperate regions, between 4500 and 6000 are threatened. Far less is known of the 155,000 species with tropical distribution, but in the light of current and anticipated destruction Raven supposes that at least a third of them will be threatened or extinct by the end of the century. Even less is known about the prospective fate of animal species, with the exception of the larger mammals and birds, many of which are even more dependent than plants on extensive and minimally disturbed areas. Here I can only refer to the vast literature on extinct and threatened animal species.

Nature reserves as evolutionary refuges?

The principal, and increasingly the only, hope for wildlife continuing to exist and to evolve under natural or near-natural conditions, with a minimum of human interference, rests on those residual habitats that are dedicated as repositories and refuges of wild biota – the nature reserves. The question is, what are the evolutionary opportunities in nature reserves?

This question must be viewed in terms of evolutionary time scales. In the long term, nature reserves can be seen as space, that is, as potential habitats for communities evolving in response to successive climatic changes, presupposing the long-term security for substantial reserves. This view may be regarded as far fetched and, in present-day circumstances, scarcely relevant because the distant future, not only of current reserves but of areas now occupied or otherwise alienated by man, cannot be perceived. However, in the shorter term, nature reserves are recognized as sites where ecosystems persist in balance with the environment, subject to processes of extinction, survival, and adaptation, hence potentially providing the material for continuing evolution. It is no doubt trite to remark, though perhaps appropriate to remind ourselves, that none but survivors become parents.

There is in fact a direct relationship between survival (i.e., fitness) and evolutionary potential because both are dependent on the level of heterozygosity. In many species, especially of animals, fitness is associated with heterozygosity, and inbreeding reduces fitness. It is generally agreed that a maximum rate of inbreeding of 1% is tolerable, and this is confirmed by experience of animal breeders. It can be shown that under conditions of unrestricted interbreeding, with equal numbers in both sexes – both rather rare conditions – a population of 50 will retain short-term fitness; in practice a number several times larger will be needed. But to maintain a genetically viable population (i.e., to avoid genetic drift and to maintain the capacity for adaptive response), a much larger population will be needed. How large? This is a critical question for the success of nature reserves.

Guidelines have been provided, which are derived either from general principles or empirically. Crow (1969) considered the dependence of heterozygosity and polymorphisms on population size and mutation rate. For example, assuming a mutation rate of 5×10^{-5} per allele per generation, a population size of 2000 would be required to maintain 30% of all loci in a heterozygous state, which is not out of order with existing information on variation in natural populations. Franklin (1980) derived a minimum population size of 500 empirically, from general considerations of quantitative variation in populations, with confirmation derived from data on variation of bristle numbers in *Drosophila*. The general conclusion on population size is that in the short term (i.e., for the maintenance of fitness), the effective population size should be of the order of hundreds; in the longer, evolutionary term, of thousands.

In practice, a number term turns into an area term, for two reasons. If a reserve is designed to protect a community rather than specific species, then, numerous population censuses being impracticable, an area specification is inevitable. Similarly, if there are specific targets of preservation such as rare or endangered species, the population size will inevitably be expressed in area terms. In the case of plants it will be determined by their geographical distribution and frequency; in the case of animals such as large predators or herbivores, the large individual space requirement will determine the area needed for survival.

Guidance on area requirements derives from the theory of island biogeography (see MacArthur and Wilson, 1967) because nature reserves tend to share the ecological characteristics of island ecosystems. Species equilibria on islands are subject to area size and migration. In nature reserves, with migration for most animal and plant species absent or minimal, the equilibrium level will be determined by the area. This area effect is extensively documented for various groups of animals through censuses of land-bridge islands (islands isolated since the last ice age), which provide comparisons with each other and the mainland. In general the number of species in equilibrium declines with island size. Similarly, area-demanding large species "cut out" on smaller islands. The minimum viable population of macropod species on offshore islands in Western

Australia was found to be between 200 and 300, and islands too small to hold this number did not have any (Main and Yadav, 1971). Only the largest islands, of more than 30 km^2, contain the large kangaroo *Macropus robustus,* the euro. Neither the grey kangaroo, *Megaleia fuliginosus,* nor the red kangaroo, *M. rufus,* occurs on any of the islands, though common on the mainland. However, even moderate immigration can reduce the effect of small area size, especially with regard to fitness (see above, in this section). This is a relevant factor for species capable of traversing large areas separating reserves, like some insects, bats, and birds, and, of course, pollen.

But for large mammals, even the largest reserves that an overpopulating rich country, let alone a poor one, is likely to safeguard, are rarely of sufficient size for a minimum population size. It has been shown that to maintain the puma, or American mountain lion (*Felis concolor*), would require an area not far short of twice the size of Yellowstone National Park, the largest in the United States. Indeed, the large mammals and many other species requiring much space or special conditions have little chance of surviving the next 500 to 5000 years, and in many instances probably much less, unless special management measures are taken that approach domestication. Similarly, large tree species, especially in tropical forests, have a low area density (in 20 common species an average of 1 per 5 acres, Whitmore, 1973) and have little chance of long-term survival. In both animals and plants, the extinction of dominant species will have profound effects on the survival of other species in their ecosystems.

It should be noted that among the many species threatened with extinction through loss or reduction of habitats are the ancestors and other relatives of many crop and horticultural species that constitute valuable resources for plant breeders. The genetic resources of present and prospective forestry species are similarly threatened, especially in tropical forests that are subject to exploitation and to replacement by other kinds of land use.

The future prospects of speciation

What then are the prospects of speciation in the centuries to come? With many of the larger animal and plant species insecurely poised on the path to extinction in threatened habitats or in nature reserves of insufficient size, new speciation of such organism, to say the least, seems unlikely. Historical evidence on minimum areas of speciation of different organisms is too episodic for generalizations, but the general trend is indicative. The smallest islands for speciation of snails are Tahiti and somewhat smaller islands, for lizards Jamaica (11,000 km²), and for small mammals Cuba and Luzon, (ca. 110,000 km²), whereas larger mammals, such as jackals, appear to require an area of the size of Madagascar (nearly 600,000 km²). Soulé (1980), who records these statistics, considers the generation of species in nature reserves unlikely for the larger vertebrates, and perhaps for many other animals and plants. In the future, evolution would be limited to the adaptation of existing species to environmental change (i.e., to prevention of extinction), with speciation, or progressive evolution in Dobzhansky's earlier-quoted term, having come to its end, at least for larger and space-demanding biota.

This need certainly not be the case for the very much larger number of biota making lesser demands on space. Here one may merely refer to the most famous case of speciation on an island archipelago, Darwin's Galapagos finches, and to the well-documented endemic insect fauna on the island of St. Helena. The remarkable fact about Darwin's finches is that they are highly differentiated – four genera with a total of 14 species are recognized – and clearly related to each other, but without any close relatives on the mainland that could be identified as a common ancestor (see Lack, 1947). The Galapagos islands range from quite small size to lengths of 15 to 25 km and one of 130 km, with widths of some 20 to 30 km, situated about 600 km from the mainland. St. Helena is about 120 km² in size, 1760 km from Africa, and 2580 km from South America. Like the Galapagos islands, it is an oceanic island and has never had land connections. Here an extensive study of the insect fauna by Basilewsky et al. (1972) established that of the 256 species of Coleoptera on the island, 159 are endemic, belonging to 17 wholly endemic genera. There is convincing evidence that this extensive evolutionary development took place entirely on the small island, commencing, as White (1978) suggests, "from a single immigrant species accidentally transported to St. Helena in the remote past." A similar, though less extensive process of speciation took place in several genera of grasshoppers, with a number of species endemic on the island.

It is an open question to what extent the historical record, scant as it is, has predictive significance for conditions of speciation that can be expected in nature reserves of different kinds, exposed to different forms of management. Yet the overall conclusion seems justified that survival and speciation are subject to the relationship between organism size and available space. I have restricted this discussion to the aspect of area size because it is of primary significance for many dominant biota, and have refrained from considering modes of speciation, though they are of obvious relevance for island-type evolution, because they are the subject of several contributions to this volume, apart from the comprehensive treatment in Michael White's recent book (White, 1978). Suffice it to say that whatever the mode adopted by any organism, speciation is likely to be favored by ecological diversity within and between reserves, in addition to area size.

Conclusions

The opportunities for species survival and for continuing evolution are rapidly changing, owing to the destruction of habitats of many species-rich ecosystems, especially in the tropics. A substantial proportion of tropical biota will be extinct or threatened by the end of the century. Habitat restrictions and species extinction or impoverishment will drastically affect the evolutionary potential. Nature reserves, it is hoped, will remain the main safeguard for survival and continuing evolution.

Nature reserves have various limitations, the most significant, and as a rule unchangeable, being size. This factor determines the carrying capacity or population size, which can be critical, especially for large animals and plants. In many organisms population size is the factor controlling heterozygosity – hence in the short term the average fitness, and in the longer term the adaptive potential of a species. The known distributions of many large species of animals – prominently the large and medium-sized vertebrates – are such as to require areas far in excess of those available in most nature reserves, which are incapable of maintaining even a fitness level of heterozygosity.

From these considerations, and from the historical record of speciation on islands of different sizes, it is deducible that for larger animals and plants restricted to nature reserves there is little if any prospect of future speciation. However, the record of island speciation shows that for less space-demanding organisms such as small birds or insects, nature reserves may offer opportunities for continuing evolution, depending on number, size, management, and ecological diversity between and within reserves.

Of the organisms adapted to man-induced ecosystems, those serving man's needs for food and other products can continue their man-controlled evolution as long as the materials for genetic adaptation and for speciation (i.e., their genetic resources) remain in existence, or, alternatively, if methods of genetic manipulation become available that render the ancient gene assemblies redundant. Other, usually less desirable, occupants of rural or urban ecosystems, including weeds and various types of predators and parasites, from mammals to bacteria, may continue their synadaptive evolutionary pathways.

If the prospect for a continuation of the evolutionary pathway for the large and dominant species appears slight, this does not reduce the reasons for attempting to keep in existence those that have so far managed to stay alive. To do so may require a concerted policy of "conservation of scarcity," in some instances involving population genetic management within and between nature reserves. I believe that this is a very real obligation we have to future generations. At the same time we must be conscious of the task of preventing other biota, though now abundant, from descending on the slippery road toward scarcity. We can call this "conservation of abundance." The difference is one of emphasis, but the objectives are essentially the same. If organisms, whether great or small, are to have a chance of a continuing existence, the need is for large and representative nature reserves *now*.

References

Basilewsky, P. et al. 1972. *La faune terrestre de l'Isle de Sainte Helène.* Deuxième partie, II. Insectes 9, Coleoptera. *Ann. Mus. R. Afr. Cent. Sér. Oct. Zool.,* No. 192.

Crow, J. F. 1969. Molecular genetics and population genetics. *Proc. XII Int. Cong. Genet.* 3:105–13.

Darwin, Charles. 1875. *The Origin of Species,* 6th ed., p. 65f. London: John Murray.

Dobzhansky, Th. 1967. *The Biology of Ultimate Concern. Perspectives in Humanism* (ed. Anshen, R.N.). New York: The New American Library.

Food and Agriculture Organization. 1976. *Forest Resources in Asia and the Far East Region.* Rome: United Nations Food and Agriculture Organization.

Frankel, O. H., and Soulé, M. 1981. *Conservation and Evolution.* Cambridge: Cambridge University Press (in press).

Franklin, I. R. 1980. Evolutionary change in small populations. In *Conservation Biology: An Evolutionary-Ecological Perspective* (ed. Soulé, M. E., and Wilcox, B. A.), pp. 135–49. Sunderland, Massachusetts: Sinauer.

Lack, D. 1947. *Darwin's Finches.* Cambridge: Cambridge University Press.

MacArthur, R. H., and Wilson E. O. 1967. *The Theory of Island Biogeography.* Princeton: Princeton University Press.

Main, A. R., and Yadav, M. 1971. Conservation of macropods in reserves in Western Australia. *Biol. Conserv.* 3:123–33.

Raven, P. H. 1976. Ethics and attitudes. In: *Conservation of Threatened Plants* (ed. Simmonds, J. B., Beyer, R. I., Brandham, P. E., Lucas, G. U., and Parry, V. T. H.), pp. 155–79. New York: Plenum Press.

Soulé, M. 1980. Thresholds for survival: maintaining fitness and evolutionary potential. In *Conservation Biology: An Evolutionary-Ecological Perspective* (ed. Soulé, M. E., and Wilcox, B. A.), pp. 151–69. Sunderland, Massachusetts: Sinauer.

White, M. J. D. 1978. *Modes of Speciation.* San Francisco: Freeman.

Whitmore, T. C. 1973. Frequency and habitat of tree species in the rain forest of Ulu Kelantan. *Gard. Bull.* 26:195–210.

Index

DATE DUE

GAYLORD   PRINTED IN U.S.A.